Benson's
Microbiological Applications

Laboratory Manual in General Microbiology

Fourteenth Edition

Alfred E. Brown
Emeritus Professor, Auburn University

Heidi R. Smith
Front Range Community College

McGraw Hill Education

BENSON'S MICROBIOLOGICAL APPLICATIONS: LABORATORY MANUAL IN GENERAL MICROBIOLOGY, CONCISE VERSION, FOURTEENTH EDITION

Published by McGraw-Hill Education, 2 Penn Plaza, New York, NY 10121. Copyright © 2017 by McGraw-Hill Education. All rights reserved. Printed in the United States of America. Previous editions © 2015, 2012, and 2009. No part of this publication may be reproduced or distributed in any form or by any means, or stored in a database or retrieval system, without the prior written consent of McGraw-Hill Education, including, but not limited to, in any network or other electronic storage or transmission, or broadcast for distance learning.

Some ancillaries, including electronic and print components, may not be available to customers outside the United States.

This book is printed on acid-free paper.

1 2 3 4 5 6 7 8 9   LMN   21 20 19 18 17 16

ISBN 978-1-259-70523-6
MHID 1-259-70523-4

Chief Product Officer, SVP Products & Markets: *G. Scott Virkler*
Vice President, General Manager, Products & Markets: *Marty Lange*
Managing Director: *Lynn Breithaupt*
Brand Manager: *Marija Magner*
Director, Product Development: *Rose Koos*
Product Developer: *Darlene M. Schueller*
Director of Digital Content: *Michael G. Koot, PhD*
Digital Product Analyst: *Jake Theobald*
Product Development Coordinator: *Andrea Eboh*
Marketing Manager: *Kristine Rellihan*
Director, Content Design & Delivery: *Linda Meehan-Avenarius*
Content Project Managers: *Mary Jane Lampe (Core) / Keri Johnson (Assessment)*
Content Licensing Specialists: *Melissa Homer (Image) / Lorraine Buczek (Text)*
Buyer: *Laura Fuller*
Designer: *Matt Backhaus*
Cover Image: *© Shutterstock/angellodeco; By R Parulan Jr./Getty Images*
Compositor: *MPS Limited, A Macmillan Company*
Typeface: *STIX MathJax Main 11/12*
Printer: *LSC Communications*

www.mhhe.com

About the Authors

Alfred Brown

Emeritus Professor of
Microbiology
Auburn University

B.S. Microbiology, California
State College, Long Beach

Ph.D. Microbiology, UCLA

Teaching Dr. Brown has been a member of the American Society for Microbiology for 50 years, and he has taught various courses in microbiology over a teaching career that spans more than 30 years. Courses have included general microbiology, medical microbiology, microbial physiology, applied and environmental microbiology, photosynthesis, microbiological methods, and graduate courses such as biomembranes. In 2008, Dr. Brown retired from the Auburn University faculty as an emeritus professor of microbiology.

Administration During his tenure at Auburn University, Dr. Brown served as the director of the University Electron Microscope Facility. He also served as the chair of the Department of Botany and Microbiology and the chair of the Department of Biological Sciences.

Research My research has focused on the physiology of the purple nonsulfur bacteria. This has involved how bacteriochlorophyll and photosynthetic membrane synthesis are coordinated. Herbicides, such as atrazine, have been used to determine the binding site for ubiquinone in photosynthetic electron transport. Binding occurs on the L-subunit, a protein in the photosynthetic reaction center. Resistance to atrazine involves a single amino acid change in the L-subunit that prevents the herbicide from binding to the protein and inhibiting electron transport. This is comparable to how atrazine inhibits electron transport in plants and how resistance to these herbicides develops. My laboratory also investigated how the sulfonylurea herbicides inhibit acetolactate synthase, a crucial enzyme in the pathway for branched-chain amino acids. Recently, I consulted for a company that manufactures roofing shingles. Because of the presence of calcium carbonate in shingles, cyanobacteria can easily grow on their surface, causing problems of contamination. My laboratory isolated various species of cyanobacteria involved in the problem and taxonomically characterized them. We also tested growth inhibitors that might be used in their control.

Dr. Brown and his wife have traveled extensively in Europe since his retirement. His three children have followed him in science, having earned doctorates in physics, chemistry and anatomy.

Heidi Smith

Front Range Community College
B.A. Biology/Pre-Medicine,
Taylor University, IN

M.S. Biology, Bowling Green
State University, OH

Heidi Smith is the lead faculty for microbiology at Front Range Community College in Fort Collins, CO, and teaches a variety of biology courses each semester including microbiology, anatomy/physiology, and biotechnology. Heidi also serves as the principal investigator on a federal grant program designed to increase student success in transfer and completion of STEM degrees at the local university. In that role, Heidi works directly with students to train them for and support them through summer undergraduate research experiences.

Student success is a strategic priority at FRCC and a personal passion of Heidi's. She continually works to develop professionally in ways that help her do a better job of reaching this important goal. During the past 7 years, Heidi has had the opportunity to collaborate with faculty throughout the United States in developing digital tools, such as SmartBook, LearnSmart Labs, and Connect, that measure and improve student learning outcomes. This collaborative experience with these tools has revolutionized her approach to teaching in both face-to-face and hybrid courses. The use of digital technology has given Heidi the ability to teach courses driven by real-time student data and with a focus on active learning and critical thinking activities.

Heidi is an active member of the American Society for Microbiology. She has presented instructional technology and best online and face-to-face teaching practices on numerous occasions at the annual conference for undergraduate educators. She also served as a member of the ASM Task Force on Curriculum Guidelines for Undergraduate Microbiology Education, assisting in the identification of core microbiology concepts as a guide to undergraduate instruction.

Off campus, Heidi spends as much time as she can enjoying the beautiful Colorado outdoors with her husband and three young children.

Contents

Indicates a LearnSmart Lab® activity is available for all or part of this exercise. For more information, visit http://www.mhhe.com /LearnSmartLabsBiology/

v

Preface

Benson's Microbiological Applications has been the gold standard of microbiology laboratory manuals for over 35 years. This manual has a number of attractive features that resulted in its adoption in universities, colleges, and community colleges for a wide variety of microbiology courses. These features include user-friendly diagrams that students can easily follow, clear instructions, and an excellent array of reliable exercises suitable for beginning or advanced microbiology courses.

In revising the lab manual for the fourteenth edition, we have tried to maintain the proven strengths of the manual and further enhance it. We have updated the introductory material in many exercises to reflect changes in scientific information and increase relevancy for students. Critical thinking questions have also been added to increase the Bloom's level of the laboratory reports. Finally, the names and biosafety levels of microorganisms used in the manual are consistent with those used by the American Type Culture Collection. This is important for those users who rely on the ATCC for a source of cultures.

Guided Tour Through a Lab Exercise

Learning Outcomes

Each exercise opens with Learning Outcomes, which list what a student should be able to do after completing the exercise.

> ### Learning Outcomes
>
> **After completing this exercise, you should be able to**
>
> 1. Prepare a thin smear of bacterial cells and stain them with a simple stain.
> 2. Understand why staining is necessary to observe bacteria with a brightfield microscope.
> 3. Observe the different morphologies of bacterial cells.

Introduction

The introduction describes the subject of the exercise or the ideas that will be investigated. It includes all of the information needed to perform the laboratory exercise. The fourteenth edition has improved its student

> The recombinant plasmid pGLO used in this exercise contains a gene of interest known as the GFP gene (figure 50.1). This gene was isolated from a marine jellyfish and introduced into a bacterial plasmid using recombinant DNA techniques. The GFP gene codes for a *green fluorescent protein* that is visible when expressed by bacterial colonies that possess the plasmid.
> In this exercise, the presence of the GFP gene on the pGLO plasmid will be verified by two molecular

> Acquisition of genetic material, in addition to mutation of genes and loss of unnecessary genes, is an important part of bacterial adaptation and evolution. In many cases, acquisition of new genes (i.e., antibiotic resistance) may mean the difference between survival and death of an organism in the environment. Thus, bacteria have developed various mechanisms, includ-

relevancy message within these introductions, explaining to students why they should care about the lab.

First and Second Periods

In many cases, instructions are presented for two or more class periods so you can proceed through an exercise in an appropriate fashion.

> ### 🕐 First Period
>
> (Inoculations and Incubation)
>
> Since six microorganisms and three kinds of media are involved in this experiment, it will be necessary for economy of time and materials to have each student work with only three organisms. The materials list for this period indicates how the organisms will

> ### 🕐 Second Period
>
> (Culture Evaluations and Spore Staining)
>
> Remove the lid from the GasPak jar. If vacuum holds the inner lid firmly in place, break the vacuum by sliding the lid to the edge. When transporting the plates and tubes to your desk, *take care not to agitate the FTM tubes*. The position of growth in the medium can be easily changed if handled carelessly.

Materials Needed

This section lists the laboratory materials that are required to complete the exercise.

> ### Materials
>
> *per student:*
> - 1 tube of nutrient broth
> - 1 petri plate of trypticase soy agar (TSA)
> - 1 sterile cotton swab
> - Sharpie marking pen
>
> *per two or more students:*
> - 1 petri plate of blood agar

Procedures

The procedures and methods provide a set of detailed instructions for accomplishing the planned laboratory activities.

> ### Procedures
>
> If your microscope has three objectives, you have three magnification options: (1) low-power, or 100× total magnification, (2) high-dry magnification, which is 400× total with a 40× objective, and (3) 1000× total magnification with a 100× oil immersion objective.
> Whether you use the low-power objective or the oil immersion objective will depend on how much

Illustrations

Illustrations provide visual instructions for performing steps in procedures or are used to identify parts of instruments or specimens.

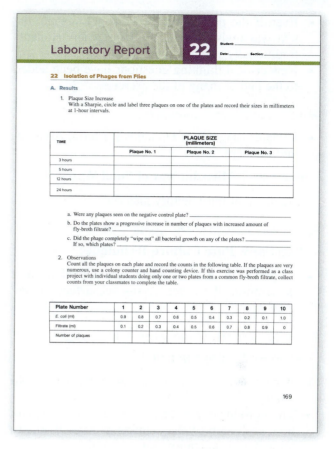

A. Results

1. Plaque Size Increase
   With a Sharpie, circle and label three plaques on one of the plates and record their sizes in millimeters at 1-hour intervals.

TIME	PLAQUE SIZE (millimeters)		
	Plaque No. 1	Plaque No. 2	Plaque No. 3
3 hours			
5 hours			
12 hours			
24 hours			

a. Were any plaques seen on the negative control plate? _____
b. Do the plates show a progressive increase in number of plaques with increased amount of fly-broth filtrate? _____
c. Did the phage completely "wipe out" all bacterial growth on any of the plates? _____ If so, which plates? _____

2. Observations
   Count all the plaques on each plate and record the counts in the following table. If the plaques are very numerous, use a colony counter and hand counting device. If this exercise was performed as a class project with individual students doing only one or two plates from a common fly-broth filtrate, collect counts from your classmates to complete the table.

Plate Number	1	2	3	4	5	6	7	8	9	10
E. coli (ml)	0.9	0.8	0.7	0.6	0.5	0.4	0.3	0.2	0.1	1.0
Filtrate (ml)	0.1	0.2	0.3	0.4	0.5	0.6	0.7	0.8	0.9	0
Number of plaques										

169

Figure 1.2 **The compound microscope.**
© Charles D. Winters/Science Source

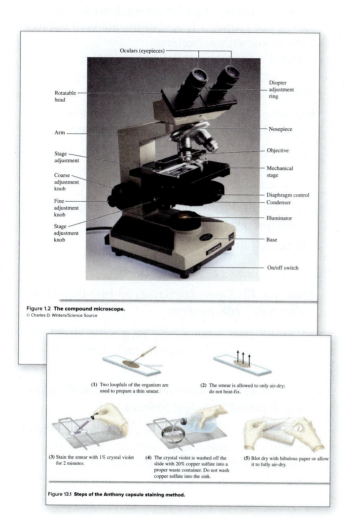

Figure 13.1 **Steps of the Anthony capsule staining method.**

Laboratory Reports

A Laboratory Report to be completed by students immediately follows most of the exercises. These Laboratory Reports are designed to guide and reinforce student learning and provide a convenient place for recording data. These reports include various types of review activities, tables for recording observations and experimental results, and questions dealing with the analysis of such data.

As a result of these activities, students will increase their skills in gathering information by observation and experimentation. By completing all of the assessments in the Laboratory Reports, students will be able to determine if they accomplished all of the learning outcomes.

Safety

In the fourteenth edition, we have followed the recent recommendations of the American Society for Microbiology concerning the biosafety levels of organisms used in the exercises in this laboratory manual. The Basic Microbiology Safety section in the introductory pages of the lab manual has been completely revised to align with ASM's *Guidelines for Biosafety in Teaching Laboratories* released in 2012. The BSL classifications of all organisms used in each exercise have been updated according to ATCC. Where possible, BSL-1 organisms replace BSL-2 organisms in many exercises. However, in some exercises, it is necessary to use BSL-2 organisms such as specific staphylococci, streptococci, and some of the *Enterobacteriaceae* due to specific tests, stains, and concepts that involve these organisms. For these exercises, we recommend that safety procedures be followed as suggested by ASM, such as the use of biosafety cabinets.

Changes to This Edition

Exercises

- Many photographs and some diagrams have been replaced, upgraded, or revised for clarity.

- Even if the procedure itself did not change, the text was revised to clarify areas where students traditionally struggle through an exercise.
- New critical thinking questions have been added to the end of many of the exercises.

Part 2 Survey of Microorganisms

- Updated diagram of Domains Bacteria, Archaea, and Eukarya.

Exercise 5 Microbiology of Pond Water

- Revision of introductory material.

Part 3 Manipulation of Microorganisms

Exercise 9 Pure Culture Techniques

- A photo of a pour plate was added with a description of surface versus subsurface colony growth.

Part 4 Staining and Observation of Microorganisms

Exercise 13 Capsular Staining

- A new photo of a capsule stain replaces the old photos.
- The introductory material was revised to explain how each step of the staining process works to allow for visualization of the capsule.

Exercise 15 Spore Staining: Two Methods

- Added a new photo of a spore stain.
- The procedures for the various staining methods were rearranged and revised for greater clarity.

Part 7 Environmental Influences and Control of Microbial Growth

Exercise 24 Effects of Oxygen on Growth

- Revised line art and photos throughout the exercise to clarify oxygen classifications.

Exercise 25 Temperature: Effects on Growth

- Added a figure depicting microbial temperature classification groups.
- Updated temperature ranges to most currently accepted upper and lower limits.

Exercise 28 Ultraviolet Light: Lethal Effects

- Replaced line art figure of UV lamp with photo of plate half covered by white card.

Part 8 Identification of Unknown Bacteria

Exercise 34 Morphological Study of an Unknown Bacterium

- New, easy-to-use descriptive chart for unknown identification.

- Added new figure of possible endospore positions in bacterial cell.

Exercise 36 Physiological Characteristics: Oxidation and Fermentation Tests

- Rearranged procedural sections and the process figures to make it easier for students to complete all of the required inoculations for both controls and unknown test bacteria.
- Added a flowchart to explain the steps of the nitrate reduction test and the correct analysis after each step.

Exercise 37 Physiological Characteristics: Hydrolytic and Degradative Reactions

- Revised figures, photos, and their associated captions to facilitate better student analysis of results for each of these tests.

Exercise 38 Physiological Characteristics: Multiple Test Media

- Removed litmus milk test.

Part 9 Miniaturized Multitest Systems

Exercise 40 *Enterobacteriaceae* Identification: The API® 20E System

- Simplified the tabulation process and enlarged the photo for easier analysis and identification by students.

Exercise 41 *Enterobacteriaceae* Identification: The EnteroPluri-*Test* System

- Completely rewrote this exercise to align with newly available test system from Becton-Dickinson due to discontinuance of the Enterotube II system.

Part 10 Applied Microbiology

Exercise 44 Bacteriological Examination of Water: Most Probable Number Determination

- Updated introduction to the exercise.
- Revised the MPN table to make it easier for students to analyze and understand their results.

Exercise 46 Reductase Test

- The introductory section to the exercise has been revised and updated.

Exercise 49 Microbiology of Alcohol Fermentation

- The introduction to the exercise has been revised, emphasizing the history of fermentation and how fermentation is a means to preserve foods.
- Added an explanation of how off-flavors produced by sulfides occur in the fermentation of

wines and how they can be monitored by a lead acetate strip placed in the fermentation flask.

Part 11 Bacterial Genetics and Biotechnology

- This section was added to provide two exercises that are appropriate for general microbiology students. These exercises can be done individually or the techniques of PCR, gel electrophoresis, and bacterial transformation can be done as a connected sequence.

Part 12 Medical Microbiology

Exercise 52 The Staphylococci: Isolation and Identification

- Upgraded photos of mannitol salt agar.

Exercise 53 The Streptococci and Enterococci: Isolation and Identification

- Upgraded photos of blood agar plate hemolysis.

Part 13 Immunology and Serology

Exercise 59 Enzyme-Linked Immunosorbent Assay (ELISA)

- Added this exercise which simulates the commonly used screening test for HIV.

Exercise 60 Blood Grouping

- Added table figure to clearly explain the ABO groups.
- Revised introductory material, adding information on the Rh factor and the use of RhoGAM in Rh-negative mothers to prevent hemolytic disease in newborns.
- Revised the procedure for the use of artificial blood available in kits that can be purchased from biological supply companies. The use of finger sticks by students to collect blood and the use of human blood in laboratories has become a serious safety issue.

We would like to thank all the people at McGraw-Hill for their tireless efforts and support with this project. They are professional and competent and always a pleasure to work with on this manual. A special and deep thanks to Darlene Schueller, our product developer. Once again, she kept the project focused, made sure we met deadlines, and made suggestions that improved the manual in many ways. Thanks as well to Marija Magner, brand manager; Mary Jane Lampe, content project manager; Kristine Rellihan, marketing manager; Matt Backhaus, designer; Keri Johnson, assessment content project manager; and Melissa Homer and Lorraine Buczek, content licensing specialists; and many who worked "behind the scenes."

Digital Tools for Your Success

Save Time with Auto-Graded Assessments. Gather Powerful Performance Data.

McGraw-Hill Connect for Benson's Microbiological Applications Laboratory Manual in General Microbiology provides assignment and assessment solutions, connecting your students with the tools and resources they'll need to achieve success.

Homework and Assessment

With **Connect for Benson's Microbiological Applications Laboratory Manual in General Microbiology,** you can deliver auto-graded assignments, quizzes, and tests online. Choose from a robust set of interactive questions and activities using high-quality art from the lab manual and animations. Assignable content is available for every Learning Outcome and is categorized according to the ASM Objectives. As an instructor, you can edit existing questions and author entirely new ones.

Instructor Resources

Customize your lecture with tools such as animations and images. An instructor's manual is also available.

Lecture Capture

McGraw-Hill Tegrity® records and distributes your class lecture with just a click of a button. Students can view this lecture material anytime, anywhere via computer or mobile device. The material is indexed as you record, so students can use keywords to find exactly what they want to study.

LearnSmart Labs® is an optional adaptive simulated lab experience that brings meaningful scientific exploration to students. Through a series of adaptive questions, LearnSmart Labs identifies a student's knowledge gaps and provides resources to quickly and efficiently close those gaps. Once students have mastered the necessary basic skills and concepts, they engage in a highly realistic simulated lab experience that allows for mistakes and the execution of the scientific method. LearnSmart Labs are currently being used as a pre-lab preparation assignment or as full lab replacement.

LearnSmart® Prep is an adaptive learning tool that prepares students for college-level work in Microbiology, and is included with Connect for Benson's Microbiological Applications purchases. LearnSmart Prep individually identifies concepts the student does not fully understand and provides learning resources to teach essential concepts so he or she enters the classroom prepared. Data-driven reports highlight areas where students are struggling.

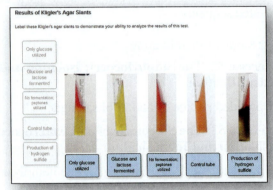

© McGraw-Hill Education/Auburn University Photographic Services

Detailed Reports

Track individual student performance—by question, by assignment, or in relation to the class overall—with detailed grade reports. Integrate grade reports easily with your Learning Management Systems (LMS).

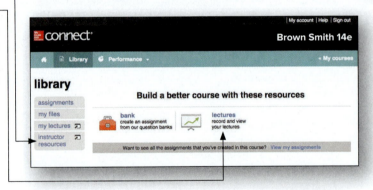

Learn more at connect.mheducation.com.

Basic Microbiology Laboratory Safety

Every student and instructor must focus on the need for safety in the microbiology laboratory. While the lab is a fascinating and exciting learning environment, there are hazards that must be acknowledged and rules that must be followed to prevent accidents and contamination with microbes. The following guidelines will provide every member of the laboratory the information required to assure a safe learning environment.

The "Biohazard" symbol must be affixed to any container or equipment used to store or transport potentially infectious materials.

Courtesy of the Centers for Disease Control.

Microbiological laboratories are special, often unique environments that may pose identifiable infectious disease risks to persons who work in or near them. Infections have been contacted in the laboratory throughout the history of microbiology. Early reports described laboratory-associated cases of typhoid, cholera, brucellosis, and tetanus, to name a few. Recent reports have documented laboratory-acquired cases in laboratory workers and health-care personnel involving *Bacillus anthracis, Bordetella pertussis, Brucella, Burkholderia pseudomallei, Campylobacter, Chlamydia,* and toxins from *Clostridium tetani, Clostridium botulinum,* and *Corynebacterium diphtheriae.* In 2011, the CDC traced an outbreak of *Salmonella* to several undergraduate microbiology laboratories, prompting further discussion about safety guidelines for the lab classroom setting.

The term "containment" is used to describe the safe methods and procedures for handling and managing microorganisms in the laboratory. An important laboratory procedure practiced by all microbiologists that will guarantee containment is **aseptic technique,** which prevents workers from contaminating themselves with microorganisms, ensures that others and the work area do not become contaminated, and also ensures that microbial cultures do not become unnecessarily contaminated with unwanted organisms. Containment involves personnel and the immediate laboratory environment. Containment also guarantees that infectious agents do not escape from the laboratory and contaminate the environment external to the lab. Containment, therefore, relies on good microbiological technique and laboratory protocol as well as the use of appropriate safety equipment.

Biosafety Levels (BSL)

The biosafety level classifications of microorganisms represent the potential of the organism to cause disease and the conditions under which the organism should be safely handled. The Centers for Disease Control classifies organisms into four levels, which take into account many factors such as virulence, pathogenicity, antibiotic resistance patterns, vaccine and treatment availability, and other factors. The four biosafety levels are described in the table on page xii.

All microorganisms used in the exercises in this manual are classified as BSL-1 or BSL-2. Note: Although some of the organisms that students will culture and work with are classified as BSL-2, these organisms may be laboratory strains that do not pose the same threat of infection as primary isolates of the same organism taken from patients in clinical samples. Hence, these laboratory strains can, in most cases, be handled using normal procedures and equipment found in the vast majority of student teaching laboratories. However, it should be emphasized that many bacteria are opportunistic pathogens, and therefore all microorganisms should be handled by observing proper techniques and precautions.

Each of the biosafety levels indicates that certain laboratory practices and techniques, safety equipment, and laboratory facilities should be used when working with organisms in that classification. Each combination is specifically appropriate for the operations performed and the documented or suspected routes of transmission of the infectious agents. In response to the *Salmonella* outbreaks in undergraduate laboratories, the American Society for Microbiology set out to define a clear set of safety practices for laboratories based on the use of BSL-1 or BSL-2 organisms. In 2013, the *Guidelines for Biosafety in Teaching Laboratories* was published in the *Journal of Microbiology Education.*

Standard Laboratory Practices (Based on ASM *Guidelines for Biosafety in Teaching Laboratories*)

BSL-1 Guidelines

Although BSL-1 organisms pose very little risk of disease for healthy students, they are still capable of causing infection under certain circumstances. These guidelines indicate the recommended best practices in the laboratory for the protection of students and the community.

Personal Protection

- It is recommended that lab coats be worn in the laboratory at all times. Lab coats can protect a student from contamination by microorganisms that he or she is working with and prevent

Biosafety Levels for Selected Infectious Agents

BIOSAFETY LEVEL (BSL)	TYPICAL RISK	ORGANISM
BSL-1	Not likely to pose a disease risk to healthy adults.	*Achromobacter denitrificans* *Alcaligenes faecalis* *Bacillus cereus* *Bacillus subtilis* *Corynebacterium pseudodiphtheriticum* *Micrococcus luteus* *Neisseria sicca* *Staphylococcus epidermidis* *Staphylococcus saprophyticus*
BSL-2	Poses a moderate risk to healthy adults; unlikely to spread throughout community; effective treatment readily available.	*Enterococcus faecalis* *Klebsiella pneumoniae* *Mycobacterium phlei* *Salmonella enterica var. Typhimurium* *Shigella flexneri* *Staphylococcus aureus* *Streptococcus pneumoniae* *Streptococcus pyogenes*
BSL-3	Can cause disease in healthy adults; may spread to community; effective treatment readily available.	*Blastomyces dermatitidis* *Chlamydia trachomatis* *Coccidioides immitis* *Coxiella burnetii* *Francisella tularensis* *Histoplasma capsulatum* *Mycobacterium bovis* *Mycobacterium tuberculosis* *Pseudomonas mallei* *Rickettsia canadensis* *Rickettsia prowazekii* *Yersinia pestis*
BSL-4	Can cause disease in healthy adults; poses a lethal risk and does not respond to vaccines or antimicrobial therapy.	Filovirus *Herpesvirus simiae* Lassa virus Marburg virus Ebola virus

contamination from stains and chemicals. At the end of the laboratory session, lab coats are usually stored in the lab in a manner prescribed by the instructor.

- You may be required to wear gloves while performing the lab exercises. They protect the hands against contamination by microorganisms and prevent the hands from coming in direct contact with stains and other reagents. This is especially important if you have open wounds.
- Wash your hands with soap and water before and after handling microorganisms.
- Safety goggles or glasses should be worn while you are performing experiments with liquid cultures, splash hazards, or while spread plating. They must also be worn when working with ultraviolet light to prevent eye damage because they block out UV rays.
- Sandals or open-toe shoes are not to be worn in the laboratory. Accidental dropping of objects or cultures could result in serious injury or infection.

- Lab coats, gloves, and safety equipment should not be worn outside of the laboratory unless properly decontaminated first.
- Students with long hair should tie the hair back to avoid accidents when working with Bunsen burners/open flames. Long hair can also be a source of contamination when working with cultures.
- Avoid wearing dangling jewelry or scarves to lab.
- If you are immune-compromised (including pregnancy) or provide care for someone who is immune-compromised, please consult with your physician about your participation in these laboratory exercises.

Laboratory Environment and Equipment

- Most importantly, read the exercise and understand the laboratory protocol before coming to laboratory. This way, you will be familiar with potential hazards in the exercise. Unless directed to do so, do not subculture any unknown organisms isolated from the environment as they could be potential pathogens.

- Students should store all books and materials not used in the laboratory in areas or receptacles designated for that purpose. Only necessary materials such as a lab notebook and the laboratory manual should be brought to the student work area. Use only institution-provided writing instruments.
- Avoid handling personal items such as cell phones and calculators while performing laboratory exercises. Students must also avoid handling contact lenses or applying makeup while in the laboratory.
- Eating, drinking (including water), gum, chewing tobacco, and smoking are not allowed in the laboratory.
- Know the location of exits and safety equipment such as the eye wash and shower stations, first aid kit, and fire extinguisher in the event of an accident that requires the use of this equipment.
- The door to the laboratory must remain closed, and only enrolled students should be allowed to enter the laboratory classroom.
- Before beginning the activities for the day, work areas should be wiped down with the disinfectant that is provided for that purpose. Likewise, when work is finished for the day, the work area should be treated with disinfectant to ensure that any contamination from the exercise performed is destroyed. To avoid contaminating the work surface, do not place contaminated pipettes, loops/needles, or swabs on the work surface.
- Always use correct labeling procedures on all containers.
- If possible, use a microincinerator or disposable loops for transferring microorganisms from one container to another. If these are unavailable, please use extreme caution when working with the open flame of a Bunsen burner. The flame is often difficult to see.
- Caution is imperative when working with alcohol and open flames. Alcohol is highly flammable, and fires can easily result when using glass rods that have been dipped in alcohol.
- Always make sure the gas is turned off before leaving the laboratory.
- Pipetting by mouth is prohibited in the lab. All pipetting must be performed with pipette aids.
- Use test tube racks when transporting cultures throughout the laboratory.
- You may be required to sign a safety agreement stating that you have been informed about safety issues and precautions and the hazardous nature of microorganisms that you may handle during the laboratory course.

BSL-2 Guidelines

BSL-2 organisms pose a moderate risk of infections, but the diseases caused by these organisms are treatable and usually not serious. Before working with these organisms, students should already show proficiency in following all of the guidelines for BSL-1 organisms. Additional precautions that should be taken when working with BSL-2 organisms include:

- Wear face shields or masks, along with proper eye wear, when working with procedures that involve a potential splash hazard. Alternatively, conduct all work with these organisms in a biological safety cabinet.
- Laboratory coats are required when working with these organisms.
- Use microincinerators or disposable loops rather than Bunsen burners.

Emergencies

- Report all spills, accidents, or injuries immediately to the laboratory instructor.
- Do not handle broken glass with your hands.
- Follow institutional policy in documenting all injuries and other emergency situations.

Disposal of Laboratory Materials

Dispose of all contaminated materials properly and in the appropriate containers. Your instructor will give you specific instructions for your laboratory classroom.

- Biohazard containers—Biohazard containers are to be lined with clearly marked biohazard bags; disposable agar plates, used gloves, and any materials such as contaminated paper towels should be discarded in these containers; no glassware, test tubes, or sharp items are to be disposed of in biohazard containers.
- Sharps containers—Sharps, needles, and Pasteur pipettes should be discarded in these containers.
- Autoclave shelf, cart, or bin—Contaminated culture tubes and glassware used to store media and other glassware should be placed in these areas for decontamination and washing.
- Trash cans—Any noncontaminated materials, paper, or trash should be discarded in these containers. Under no circumstances should laboratory waste be disposed of in trash cans.
- Slides and broken glass may be disposed of in a sharps container, a beaker filled with disinfectant, or a labeled cardboard box. Listen carefully to your instructor's directions for these items.

Microorganisms Used or Isolated in the Lab Exercises in This Manual

ORGANISM	GRAM STAIN AND MORPHOLOGY	HABITAT	BSL	LAB EXERCISE
Alcaligenes faecalis ATCC 8750	Negative rod	Decomposing organic material, feces	1	26, 39
Bacillus coagulans ATCC 7050	Positive rod	Spoiled food, silage	1	48
Bacillus megaterium ATCC 14581	Positive rod	Soil, water	1	11, 12, 14, 15, 28, 47
Bacillus subtilis ATCC 23857	Positive rod	Soil, decomposing organic matter	1	24, 37
Candida glabrata ATCC 200918	Yeast	Human oral cavity	1	26
Chromobacterium violaceum ATCC 12472	Negative rod	Soil and water; opportunistic pathogen in humans	2	9
Citrobacter freundii ATCC 8090	Negative rod	Humans, animals, soil water; sewage opportunistic pathogen	1	54
Clostridium beijerinckii ATCC 25752	Positive rod	Soil	1	24
Clostridium sporogenes ATCC 3584	Positive rod	Soil, animal feces	1	24, 48
Corynebacterium xerosis ATCC 373	Positive rods, club-shaped	Conjunctiva, skin	1	11
Enterobacter aerogenes ATCC 13048	Negative rods	Feces of humans and animals	1	24, 36, 39, 44
Enterococcus faecalis ATCC 19433	Positive cocci in pairs, short chains	Water, sewage, soil, dairy products	2	24, 39, 53, 58
Enterococcus faecium ATCC 19434	Positive cocci in pairs, short chains	Feces of humans and animals	2	53, 58
Escherichia coli ATCC 11775	Negative rods	Sewage, intestinal tract of warm-blooded animals	1	8, 9, 14, 19, 21, 22, 24, 25, 26, 27, 29, 31, 36, 37, 38, 39, 44, 47, 48, 50, 51
Geobacillus stearothermophilus ATCC 12980	Gram-positive rods	Soil, spoiled food	1	25, 48
Halobacterium salinarium ATCC 33170	Gives gram-negative reaction; rods	Salted fish, hides, meats	1	27
Klebsiella pneumoniae ATCC 13883	Negative rods	Intestinal tract of humans; respiratory and intestinal pathogen in humans	2	13, 39
Streptococcus lactis ATCC 19435	Positive cocci in chains	Milk and milk products	1	11

Microorganisms Used or Isolated in the Lab Exercises in This Manual (continued)

ORGANISM	GRAM STAIN AND MORPHOLOGY	HABITAT	BSL	LAB EXERCISE
Micrococcus luteus ATCC 12698	Positive cocci that occur in pairs	Mammalian skin	1	9, 17, 29, 39
Moraxella catarrhalis ATCC 25238	Negative cocci that often occur in pairs with flattened sides	Pharynx of humans	1	14
Mycobacterium smegmatis ATCC 19420	Positive rods; may be Y-shaped or branched	Smegma of humans	1	16
Proteus vulgaris ATCC 29905	Negative rods	Intestines of humans, and animals; soil and polluted waters	2	17, 31, 37, 38, 39, 54
Pseudomonas aeruginosa ATCC 10145	Negative rods	Soil and water; opportunistic pathogen in humans	2	14, 31, 32, 36, 39
Saccharomyces cerevisiae ATCC 18824	Yeast	Fruit, used in beer, wine, and bread	1	26
Salmonella enterica subsp. enterica serovar *Typhimurium* ATCC 700720	Negative rods	Most frequent agent of *Salmonella* gastroenteritis in humans	2	39, 54, 56
Serratia marcescens ATCC 13880	Negative rods	Opportunistic pathogen in humans	1	9, 25, 39, 55
Shigella flexneri ATCC 29903	Negative rods	Pathogen of humans	2	54
Staphylococcus aureus ATCC 12600	Positive cocci, irregular clusters	Skin, nose, GI tract of humans, pathogen	2	11, 14, 23, 24, 29, 31, 32, 36, 37, 38, 47, 52, 53, 57
Staphylococcus epidermidis ATCC 14990	Positive cocci that occur in pairs and tetrads	Human skin, animals; opportunistic pathogen	1	12, 14, 16, 26, 27, 28, 39, 52
Staphylococcus saprophyticus ATCC 15305	Positive cocci that occur singly and in pairs	Human skin; opportunistic pathogen in the urinary tract	1	52
Streptococcus agalactiae ATCC 13813	Positive cocci; occurs in long chains	Upper respiratory and vaginal tract of humans, cattle; pathogen	2	53, 58
Streptococcus bovis ATCC 33317	Positive cocci; pairs and chains	Cattle, sheep, pigs; occasional pathogen in humans	1	53, 58
Streptococcus dysagalactiae subsp. equisimilis ATCC 12394	Positive cocci in chains	Mastitis in cattle	2	53

Microorganisms Used or Isolated in the Lab Exercises in This Manual (continued)

ORGANISM	GRAM STAIN AND MORPHOLOGY	HABITAT	BSL	LAB EXERCISE
Streptococcus mitis ATCC 49456	Positive cocci in pairs and chains	Oral cavity of humans	2	53
Streptococcus mutans ATCC 25175D-5	Positive cocci in pairs and chains	Tooth surface of humans, causes dental caries	1	53
Streptococcus pneumoniae ATCC 33400	Positive cocci in pairs	Human pathogen	2	53
Streptococcus pyogenes ATCC 12344	Positive cocci in chains	Human respiratory tract; pathogen	2	53, 58
Streptococcus salivarius ATCC 19258	Positive cocci in short and long chains	Tongue and saliva	1	53
Thermoanaerobacterium thermosaccharolyticum ATCC 7956	Negative rods; single cells or pairs	Soil, spoiled canned foods	1	48

Microscopy

Although there are many kinds of microscopes available to the microbiologist today, only three types will be described here for our use: the brightfield, and phase-contrast microscopes. If you have had extensive exposure to microscopy in previous courses, this unit may not be of great value to you; however, if the study of microorganisms is a new field of study for you, there is a great deal of information that you need to acquire about the proper use of these instruments.

Microscopes in a college laboratory represent a considerable investment and require special care to prevent damage to the lenses and mechanical parts. A microscope may be used by several people during the day and moved from the work area to storage, which results in a much greater chance for damage to the instrument than if the microscope were used by only a single person.

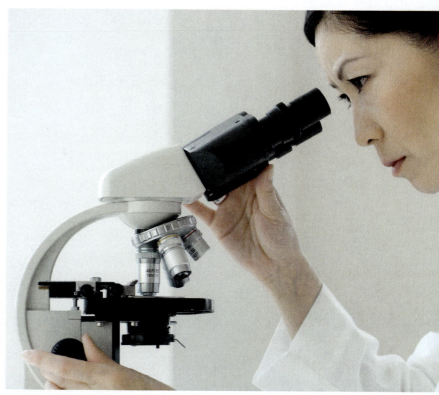

© JGI/Blend Images LCC RF

The complexity of some of the more expensive microscopes also requires that certain adjustments be made periodically. Knowing how to make these adjustments to get the equipment to perform properly is very important. An attempt is made in the four exercises of this unit to provide the necessary assistance for getting the most out of the equipment.

Microscopy should be as fascinating to the beginner as it is to the professional of long standing; however, only with intelligent understanding can the beginner approach the achievement that occurs with years of experience.

Brightfield Microscopy

Learning Outcomes

After completing this exercise, you should be able to

1. Identify the basic components of a brightfield microscope and understand the function of each component in proper specimen observation.

2. Examine a specimen using the low-power, high-dry, and oil immersion lenses.

3. Understand the proper use, care, and storage of a microscope.

A microscope that allows light rays to pass directly to the eye without being deflected by an intervening opaque plate in the condenser is called a ***brightfield microscope.*** This is the conventional type of instrument encountered by students in beginning courses in biology; it is also the first type to be used in this laboratory.

All brightfield microscopes have certain things in common, yet they differ somewhat in mechanical operation. Similarities and differences of various makes are discussed in this exercise so that you will know how to use the instrument that is available to you. Before attending the first laboratory session in which the microscope is used, read over this exercise and answer all the questions on the Laboratory Report. Your instructor may require that the Laboratory Report be handed in prior to doing any laboratory work.

Care of the Instrument

Microscopes represent considerable investment and can be damaged easily if certain precautions are not observed. The following suggestions cover most hazards.

Transport When carrying your microscope from one part of the room to another, use both hands to hold the instrument, as illustrated in figure 1.1. If it is carried with only one hand and allowed to dangle at your side, there is always the danger of collision with furniture or some other object. And, *under no circumstances should one attempt to carry two microscopes at one time.*

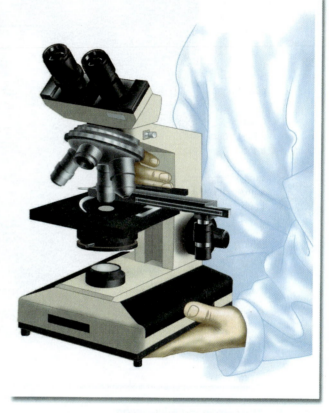

Figure 1.1 The microscope should be held firmly with both hands while being carried.

Clutter Keep your workstation uncluttered while doing microscopy. Keep unnecessary books and other materials away from your work area. A clear work area promotes efficiency and results in fewer accidents.

Electric Cord Microscopes have been known to tumble off of tabletops when students have entangled a foot in a dangling electric cord. Don't let the electric cord on your microscope dangle in such a way as to risk foot entanglement.

Lens Care At the beginning of each laboratory period, check the lenses to make sure they are clean. At the end of each lab session, be sure to wipe any immersion oil off the immersion lens if it has been used. More specifics about lens care are provided later in this exercise.

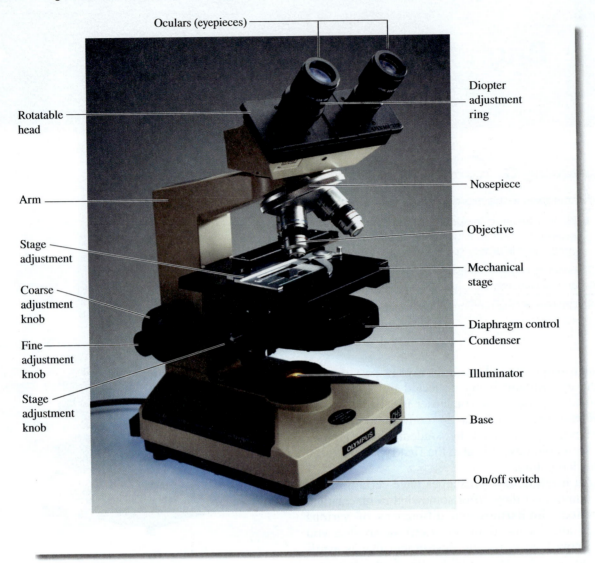

Oculars (eyepieces)

Diopter adjustment ring

Rotatable head

Arm

Nosepiece

Objective

Stage adjustment

Mechanical stage

Coarse adjustment knob

Diaphragm control

Condenser

Fine adjustment knob

Illuminator

Stage adjustment knob

Base

On/off switch

Figure 1.2 The compound microscope.
© Charles D. Winters/Science Source

Dust Protection In most laboratories dustcovers are used to protect the instruments during storage. If one is available, place it over the microscope at the end of the period.

Components

Light Source In the base of most microscopes is positioned some kind of light source. Ideally, the lamp should have a **light intensity control** to vary the intensity of light. The microscope in figure 1.2 has a knurled wheel on the right side of its base to regulate the voltage supplied to the lightbulb.

Most microscopes have some provision for reducing light intensity with a **neutral density filter.** Such

a filter is often needed to reduce the intensity of light below the lower limit allowed by the voltage control. On microscopes such as the Olympus CH2, one can simply place a neutral density filter over the light source in the base. On some microscopes a filter is built into the base.

Lens Systems All compound microscopes have three lens systems: the oculars, the objectives, and the condenser. Figure 1.3 illustrates the light path through these three systems.

The **ocular,** or eyepiece, is a complex piece, located at the top of the instrument, that consists of two or more internal lenses and usually has a magnification of 10×. Most modern microscopes (figure 1.2) have two ocular (binocular) lenses.

from the lamp to the slide being studied. Unlike the ocular and objective lenses, the condenser lens does not affect the magnifying power of the compound microscope. The condenser can be moved up and down by a knob under the stage. A **diaphragm** within the condenser regulates the amount of light that reaches the slide. Microscopes that lack a voltage control on the light source rely entirely on the diaphragm for controlling light intensity. On the Olympus CH2 microscope in figure 1.2, the diaphragm is controlled by turning a knurled ring. On some microscopes, a diaphragm lever is present. Figure 1.2 illustrates the location of the condenser and diaphragm.

Focusing Knobs The concentrically arranged **coarse adjustment** and **fine adjustment knobs** on the side of the microscope are used for bringing objects into focus when studying an object on a slide. On some microscopes, these knobs are not positioned concentrically as shown here.

Ocular Adjustments On binocular microscopes, one must be able to change the distance between the oculars and to make diopter changes for eye differences. On most microscopes, the interocular distance is changed by simply pulling apart or pushing together the oculars.

To make diopter adjustments, one focuses first with the right eye only. Without touching the focusing knobs, diopter adjustments are then made on the left eye by turning the knurled **diopter adjustment ring** (figure 1.2) on the left ocular until a sharp image is seen. One should now be able to see sharp images with both eyes.

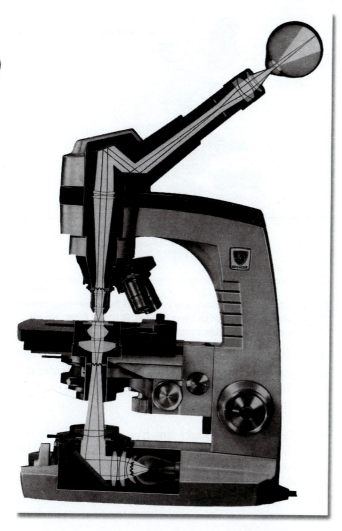

Figure 1.3 **The light pathway of a microscope.**

Three or more **objectives** are usually present. Note that they are attached to a rotatable **nosepiece,** which makes it possible to move them into position over a slide. Objectives on most laboratory microscopes have magnifications of 10×, 40×, and 100×, designated as **low-power, high-dry,** and **oil immersion,** respectively. Some microscopes will have a fourth objective for rapid scanning of microscopic fields that is only 4×.

The total magnification of a compound microscope is determined by multiplying the power of the ocular lens times the power of the objective lens used. Thus, the magnification of a microscope in which the oil immersion lens is being used is:

$$10 \times 100 = 1000$$

The object is now magnified 1000 times its actual size. The third lens system is the **condenser,** which is located under the stage. It collects and directs the light

Resolution

It would appear that the magnification of a microscope is only limited by the magnifying power of a lens system. However, in reality the limit for most light microscopes is 1000×, which is set by an intrinsic property of lenses called **resolving power.** The resolving power of a lens is its ability to completely separate two objects in a microscopic field. The resolving power is given by the formula $d = 0.5\ \lambda/NA$. The limit of resolution, d, or the distance between the two objects, is a function of two properties: the wavelength of the light used to observe a specimen, λ, and a property of lenses called the **numerical aperture,** or NA. Numerical aperture is a mathematical expression that describes how the condenser lens concentrates and focuses the light rays from the light source. Its value is maximized when the light rays are focused into a cone of light that then passes through the specimen into the objective lens. However, because some light is refracted or bent as it passes from glass into air, the refracted light rays are lost, and as a result the

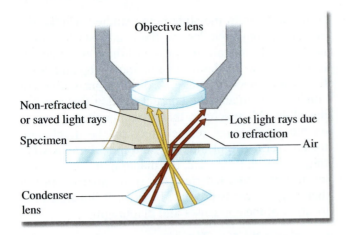

Figure 1.4 **Immersion oil, having the same refractive index as glass, prevents light loss due to refraction.**

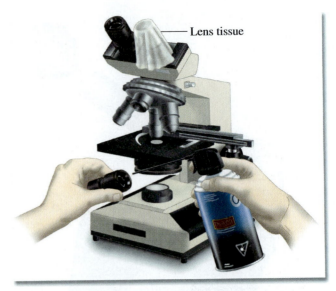

Figure 1.5 **When oculars are removed for cleaning, cover the ocular opening with lens tissue. A blast from an air syringe or gas canister removes dust and lint.**

numerical aperture is diminished (figure 1.4). The greater the loss of refracted light, the lower the numerical aperture. The final result is that the resolving power is greatly reduced.

For any light microscope, the limit of resolution is about 0.2 μm. This means that two objects closer than 0.2 μm would not be seen as two distinct objects. Because bacterial cells are about 1 μm, the cells can be resolved by the light microscope, but that is not the case for internal structures in bacterial cells that are smaller than 0.2 μm.

In order to maximize the resolving power from a lens system, the following should be considered:

- A **blue filter** should be placed over the light source because the shorter wavelength of the resulting light will provide maximum resolution.
- The condenser should be kept at the highest position that allows the maximum amount of light to enter the objective lens and therefore limit the amount of light lost due to refraction.
- The diaphragm should not be stopped down too much. While closing the diaphragm improves the contrast, it also reduces the numerical aperture.
- **Immersion oil** should be used between the slide and the 100× objective lens. This is a special oil that has the same refractive index as glass. When placed between the specimen and objective lens, the oil forms a continuous lens system that limits the loss of light due to refraction.

The bottom line is that for magnification to increase, resolution must also increase. Thus, a greater magnification cannot be achieved simply by adding a stronger ocular lens.

Lens Care

Keeping the lenses of your microscope clean is a constant concern. Unless all lenses are kept free of

dust, oil, and other contaminants, they cannot achieve the degree of resolution that is intended. Consider the following suggestions for cleaning the various lens components:

Cleaning Tissues Only lint-free, optically safe tissues should be used to clean lenses. Tissues free of abrasive grit fall in this category. Booklets of lens tissue are most widely used for this purpose. Although several types of boxed tissues are also safe, *use only the type of tissue that is recommended by your instructor* (figure 1.5).

Solvents Various liquids can be used for cleaning microscope lenses. Green soap with warm water works very well. Xylene is universally acceptable. Alcohol and acetone are also recommended, but often with some reservations. Acetone is a powerful solvent that could possibly dissolve the lens mounting cement in some objective lenses if it were used too liberally. When it is used it should be used sparingly. Your instructor will inform you as to what solvents can be used on the lenses of your microscope.

Oculars The best way to determine if your eyepiece is clean is to rotate it between the thumb and forefinger as you look through the microscope. A rotating pattern will be evidence of dirt.

If cleaning the top lens of the ocular with lens tissue fails to remove the debris, one should try cleaning the lower lens with lens tissue and blowing off any excess lint with an air syringe or gas canister. *Whenever the ocular is removed from the microscope, it is imperative that a piece of lens tissue be placed over the open end of the microscope as illustrated in figure 1.5.*

Objectives Objective lenses often become soiled by materials from slides or fingers. A piece of lens tissue moistened with green soap and water, or one of the acceptable solvents mentioned previously, will usually remove whatever is on the lens. Sometimes a cotton swab with a solvent will work better than lens tissue. At any time that the image on the slide is unclear or cloudy, assume at once that the objective you are using is soiled.

Condenser Dust often accumulates on the top surface of the condenser; thus, wiping it off occasionally with lens tissue is desirable.

Procedures

If your microscope has three objectives, you have three magnification options: (1) low-power, or 100× total magnification, (2) high-dry magnification, which is 400× total with a 40× objective, and (3) 1000× total magnification with a 100× oil immersion objective.

Whether you use the low-power objective or the oil immersion objective will depend on how much magnification is necessary. Generally speaking, however, it is best to start with the low-power objective and progress to the higher magnifications as your study progresses. Consider the following suggestions for setting up your microscope and making microscopic observations.

Low-Power Examination The main reason for starting with the low-power objective is to enable you to explore the slide to look for the object you are planning to study. Once you have found what you are looking for, you can proceed to higher magnifications. Use the following steps when exploring a slide with the low-power objective:

1. Position the slide on the stage with the material to be studied on the *upper* surface of the slide. Figure 1.6 illustrates how the slide must be held in place by the mechanical stage retainer lever.
2. Turn on the light source, using a *minimum* amount of voltage. If necessary, reposition the slide so that the stained material on the slide is in the *exact center* of the light source.
3. Check the condenser to see that it has been raised to its highest point.
4. If the low-power objective is not directly over the center of the stage, rotate it into position. Be sure that as you rotate the objective into position it clicks into its locked position.
5. Turn the coarse adjustment knob to lower the objective until it stops. A built-in stop will prevent the objective from touching the slide.
6. While looking down through the ocular (or oculars), bring the object into focus by turning the fine adjustment focusing knob. Don't readjust the

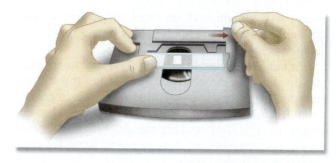

Figure 1.6 The slide must be properly positioned as the retainer lever is moved to the right.

coarse adjustment knob. If you are using a binocular microscope, it will also be necessary to adjust the interocular distance and diopter adjustment to match your eyes.

7. For optimal viewing, it is necessary to focus the condenser and adjust it for maximum illumination. This procedure should be performed each time the objective lens is changed. Raise the iris diaphragm to its highest position. Close the iris diaphragm until the edges of the diaphragm image appear fuzzy. Lower the condenser using its adjustment knob until the edges of the diaphragm are brought into sharp focus. You should now clearly see the sides of the diaphragm expand beyond the field of view. Refocus the specimen using the fine adjustment. Note that as you close the iris diaphragm to reduce the light intensity, the contrast improves and the depth of field increases. **Depth of field** is defined as the range of distance in front of and behind a focused image within which other objects will appear clear and sharply defined.
8. Once an image is visible, move the slide about to search out what you are looking for. The slide is moved by turning the knobs that move the mechanical stage.
9. Check the cleanliness of the ocular, using the procedure outlined earlier.
10. Once you have identified the structures to be studied and wish to increase the magnification, you may proceed to either high-dry or oil immersion magnification. However, before changing objectives, *be sure to center the object you wish to observe.*

High-Dry Examination To proceed from low-power to high-dry magnification, all that is necessary is to rotate the high-dry objective into position and open up the diaphragm somewhat. It may be necessary to make a minor adjustment with the fine adjustment knob to sharpen up the image, but *the coarse adjustment knob should not be touched.*

Good quality modern microscopes are usually both **parfocal** and **parcentral.** This means that the image will remain both centered and in focus when changing from a lower-power to a higher-power objective lens.

When increasing the lighting, be sure to open up the diaphragm first instead of increasing the voltage on your lamp; the reason is that *lamp life is greatly extended when used at low voltage.* If the field is not bright enough after opening the diaphragm, feel free to increase the voltage. A final point: Keep the condenser at its highest point.

Oil Immersion Techniques The oil immersion lens derives its name from the fact that a special mineral oil is interposed between the specimen and the 100× objective lens. As stated previously, this reduces light refraction and maximizes the numerical aperture to improve the resolution. The use of oil in this way enhances the resolving power of the microscope. Figure 1.4 reveals this phenomenon.

With parfocal objectives one can go directly to oil immersion from either low-power or high-dry. On some microscopes, however, going from low-power to high-power and then to oil immersion is better. Once the microscope has been brought into focus at one magnification, the oil immersion lens can be rotated into position without fear of striking the slide.

Before rotating the oil immersion lens into position, however, a drop of immersion oil must be placed on the slide. An oil immersion lens should never be used without oil. Incidentally, if the oil appears cloudy, it should be discarded.

When using the oil immersion lens, more light is necessary to adequately visualize an image. Opening the diaphragm increases the resolving power of the microscope at higher magnifications. Thus, the iris diaphragm must be opened wider when using the oil immersion lens. Also, do not forget to refocus the condenser when moving from lower-power to higher-power objectives. Some microscopes also employ blue or green filters on the lamp housing to enhance resolving power.

Since the oil immersion lens will be used extensively in all bacteriological studies, it is of paramount importance that you learn how to use this lens properly. Using this lens takes a little practice due to the difficulties usually encountered in manipulating the lighting. It is important for all beginning students to appreciate that the working distance of a lens, the distance between the lens and microscope slide, decreases significantly as the magnification of the lens increases (table 1.1). Hence, the potential for damage to the oil immersion lens because of a collision with the microscope slide is very great. A final comment of importance: At the end of the laboratory period remove all immersion oil from the lens tip with lens tissue.

Table 1.1 Relationship of Working Distance to Magnification

LENS	MAGNIFICATION	FOCAL LENGTH (mm)	WORKING DISTANCE (mm)
Low-power	10×	16.0	7.7
High-dry	40×	4.0	0.3
Oil immersion	100×	1.8	0.12

Putting It Away

When you take a microscope from the cabinet at the beginning of the period, you expect it to be clean and in proper working condition. The next person to use the instrument after you have used it will expect the same consideration. A few moments of care at the end of the period will ensure these conditions. Check over the following list of items at the end of each period before you return the microscope to the cabinet.

1. Remove the slide from the stage.
2. If immersion oil has been used, wipe it off the lens and stage with lens tissue. Also, make sure that no immersion oil is on the 40× objective. This lens often becomes contaminated with oil as a result of mistakes made by beginning students. (Do not wipe oil off slides you wish to keep. Simply put them into a slide box and let the oil drain off.)
3. Rotate the low-power objective into position.
4. If the microscope has been inclined, return it to an erect position.
5. If the microscope has a built-in movable lamp, raise the lamp to its highest position.
6. If the microscope has a long attached electric cord, wrap it around the base.
7. Adjust the mechanical stage so that it does not project too far on either side.
8. Replace the dustcover.
9. If the microscope has a separate transformer, return it to its designated place.
10. Return the microscope to its correct place in the cabinet.

Laboratory Report

Before the microscope is to be used in the laboratory, answer all the questions in Laboratory Report 1. Preparation on your part prior to going to the laboratory will greatly facilitate your understanding. Your instructor may wish to collect this report at the *beginning of the period* on the first day that the microscope is to be used in class.

Laboratory Report

Student: _____

Date: _____ Section: _____

1

1 Brightfield Microscopy

A. Short-Answer Questions

1. Describe the position of your hands when carrying the microscope to and from your laboratory bench.

    _____

    _____

2. Differentiate between the limit of resolution of the typical light microscope and that of the unaided human eye.

    _____

    _____

3. (a) What two adjustments can be made to the condenser? (b) What effect do these adjustments have on the image?

    _____

    _____

4. Why are condenser adjustments generally preferred over the use of the light intensity control?

    _____

    _____

5. When using the oil immersion lens, what four procedures can be implemented to achieve the maximum resolution?

    _____

    _____

    _____

6. Why is it advisable to start first with the low-power lens when viewing a slide?

    _____

    _____

7. Why is it necessary to use oil in conjunction with the oil immersion lens and not with the other objectives?

    _____

    _____

8. What is the relationship between the working distance of an objective lens and its magnification power?

    _____

    _____

B. Matching Questions

Match the lens (condenser, high-dry, low-power, ocular, or oil immersion) to its description. Choices may be used more than once.

1. This objective lens provides the highest magnification.
2. This objective lens provides the second-highest magnification.
3. This objective lens provides the lowest magnification.
4. This objective lens has the shortest working distance.
5. The coarse focus knob should be adjusted only when using this objective lens.
6. This lens collects and focuses light from the lamp onto the specimen on the slide.
7. This lens, also known as the eyepiece, often comes in pairs.
8. Diopter adjustments can be made to this lens.
9. A diaphragm is used to regulate light passing through this lens.

C. True-False

1. Acetone is the safest solvent for cleaning an objective lens.
2. Only lint-free, optically safe tissue should be used to wipe off microscope lenses.
3. The total magnification capability of a light microscope is only limited by the magnifying power of the lens system.
4. The coarse focus knob can be used to adjust the focus when using any of the objective lenses.
5. Once focus is achieved at one magnification, a higher-power objective lens can be rotated into position without fear of striking the slide.

D. Multiple Choice

Select the answer that best completes the following statements.

1. The resolving power of a microscope is a function of
   a. the magnifying power of the lenses.
   b. the numerical aperture of the lenses.
   c. the wavelength of light.
   d. Both (a) and (b) are correct.
   e. Both (b) and (c) are correct.

2. The coarse and fine focus knobs adjust the distance between
   a. the objective and ocular lenses.
   b. the ocular lenses.
   c. the ocular lenses and your eyes.
   d. the stage and the condenser lens.
   e. the stage and the objective lens.

3. A microscope that maintains focus when the objective magnification is increased is called
   a. binocular.
   b. myopic.
   c. parfocal.
   d. refractive.
   e. resolute.

ANSWERS

Matching

1. _____
2. _____
3. _____
4. _____
5. _____
6. _____
7. _____
8. _____
9. _____

True-False

1. _____
2. _____
3. _____
4. _____
5. _____

Multiple Choice

1. _____
2. _____
3. _____

4. The total magnification achieved when using a 100× oil immersion lens with 10× binocular eyepieces is
   a. 10×.
   b. 100×.
   c. 200×.
   d. 1000×.
   e. 2000×.

5. The most useful adjustment for increasing image contrast in low-power magnification is
   a. closing down the diaphragm.
   b. closing one eye.
   c. opening up the diaphragm.
   d. placing a drop of oil on the slide.
   e. using a blue filter.

6. Before the oil immersion lens is rotated into place, you should
   a. center the object of interest in the preceding lens.
   b. lower the stage with use of the coarse focus adjustment knob.
   c. place a drop of oil on the slide.
   d. Both (a) and (c) are correct.
   e. All are correct.

4. _____

5. _____

6. _____

Darkfield Microscopy

Delicate transparent living organisms can be more easily observed with darkfield microscopy than with conventional brightfield microscopy. This method is particularly useful when one is attempting to identify spirochaetes in an exudate from a syphilitic lesion. Figure 2.1 illustrates the appearance of these organisms under such illumination. This effect may be produced by placing a darkfield stop below the regular condenser or by replacing the condenser with a specially constructed one.

Another application of darkfield microscopy is in the fluorescence microscope. Although fluorescence may be seen without a dark field, it is greatly enhanced with this application.

To achieve the darkfield effect it is necessary to alter the light rays that approach the objective in such a way that only oblique rays strike the objects being viewed. The obliquity of the rays must be so extreme that if no objects are in the field, the background is completely light-free. Objects in the field become brightly illuminated by the rays that are reflected up through the lens system of the microscope.

Although there are several different methods for producing a dark field, only two devices will be described here: the star diaphragm and the cardioid condenser. The availability of equipment will determine the method to be used in this laboratory.

The Star Diaphragm

One of the simplest ways to produce the darkfield effect is to insert a star diaphragm into the filter slot of the condenser housing, as shown in figure 2.2. This device has an opaque disk in the center that blocks the central rays of light. Figure 2.3 reveals the effect of this stop on the light rays passing through the condenser. If such a device is not available, one can be made by cutting round disks of opaque paper of different sizes that are cemented to transparent celluloid disks that will fit into the slot. If the microscope normally has a diffusion disk in this slot, it is best to replace it with rigid clear celluloid or glass.

An interesting modification of this technique is to use colored celluloid stops instead of opaque paper. Backgrounds of blue, red, or any color can be produced in this way.

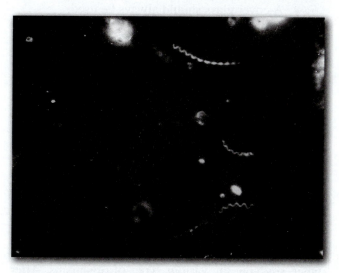

Figure 2.1 Darkfield image of *Treponema pallidum*, the bacterium that causes syphilis.
Source: Susan Lindsley/Centers for Disease Control and Prevention

Figure 2.2 The insertion of a star diaphragm into the filter slot of the condenser will produce a dark field suitable for low magnifications.
© McGraw-Hill Education/Auburn University Photographics Services

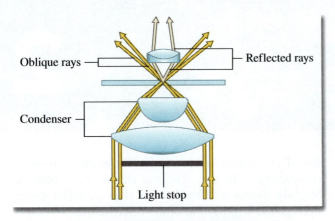

Figure 2.3 The star diaphragm allows only peripheral light rays to pass up through the condenser. This method requires maximum illumination.

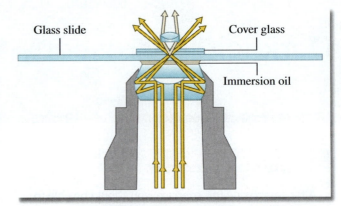

Figure 2.4 A cardioid condenser provides greater light concentration for oblique illumination than the star diaphragm.

In setting up this type of darkfield illumination, it is necessary to keep these points in mind:

1. Limit this technique to the study of large organisms that can be seen easily with low-power magnification. *Good resolution with higher-powered objectives is difficult with this method.*
2. Keep the diaphragm wide open and use as much light as possible. If the microscope has a voltage regulator, you will find that the higher voltages will produce better results.
3. Be sure to center the stop as precisely as possible.
4. Move the condenser up and down to produce the best effects.

The Cardioid Condenser

The difficulty that results from using the star diaphragm or opaque paper disks with high-dry and oil immersion objectives is that the oblique rays are not as carefully metered as is necessary for the higher magnifications. Special condensers such as the cardioid or paraboloid types must be used. Since the cardioid type is the most frequently used type, its use will be described here.

Figure 2.4 illustrates the light path through such a condenser. Note that the light rays entering the lower element of the condenser are reflected first off a convex mirrored surface and then off a second concave surface to produce the desired oblique rays of light. Once the condenser has been installed in the microscope, the following steps should be followed to produce ideal illumination.

Materials

- slides and cover glasses of excellent quality (slides of 1.15–1.25 mm thickness and No. 1 cover glasses)

1. Adjust the upper surface of the condenser to a height just below stage level.
2. Place a clear glass slide in position over the condenser.
3. Focus the 10× objective on the top of the condenser until a bright ring comes into focus.

4. Center the bright ring so that it is concentric with the field edge by adjusting the centering screws on the darkfield condenser. If the condenser has a light source built into it, it will also be necessary to center it as well to achieve even illumination.
5. Remove the clear glass slide.
6. If a funnel stop is available for the oil immersion objective lens, remove the oil immersion objective and insert the funnel stop. (This stop serves to reduce the numerical aperture of the oil immersion objective to a value that is less than that of the condenser.)
7. Place a drop of immersion oil on the upper surface of the condenser and place the slide on top of the oil. The following preconditions in slide usage must be adhered to:

   - Slides and cover glasses should be optically perfect. Scratches and imperfections will cause annoying diffractions of light rays.
   - Slides and cover glasses must be free of dirt or grease of any kind.
   - A cover glass should always be used.

8. If the oil immersion lens is to be used, place a drop of oil on the cover glass.
9. If the field does not appear dark and lacks contrast, return to the 10× objective and check the ring concentricity and light source centration. If contrast is still lacking after these adjustments, the specimen is probably too thick.
10. If sharp focus is difficult to achieve under oil immersion, try using a thinner cover glass and adding more oil to the top of the cover glass and bottom of the slide.

Laboratory Report

This exercise may be used in conjunction with Part 2 when studying the various types of organisms. After reading over this exercise and doing any special assignments made by your instructor, answer the questions in Laboratory Report 2 about darkfield microscopy.

2 Darkfield Microscopy

A. Short-Answer Questions

1. For which types of specimens is darkfield microscopy preferred over brightfield microscopy?

   _____

   _____

   _____

2. If a darkfield condenser causes all light rays to bypass the objective, where does the light come from that makes an object visible in a dark field?

   _____

   _____

   _____

3. What advantage does a cardioid condenser have over a star diaphragm?

   _____

   _____

   _____

4. If accessories for darkfield microscopy were not available, how would you construct a simple star diaphragm?

   _____

   _____

   _____

Phase-Contrast Microscopy

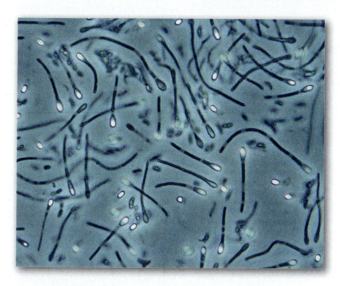

Figure 3.1 Phase-contrast image of an endospore-forming bacterium.
© Stephen Durr

If one tries to observe cells without the benefit of staining, very little contrast or detail can be seen because the cells appear transparent against the aqueous medium in which they are usually suspended. Staining increases the contrast between the cell and its surrounding medium, allowing the observer to see more cellular detail, including some inclusions and various organelles. However, staining usually results in cell death, which means we are unable to observe living cells or their activities, and staining can also lead to undesirable artifacts.

A microscope that is able to differentiate the transparent protoplasmic structures and enhance the contrast between a cell and its surroundings, without the necessity of staining, is the **phase-contrast microscope.** This microscope was developed by the Dutch physicist Frits Zernike in the 1930s. For his discovery of phase-contrast microscopy, he was awarded the Nobel Prize in 1953. Today it is the microscope of choice for viewing living cells and their activities such as motility. Figure 3.1 illustrates a phase-contrast image of an endospore-forming bacterium. In this exercise, you will learn to use the phase-contrast microscope and observe the activities of living cells.

To understand how a phase-contrast microscope works, it is necessary to review some of the physical properties of light and how it interacts with matter such as biological material. Light energy can be represented as a waveform that has both an amplitude and a characteristic wavelength (illustration 1, figure 3.2). Some objects can reduce the amplitude of a light wave, and they would appear as dark objects in a microscope. In contrast, light can pass through matter without affecting the amplitude, and these objects would

appear transparent in a microscope. However, as light passes through some of the transparent objects, it can be slowed down by $\frac{1}{4}$ wavelength, resulting in a **phase shift** of the light's wavelength (illustration 2, figure 3.2). For a cell, the phase shift without a reduction in amplitude results in the cell having a different refractive index than its surroundings. However, the phase shifts caused by biological material are usually too small to be seen as contrast differences in a brightfield microscope. Therefore, in a brightfield microscope, cells appear transparent rather than opaque against their surroundings. Since biological material lacks any appreciable contrast, it becomes necessary to stain cells with various dyes in order to study them. However, Zernike took advantage of the $\frac{1}{4}$ wavelength phase shift to enhance the small contrast differences in the various components that comprise a cell, making them visible in his microscope. This involved manipulating the light rays that were shifted and those that were unchanged as they emerged from biological material.

Two Types of Light Rays

Light rays passing through a transparent object emerge as either direct or diffracted rays. Those rays that pass straight through unaffected by the medium are called

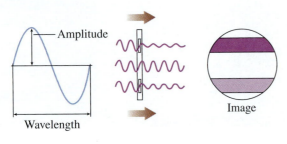

AMPLITUDE OBJECTS

(1) The extent to which the amplitude of light rays is diminished determines the darkness of an object in a microscopic field.

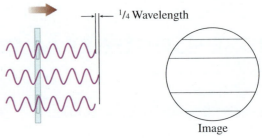

PHASE OBJECTS

(2) Note that the retardation of light rays without amplitude diminution results in transparent phase objects.

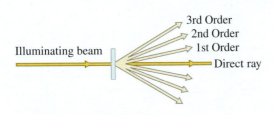

DIRECT AND DIFFRACTED RAYS

(3) A light ray passing through a slit or transparent object emerges as a direct ray with several orders of diffracted rays. The diffracted rays are $\frac{1}{4}$ wavelength out of phase with the direct ray.

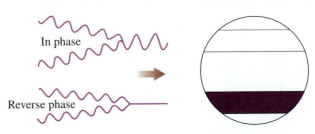

COINCIDENCE AND INTERFERENCE

(4) Note that when two light rays are in phase they will unite to produce amplitude summation. Light rays in reverse phase, however, cancel each other (interference) to produce dark objects.

Figure 3.2 **The utilization of light rays in phase-contrast microscopy.**

direct rays. They are unaltered in amplitude and phase. The rays that are bent because they are retarded by the medium (due to density differences) emerge from the object as **diffracted rays.** It is these specific light rays that are retarded $\frac{1}{4}$ wavelength. Illustration 3, figure 3.2, shows these two types of light rays.

If the direct and diffracted light waves are brought into exact phase with each other, the result is **coincidence** with the resultant amplitude of the converged waves being the sum of the two waves. This increase in amplitude will produce increased brightness of the object in the field. In contrast, if two light waves of equal amplitude are in reverse phase ($\frac{1}{2}$ wavelength off), their amplitudes will cancel each other to produce a dark object. This is called **interference.** Illustration 4, figure 3.2, shows these two conditions.

Phase-Contrast Microscope

In constructing his first phase-contrast microscope, Zernike experimented with various configurations of diaphragms and various materials that could be used to retard or advance the direct light rays. Figure 3.3 illustrates the optical system of a typical modern phase-contrast microscope. It differs from a conventional brightfield microscope by having (1) a different type of diaphragm and (2) a phase plate.

The diaphragm consists of an **annular stop** that allows only a hollow cone of light rays to pass through the condenser to the specimen on the slide. The phase plate is a special optical disk located on the objective lens near its rear focal plane. It has a phase ring that advances or retards the direct light rays $\frac{1}{4}$ wavelength.

Note in figure 3.3 that the direct rays converge on the phase ring to be advanced or retarded $\frac{1}{4}$ wavelength. These rays emerge as solid lines from the object on the slide. This ring on the phase plate is coated with a material that will produce the desired phase shift. The diffracted rays, on the other hand, which have already been retarded $\frac{1}{4}$ wavelength by the phase object on the slide, completely miss the phase ring and are not affected by the phase plate. It should be clear, then, that depending on the type of phase-contrast microscope, the convergence of diffracted and direct rays on the image plane will result in either a brighter image (*amplitude summation*) or a darker image (*amplitude interference* or *reverse phase*). The former is referred to as ***bright-phase*** microscopy;

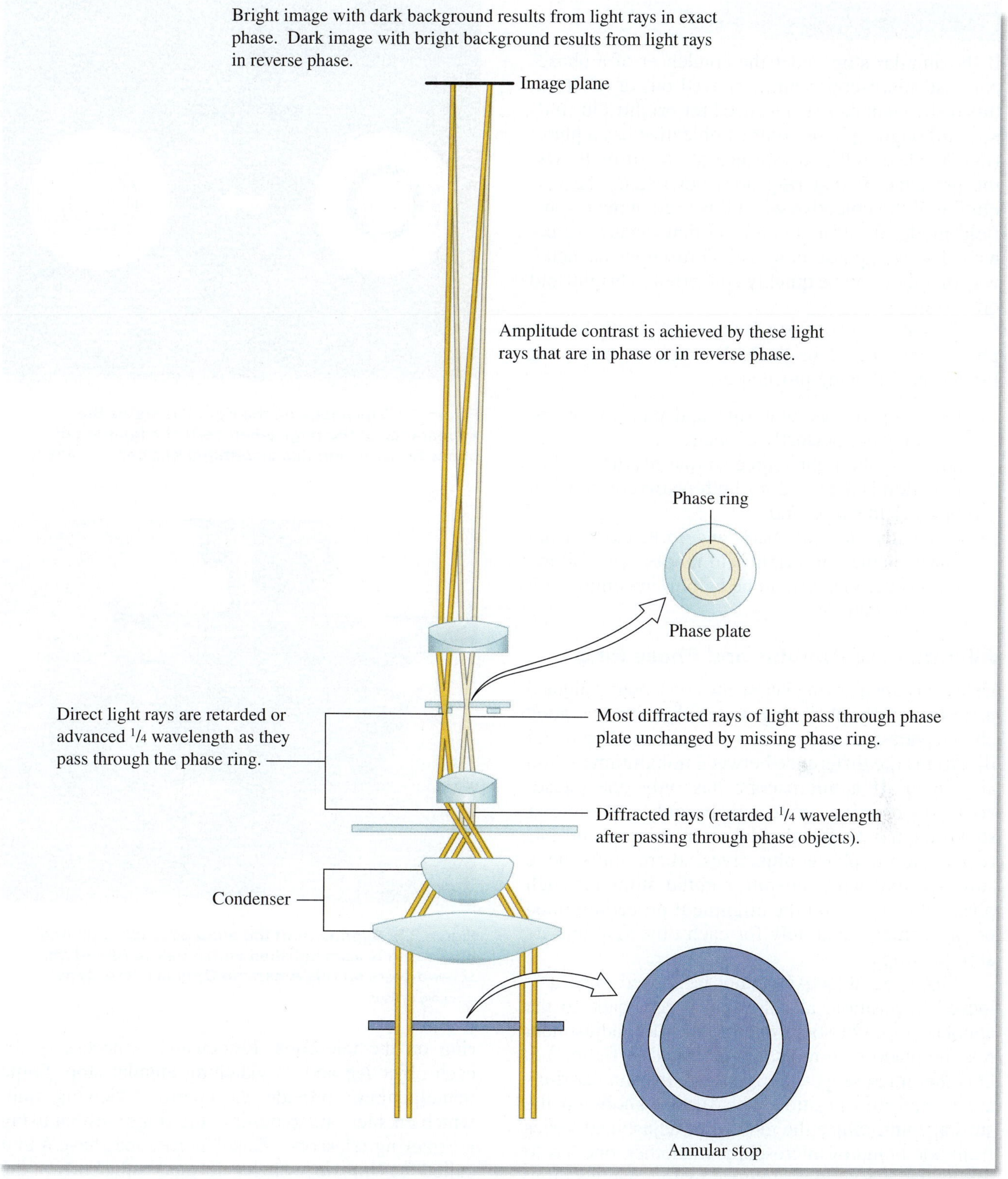

Bright image with dark background results from light rays in exact phase. Dark image with bright background results from light rays in reverse phase.

Image plane

Amplitude contrast is achieved by these light rays that are in phase or in reverse phase.

Phase ring

Phase plate

Direct light rays are retarded or advanced $1/4$ wavelength as they pass through the phase ring.

Most diffracted rays of light pass through phase plate unchanged by missing phase ring.

Diffracted rays (retarded $1/4$ wavelength after passing through phase objects).

Condenser

Annular stop

Figure 3.3 **The optical system of a phase-contrast microscope.**

the latter as **_dark-phase_** microscopy. The apparent brightness or darkness, incidentally, is proportional to the square of the amplitude; thus, the image will be four times as bright or dark as one seen through a brightfield microscope.

It should be added here that the phase plates of some microscopes have coatings to change the phase of the diffracted rays. In any event, the end result will be the same: to achieve coincidence or interference of direct and diffracted rays.

Microscope Adjustments

If the annular stop under the condenser of a phase-contrast microscope can be moved out of position, this instrument can also be used for brightfield studies. Although a phase-contrast objective has a phase ring attached to the top surface of one of its lenses, the presence of that ring does not impair the resolution of the objective when it is used in the brightfield mode. It is for this reason that manufacturers have designed phase-contrast microscopes in such a way that they can be quickly converted to brightfield operation.

To make a microscope function efficiently in both phase-contrast and brightfield situations, one must master the following procedures:

- lining up the annular ring and phase rings so that they are perfectly concentric,
- adjusting the light source so that maximum illumination is achieved for both phase-contrast and brightfield usage, and
- being able to shift back and forth easily from phase-contrast to brightfield modes. The following suggestions should be helpful in coping with these problems.

Alignment of Annulus and Phase Ring

Unless the annular ring below the condenser is aligned perfectly with the phase ring in each objective lens, good phase-contrast imagery cannot be achieved. Figure 3.4 illustrates the difference between nonalignment and alignment. If a microscope has only one phase-contrast objective, there will be only one annular stop that has to be aligned. If a microscope has two or more phase objectives, there must be a substage unit with separate annular stops for each phase objective, and the alignment procedure must be performed separately for each objective and its annular stop.

Since the objective cannot be moved once it is locked in position, all adjustments are made to the annular stop. On some microscopes the adjustment may be made with tools, as illustrated in figure 3.5. On other microscopes, in figure 3.6, the annular rings are moved into position with special knobs on the substage unit. Since the method of adjustment varies from one brand of microscope to another, one has to follow the instructions provided by the manufacturer. Once the adjustments have been made, they are rigidly set and needn't be changed unless someone inadvertently disturbs them.

To observe ring alignment, one can replace the eyepiece with a **centering telescope,** as shown in figure 3.7. With this unit in place, the two rings can be brought into sharp focus by rotating the focusing

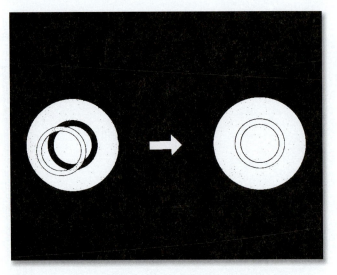

Figure 3.4 **The image on the right illustrates the appearance of the rings when perfect alignment of phase ring and annulus diaphragm has been achieved.**

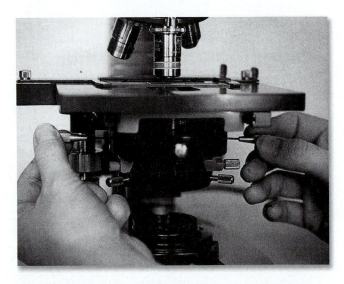

Figure 3.5 **Alignment of the annulus diaphragm and phase ring is accomplished with a pair of Allen-type screwdrivers on this American Optical microscope.**
© Harold Benson

ring on the telescope. Refocusing is necessary for each objective and its matching annular stop. Some manufacturers provide an aperture viewing unit, which enables one to observe the rings without using a centering telescope. Zeiss microscopes have a unit called the *Optovar,* which is located in the body of the microscope.

Light Source Adjustment

For both brightfield and phase-contrast modes, it is essential that optimum lighting be achieved. For multiple-phase objective microscopes, however, there are many more adjustments that need to be made.

Figure 3.6 Alignment of the annulus and phase ring is achieved by adjusting the two knobs as shown.
© McGraw-Hill Education/Auburn University Photographics Services

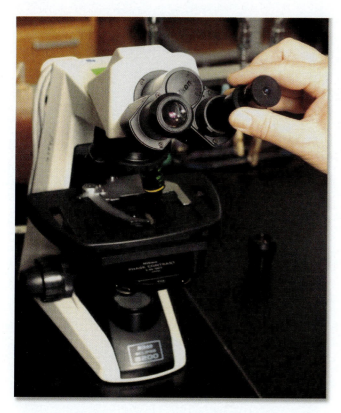

Figure 3.7 If the ocular of a phase-contrast microscope is replaced with a centering telescope, the orientation of the phase ring and annular ring can be viewed.
© McGraw-Hill Education/Auburn University Photographics Services

A few suggestions that highlight some of the problems and solutions follow:

1. Since blue light provides better images for both phase-contrast and brightfield modes, make certain that a blue filter is placed in the filter holder that is positioned in the light path. If the microscope has no filter holder, placing the filter over the light source on the base will help.

2. Brightness of field under phase-contrast is controlled by adjusting the voltage or the iris diaphragm on the base. Considerably more light is required for phase-contrast than for brightfield since so much light is blocked out by the annular stop.

3. The evenness of illumination on some microscopes, seen on these pages, can be adjusted by removing the lamp housing from the microscope and focusing the light spot on a piece of translucent white paper. For the detailed steps in this procedure, one should consult the instruction manual that comes with the microscope. Light source adjustments of this nature are not necessary for the simpler types of microscopes.

Working Procedures

Once the light source is correct and the phase elements are centered you are finally ready to examine slide preparations. Keep in mind that from now on most of the adjustments described earlier should not be altered; however, if misalignment has occurred due to mishandling, it will be necessary to refer back to the alignment procedures. The following guidelines should be adhered to in all phase-contrast studies:

- Use only optically perfect slides and cover glasses (no bubbles or striae in the glass).
- Be sure that slides and cover glasses are completely free of grease or chemicals.
- Use wet mount slides instead of hanging drop preparations. The latter leave much to be desired. Culture broths containing bacteria or protozoan suspensions are ideal for wet mounts.
- In general, limit observations to living cells. In most instances, stained slides are not satisfactory.

The first time you use phase-contrast optics to examine a wet mount, follow these suggestions:

1. Place the wet mount slide on the stage and bring the material into focus, *using brightfield optics* at low-power magnification.

2. Once the image is in focus, switch to phase optics at the same magnification. Remember, it is necessary to place in position the matching annular stop.

3. Adjust the light intensity, first with the base diaphragm and then with the voltage regulator. In

most instances, you will need to increase the amount of light for phase-contrast.

4. Switch to higher magnifications, much in the same way you do for brightfield optics, except that you have to rotate a matching annular stop into position.

5. If an oil immersion phase objective is used, add immersion oil to the top of the condenser as well as to the top of the cover glass.

6. Don't be disturbed by the "halo effect" that you observe with phase optics. Halos are normal.

Laboratory Report

This exercise may be used in conjunction with Part 2 in studying various types of organisms. Organelles in protozoans and algae will show up more distinctly than with brightfield optics. After reading this exercise and doing any special assignments made by your instructor, answer the questions in Laboratory Report 3.

Laboratory Report

3

3 Phase-Contrast Microscopy

A. Short-Answer Questions

1. Staining of cells is often performed to enhance images acquired by brightfield microscopy. Phase-contrast microscopy does not require cell staining. Why is this advantageous?

    _____

    _____

2. As light passes through a transparent object, how are direct and diffracted light rays produced? How much phase shift occurs?

    _____

    _____

3. How do coincidence and interference of light rays differ? What type of image does each produce? How does that contribute to a sharper image?

    _____

    _____

    _____

4. Differentiate between bright-phase and dark-phase microscopy in terms of phase shift.

    _____

    _____

    _____

5. Which two items can be used to check the alignment of the annulus and phase ring?

    _____

    _____

B. Multiple Choice

Select the answer that best completes the following statements.

1. A phase-contrast microscope differs from a brightfield microscope by having a
    a. blue filter in the ocular lens.
    b. diaphragm with an annular stop.
    c. phase plate in the objective lens.
    d. Both (b) and (c) are correct.
    e. All are correct.

ANSWERS

Multiple Choice

1. _____

2. Which of the following would be best observed for a bacterial cell using phase-contrast microscopy?
   a. Motility of cells
   b. Bacterial nucleoid
   c. Cell wall
   d. Glycocalyx

3. Amplitude summation occurs in phase-contrast optics when both direct and diffracted rays are
   a. in phase.
   b. in reverse phase.
   c. off $\frac{1}{4}$ wavelength.
   d. None of these is correct.

4. The phase-contrast microscope is best suited for observing
   a. living organisms in an uncovered drop on a slide.
   b. stained slides with cover glasses.
   c. living organisms in hanging drop slide preparations.
   d. living organisms on a slide with a cover glass.

2. _____

3. _____

4. _____

Microscopic Measurements

Learning Outcomes

After completing this exercise, you should be able to

1. Calibrate an ocular micrometer using a stage micrometer.

2. Use an ocular micrometer to accurately measure the dimensions of stained cells.

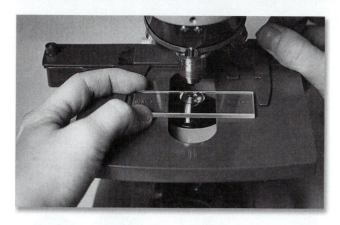

Figure 4.2 Stage micrometer is positioned by centering the small glass disk over the light source.
© Harold Benson

Calibration Procedure

The distance between the lines of an ocular micrometer is an arbitrary value that has meaning only if the ocular micrometer is calibrated for the objective that is being used. A **stage micrometer** (figure 4.2), also known as an *objective micrometer,* has lines inscribed on it that are exactly 0.01 mm (10 μm) apart. Illustration (c), figure 4.4, shows these graduations.

To calibrate the ocular micrometer for a given objective, it is necessary to superimpose the two scales and determine how many of the ocular graduations coincide with one graduation on the scale of the stage micrometer. Illustration (a) in figure 4.4 shows how the two scales appear when they are properly aligned in the microscopic field. In this case, seven ocular divisions match up with one stage micrometer division of 0.01 mm to give an ocular value of 0.01/7, or 0.00143 mm. Since there are 1000 micrometers in 1 millimeter, these divisions are 1.43 μm apart.

With this information known, the stage micrometer is replaced with a slide of organisms (figure 4.3) to be measured. Illustration (d), figure 4.4, shows how a field of microorganisms might appear with the ocular micrometer in the eyepiece. To determine the size of an organism, then, it is a simple matter to count the graduations and multiply this number by the known distance between the graduations. When calibrating the objectives of a microscope, proceed as follows.

Figure 4.1 The ocular micrometer with retaining ring is inserted into the base of the eyepiece.
© Harold Benson

With an ocular micrometer properly installed in the eyepiece of your microscope, it is a simple matter to measure the size of microorganisms that are seen in the microscopic field. An **ocular micrometer** consists of a circular disk of glass that has graduations engraved on its upper surface. These graduations appear as shown in illustration (b), figure 4.4. On some microscopes one has to disassemble the ocular so that the disk can be placed on a shelf in the ocular tube between the two lenses. On most microscopes, however, the ocular micrometer is simply inserted into the bottom of the ocular, as shown in figure 4.1. Before one can use the micrometer, it is necessary to calibrate it for each of the objectives by using a stage micrometer.

The principal purpose of this exercise is to show you how to calibrate an ocular micrometer for the various objectives on your microscope.

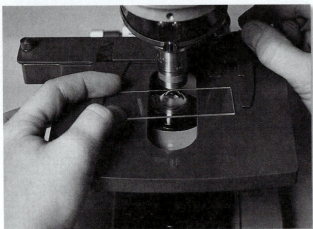

Figure 4.3 After calibration is completed, the stage micrometer is replaced with a slide for measurements.
© Harold Benson

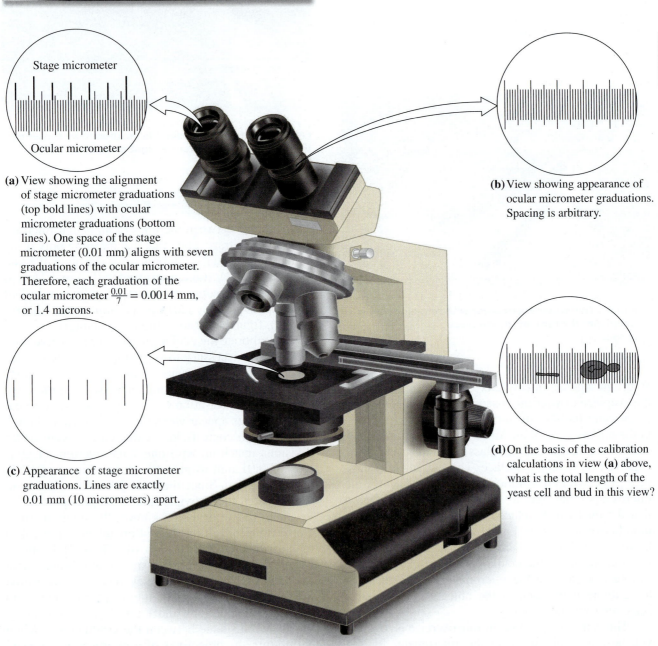

Stage micrometer

Ocular micrometer

(a) View showing the alignment of stage micrometer graduations (top bold lines) with ocular micrometer graduations (bottom lines). One space of the stage micrometer (0.01 mm) aligns with seven graduations of the ocular micrometer. Therefore, each graduation of the ocular micrometer $\frac{0.01}{7} = 0.0014$ mm, or 1.4 microns.

(b) View showing appearance of ocular micrometer graduations. Spacing is arbitrary.

(c) Appearance of stage micrometer graduations. Lines are exactly 0.01 mm (10 micrometers) apart.

(d) On the basis of the calibration calculations in view **(a)** above, what is the total length of the yeast cell and bud in this view?

Figure 4.4 Calibration of ocular micrometer.

Materials

- ocular micrometer or eyepiece that contains a micrometer disk
- stage micrometer

1. If eyepieces are available that contain ocular micrometers, replace the eyepiece in your microscope with one of them. If it is necessary to insert an ocular micrometer in your eyepiece, find out from your instructor whether it is to be inserted below the bottom lens or placed between the two lenses within the eyepiece. In either case, great care must be taken to avoid dropping the eyepiece or reassembling the lenses incorrectly. *Only with your instructor's prior approval shall eyepieces be disassembled.* Be sure that the graduations are on the upper surface of the glass disk.
2. Place the stage micrometer on the microscope stage and center it exactly over the light source.
3. With the low-power (10×) objective in position, bring the graduations of the stage micrometer into focus, *using the coarse adjustment knob. Reduce the lighting.* Note: If the microscope has an automatic stop, do not use it as you normally would for regular microscope slides. The stage micrometer slide is too thick to allow it to function properly.
4. Rotate the eyepiece until the graduations of the ocular micrometer lie parallel to the lines of the stage micrometer.
5. If the **low-power objective** is the objective to be calibrated, proceed to step 8.
6. If the **high-dry objective** is to be calibrated, swing it into position and proceed to step 8.
7. If the **oil immersion lens** is to be calibrated, place a drop of immersion oil on the stage micrometer, swing the oil immersion lens into position, and bring the lines into focus; then, proceed to the next step.
8. Move the stage micrometer laterally until the lines at one end coincide. Then look for another line on the ocular micrometer that coincides *exactly* with one on the stage micrometer. Occasionally, one stage micrometer division will include an even number of ocular divisions, as shown in illustration (a) of figure 4.4. In most instances, however, several stage graduations will be involved. In this case, divide the number of stage micrometer divisions by the number of ocular divisions that coincide. The figure you get will be that part of a stage micrometer division that is seen in an ocular division. This value must then be multiplied by 0.01 mm to get the amount of each ocular division.

Example: 3 divisions of the stage micrometer line up with 20 divisions of the ocular micrometer.

$$\text{Each ocular division} = \tfrac{3}{20} \times 0.01$$
$$= 0.0015 \text{ mm}$$
$$= 1.5 \text{ }\mu\text{m}$$

Replace the stage micrometer with slides of organisms to be measured.

Measuring Assignments

Organisms such as protozoa, algae, fungi, and bacteria in the next few exercises may need to be measured. If your instructor requires that measurements be made, you will be referred to this exercise.

Later on you will be working with unknowns. In some cases measurements of the unknown organisms will be pertinent to identification.

If trial measurements are to be made at this time, your instructor will make appropriate assignments. **Important:** Remove the ocular micrometer from your microscope at the end of the laboratory period.

Laboratory Report

Answer the questions in Laboratory Report 4.

Laboratory Report

4 Microscopic Measurements

A. Short-Answer Questions

1. How do the graduations differ between ocular and stage micrometers?

   _____

   _____

2. If 13 ocular divisions line up with two divisions of the stage micrometer, what is the diameter (μm) of a cell that spans 16 ocular divisions?

   _____

   _____

3. Why must the entire calibration procedure be performed for each objective?

   _____

   _____

Survey of Microorganisms

Microorganisms abound in the environment. Eukaryotic microbes such as protozoa, algae, diatoms, and amoebas are plentiful in ponds and lakes. Bacteria are found associated with animals, occur abundantly in the soil and in water systems, and have even been isolated from core samples taken from deep within the earth's crust. Bacteria are also present in the air where they are distributed by convection currents that transport them from other environments. The Archaea, modern-day relatives of early microorganisms, occupy some of the most extreme environments such as acidic-volcanic hot springs, anaerobic environments devoid of any oxygen, and lakes

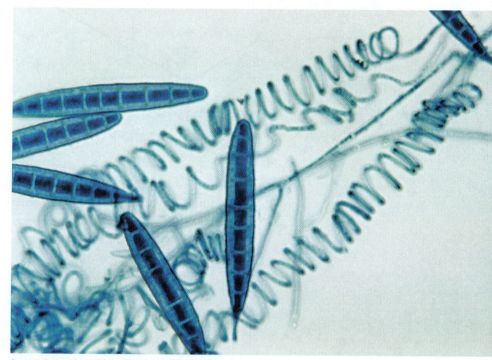

Source: Dr. Lucille K. Georg/Centers for Disease Control

and salt marshes excessively high in sodium chloride. Cyanobacteria are photosynthetic prokaryotes that can be found growing in ponds and lakes, on limestone rocks, and even on the shingles that protect the roofs of our homes. Fungi are a very diverse group of microorganisms that are found in most common environments. For example, they degrade complex molecules in the soil, thus contributing to its fertility. Sometimes, however, they can be nuisance organisms; they form mildew in our bathroom showers and their spores cause allergies. The best way of describing the distribution of microorganisms is to say that they are ubiquitous, or found everywhere.

Intriguing questions to biologists are how are the various organisms related to one another and where do the individual organisms fit in an evolutionary scheme? Molecular biology techniques have provided a means to analyze the genetic relatedness of the organisms that comprise the biological world and determine where the various organisms fit into an evolutionary scheme. By comparing the sequence of ribosomal RNA molecules, coupled with biochemical data, investigators have developed a phylogenetic tree that illustrates the current thinking on the placement of the various organisms into such a scheme. This evolutionary scheme divides the biological world into three domains.

Domain Bacteria   These organisms have a prokaryotic cell structure. They lack organelles such as mitochondria and chloroplasts, are devoid of an organized nucleus with a nuclear membrane, and possess 70S ribosomes that are inhibited by many broad-spectrum antibiotics. The vast majority of organisms are enclosed in a cell wall composed of peptidoglycan. The bacteria and cyanobacteria are members of this domain.

Bacteria   **Archaea**   **Eukarya**

Domain Eukarya   Organisms in this domain have a eukaryotic cell structure. They contain membrane-bound organelles such as mitochondria and chloroplasts, an organized nucleus enclosed in a nuclear membrane, and 80S ribosomes that are not inhibited by broad-spectrum antibiotics. Plants, animals, and microorganisms such as protozoa, algae, and fungi belong in this domain. Plants have cell walls composed of cellulose and fungi have cell walls composed of chitin. In contrast, animal cells lack a cell wall structure.

Domain Archaea   The Archaea exhibit the characteristics of both the Bacteria and Eukarya. These organisms are considered to be the relatives of ancient microbes that existed during Archaean times. Like their bacterial counterparts, they possess a simple cell structure that lacks organelles and an organized nucleus. They have 70S ribosomes like bacteria, but the protein makeup and morphology of their ribosomes are more similar to eukaryotic ribosomes. Like eukaryotes, the ribosomes in Archaea are not sensitive to antibiotics. They have a cell wall but its structure is not composed of peptidoglycan. The principal habitats of these organisms are extreme environments suchas volcanic hot springs, environments with excessively high salt, and environments devoid of oxygen. Thus, they are referred to as "extremophiles." The acido-thermophiles, the halobacteria, and the methanogens (methane bacteria) are examples of the Archaea.

In the exercises of Part 2, you will have the opportunity to study some of these organisms. In pond water, you may see amoebas, protozoa, various algae, diatoms, and cyanobacteria. You will sample for the presence of bacteria by exposing growth media to various environments. The fungi will be studied by looking at cultures and preparing slides of these organisms. Because the Archaea occur in extreme conditions and also require specialized culture techniques, it is unlikely that you will encounter any of these organisms.

MICROBIOLOGY

Microbiology of Pond Water—Protists, Algae, and Cyanobacteria

Learning Outcomes

After completing this exercise, you should be able to

1. Prepare wet mounts of samples from aquatic environments.

2. Identify various eukaryotic microorganisms belonging to the algae and protozoa groups, and differentiate them from the prokaryotic cyanobacteria.

3. Differentiate between the prokaryotic cyanobacteria and the eukaryotic algae.

In this exercise you will study the diverse organisms that can occur in pond water. This will include eukaryotic organisms classified as algae and the protozoa, as well as the prokaryotic cyanobacteria. Illustrations and descriptions are provided to assist you in identifying some of these organisms. In addition, you may also observe small nematodes, insect larvae, microcrustaceans, rotifers, and other invertebrates. Your instructor may supply you with additional materials to identify these organisms, as they are not covered in this manual.

The purpose of this exercise is to familiarize you with the diversity of organisms that occur in a pond habitat. Most of your identifications will involve comparisons of morphological characteristics and differences. To study these organisms, you will make wet mounts of various samples of pond water. This is achieved by pipetting a drop of pond water onto a microscope slide and covering the drop with a coverslip. To achieve the best results, use the following guidelines:

1. Use a clean slide and coverslip.
2. Insert the pipette into the bottom of the sample bottle to obtain the maximum number of organisms. Fewer will occur at mid-depth.
3. Remove filamentous algae using forceps. Avoid using too much material.
4. Focus first with the low-power objective. Reduce the illumination using the iris diaphragm to provide better contrast.

5. When you find an organism of interest, switch to the high-dry objective and adjust the illumination appropriately.
6. Refer to figures 5.1 to 5.7 and the accompanying text to help in your identification.
7. Record your observations on the Laboratory Report.

Materials

- bottles of pond water
- microscope slides and cover glasses
- Pasteur pipettes, pipette aids, and forceps
- marking pen
- additional reference books
- prepared slides of protozoa, amoeba, and algae

Survey of Organisms

You will likely see a variety of eukaryotic microorganisms as well as possibly cyanobacteria in pond water used in this exercise. You will identify and categorize these organisms based on differences in morphological characteristics. Traditionally, morphology was a primary means for constructing formal classification schemes for these organisms. However, true evolutionary relationships cannot be determined by simply observing morphological characteristics. Nowadays molecular genetics is used to more clearly establish the taxonomic and evolutionary relationships among these organisms. Observable morphological differences do exist between these organisms, and therefore you can use an informal system based on morphology to characterize and categorize the various organisms that you observe. Table 5.1 will help you understand some of the major morphological groups of the organisms that you may see. **Please keep in mind that these are not formally recognized taxonomic groups,** but they are useful for identifying and categorizing organisms in the laboratory based on specific physical traits.

Eukaryotes

Protozoa

The Protozoa are protists, which means "first animals." They were some of the original microorganisms observed by Leeuwenhoek over 300 years ago

Table 5.1   **Classification of Organisms**

PROKARYOTES (DOMAIN BACTERIA)	EUKARYOTES (DOMAIN EUKARYA)

Cyanobacteria (figure 5.7)	Protozoa
	    Diplomonads and Parabasalids
	    Euglenozoans
	        Euglenids (figure 5.2, illustrations 1, 2)
	        Kinetoplastids
	    Alveolates
	        Ciliates
	        Dinoflagellates (figure 5.3, illustrations 10, 11)
	        Apicomplexans
	    Stramenophiles
	        Diatoms (figure 5.6)
	        Golden algae (figure 5.3, illustrations 1–4)
	    Amoebozoans
	        Gymnamoebas (figure 5.1, illustration 6)
	        Entamoebas
	Algae
	    Red and green algae
	        Unicellular red algae
	        Unicellular green algae (figures 5.2, 5.3, 5.4)

when he studied samples of rain water with his simple microscopes. Protozoa were initially classified as unicellular, colorless organisms, lacking a cell wall and were further subdivided based on their modes of motility. However, the application of modern genetic analysis has shown that these organisms are a diverse group that have little relationship to one another. They are mostly heterotrophic, unicellular organisms that lack a rigid cellulose cell wall characteristic of algae or a chitinous cell wall characteristic of fungi. Their specialized structures for motility such as cilia, flagella, or pseudopodia were important in previous classification schemes, but genetic analysis subdivides them into the following groups with the accompanying characteristics:

Diplomonads and Parabasalids   These organisms contain two nuclei and organelles called **mitosomes,** which are reduced mitochondria that lack electron transport components and Krebs cycle constituents. *Giardia lamblia* is a diplomonad that causes giardiasis, a diarrheal disease acquired by campers drinking contaminated water from lakes and streams. The parabasalids are characterized by the presence of a **parabasal body** that is associated with the Golgi apparatus. They lack mitochondria but instead contain hydrogenosomes that carry out anaerobic respiration, releasing hydrogen gas. Some inhabit the guts of termites where they digest cellulose for the insect. They are also serious pathogens of vertebrates, causing infections in the intestinal or urogenital tracts. *Trichomonas vaginalis* is a parabasalid that causes a sexually transmitted disease in humans.

Euglenozoans   These organisms are a diverse group that are unicellular and contain a unique crystalline rod associated with their flagellum. Some are pathogens while others are free living. They are comprised of the kinetoplastids and the euglenids.

- **Kinetoplastids**
  The kinetoplastids have a single large mitochondrion in the cell that contains a kinetoplast, which is a large mass of DNA. Some are found in aquatic habitats while others are very serious pathogens in humans. The trypanosomes are responsible for several diseases in humans, such as African sleeping sickness caused by *Trypanosoma brucei*, Chagas' disease caused by *Trypanosoma cruzi*, and leishmaniasis caused by *Leishmania major*. Trypanosomes are crescent-shaped and possess a single flagellum that originates at a basal body and is enclosed by part of the cytoplasmic membrane, thus forming an undulating structure over the cell body.

- **Euglenids**
  The euglenids can grow either heterotrophically or phototrophically because they contain chloroplasts. They are nonpathogens that occur primarily in aquatic habitats, where many subsist on bacteria. Food is taken in by the process of **phagocytosis** when the cytoplasmic membrane surrounds the food and brings it into the cell for digestion. *Euglena* (figure 5.2, illustrations 1 and 2) is a member of this group.

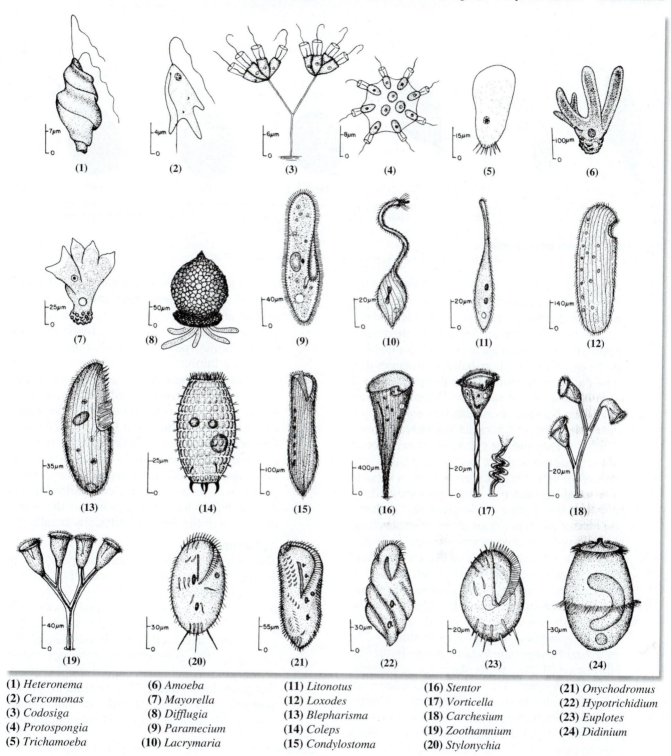

Figure 5.1 Protozoa.

(1) *Heteronema*
(2) *Cercomonas*
(3) *Codosiga*
(4) *Protospongia*
(5) *Trichamoeba*
(6) *Amoeba*
(7) *Mayorella*
(8) *Difflugia*
(9) *Paramecium*
(10) *Lacrymaria*
(11) *Litonotus*
(12) *Loxodes*
(13) *Blepharisma*
(14) *Coleps*
(15) *Condylostoma*
(16) *Stentor*
(17) *Vorticella*
(18) *Carchesium*
(19) *Zoothamnium*
(20) *Stylonychia*
(21) *Onychodromus*
(22) *Hypotrichidium*
(23) *Euplotes*
(24) *Didinium*

Alveolates The alveolates are comprised of the ciliates, the dinoflagellates, and the apicomplexans. Members of this group contain sacs called **alveoli** associated with the cytoplasmic membrane that may function in maintaining the osmotic balance of the cell. The apicomplexans are obligate parasites.

- **Ciliates**
Ciliates are characterized by having two kinds of nuclei, *micronuclei* and *macronuclei*. Micronuclei genes function in sexual reproduction, whereas genes encoding for cellular functions are associated with the macronuclei. An example of a

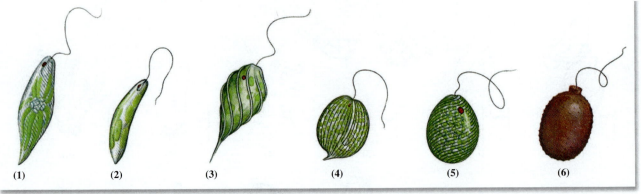

(1) *Euglena* (700×)   **(2)** *Euglena* (700×)   **(3)** *Phacus* (1000×)   **(4)** *Phacus* (350×)   **(5)** *Lepocinclis* (350×)   **(6)** *Trachelomonas* (1000×)

Figure 5.2 **Flagellated euglenids.**
(Courtesy of the U.S. Environmental Protection Agency, Office of Research & Development, Cincinnati, OH 45268)

ciliate is *Paramecium* (figure 5.1, illustration 9). Reproduction occurs by the process of conjugation, in which two *Paramecium* cells fuse and exchange their micronuclei. The ciliates are usually covered with cilia that are responsible for motility of the cell. Cilia are short, hairlike structures that have the same structure as flagella, that is, a 9 + 2 arrangement of microtubules. Motility occurs when cilia beat in a coordinated fashion to propel the cell forward or backward. Cilia also function in digestion. They line the cytostome or oral groove of the organism to direct food such as bacteria to the cell mouth, where it is enclosed in a vacuole and taken into the cell by phagocytosis. Many *Paramecium* harbor endosymbiotic prokaryotes in their cytoplasm or in the macronucleus; for example, methanogens produce hydrogen gas in the hydrogenosome of the *Paramecium*. Some species can attach to solid surfaces—for example, *Vorticella* and *Zoothamnium* (figure 5.1, illustrations 17 and 19). Ciliates can be pigmented: *Stentor* (figure 5.1, illustration 16), blue; *Blepharisma* (figure 5.1, illustration 13), pink; and *Paramecium busaria*, green (figure 5.1, illustration 9). The green color is due to the presence of endosymbiotic algal cells. The ciliates are important inhabitants of the forestomach of ruminant animals such as cattle. In the rumen (stomach), they degrade cellulose and starch, which can be used by the animal for nutrition. They also feed on rumen bacteria to maintain their numbers in the rumen. The ciliate *Balantidium coli* is an intestinal pathogen of domestic animals that occasionally infects humans. It causes an intestinal disease similar to *Entamoeba histolytica*.

- **Dinoflagellates**
  The dinoflagellates (figure 5.3, illustrations 10, 11) are characterized by the presence of two flagella of different lengths. The flagella encircle the cell and when they beat, they cause the cell to whirl or spin. These organisms occur in both marine and freshwater habitats. *Gonyaulax* is a dinoflagellate that occurs in marine coastal waters. Some species are bioluminescent and can cause luminescence when waves break on the shore or boats move through the water. The organisms produce "red tides" caused by xanthophyll in the cells. Blooms of these organisms can occur naturally or by increased pollution in coastal waters. These blooms can result in fish kills and poisoning in humans when contaminated shellfish are consumed. *Gonyaulax* produces a neurotoxin, called **saxitoxin,** that is responsible for the symptoms of dizziness, numbness of the lips, and difficulty breathing.

- **Apicomplexans**
  The apicomplexans are obligate parasites in animals and humans. In humans they are the cause of malaria (*Plasmodium* spp.), toxoplasmosis (*Toxoplasma gondii*), and coccidiosis (*Eimeria*). They produce resting stages called *sporozoites* that facilitate transmission of the pathogen. The cells contain structures called *apicoplasts*, which are vestiges of chloroplasts that have degenerated. They contain no photosynthetic pigments, but they do have some genes that encode for fatty acid and heme synthesis.

Stramenophiles   The stramenophiles are composed of the oomycetes (water molds), the diatoms, the golden algae, and the multicellular paeophytes, which includes the seaweeds. The unicelluar stramenophiles have flagella with short, hairlike extensions. The name is derived from this feature. *Stramen* is Latin for straw and *pilos* means hair.

- **Diatoms**
  Diatoms are phototrophic freshwater and marine organisms that are unique in the biological

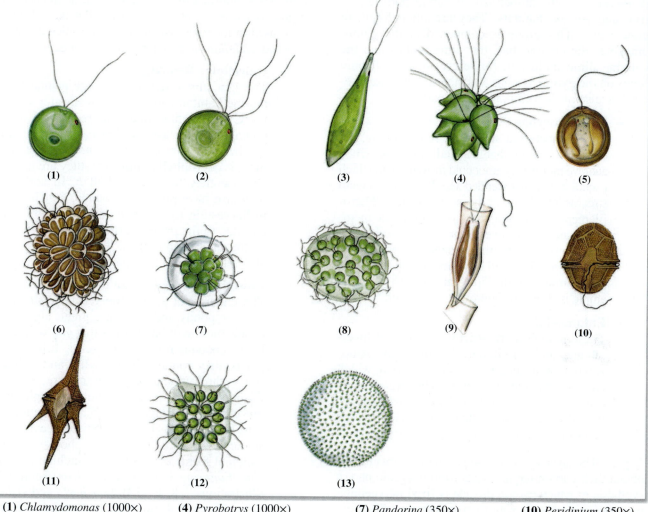

Figure 5.3 Flagellated green algae.

(1) *Chlamydomonas* (1000×)
(2) *Carteria* (1500×)
(3) *Chlorogonium* (1000×)
(4) *Pyrobotrys* (1000×)
(5) *Chrysococcus* (3000×)
(6) *Synura* (350×)
(7) *Pandorina* (350×)
(8) *Eudorina* (175×)
(9) *Dinobryon* (1000×)
(10) *Peridinium* (350×)
(11) *Ceratium* (175×)
(12) *Gonium* (350×)
(13) *Volvox* (100×)

(Courtesy of the U.S. Environmental Protection Agency, Office of Research & Development, Cincinnati, OH 45268)

world because they synthesize a cell wall composed of silica. The external portion of the wall is called a *frustule* and is very diverse in its shape. The frustule consists of two halves, the epitheca and the hypotheca. It is analogous to a box with a lid, with the epitheca fitting over the hypotheca. Pores in the frustule called *areolae* function as passageways for gases and nutrients. The frustules remain after the death of the organism and are referred to as diatomaceous earth, which is used in many applications such as a polishing abrasive in toothpaste. They do not decay when the organism dies, and hence they are a preserved fossil record of diatoms. Based on fossil records of the frustules, it is estimated that diatoms were present on the earth about 200 million years ago. *Nitzschia* is a common diatom (figure 5.6, illustration 7).

- **Golden Algae**
The *chrysophytes* or golden (brown) algae are classified with the stramenophiles because genetic analysis has determined that they are more closely related to the diatoms and oomycetes (water molds) than to the other unicellular algae. They are inhabitants of fresh and marine waters. Some species are also chemoorganotrophs, which means they can derive food from the transport of organic compounds across the cytoplasmic membrane instead of carrying out photosynthesis. They are motile by means of two flagella. Their gold or brown color is the result of the carotenoid, *fucoxanthin*. They also contain chlorophyll *c* rather than the chlorophyll *a* found in other algae. *Dinobryon* is a colonial golden alga that occurs in fresh water (figure 5.3, illustration 9).

Amoebozoa The amoebozoa occur in both terrestrial and aquatic habitats. They are often found in pond water. This group is composed of the gymnamoebas, the entamoebas, the slime molds, and the cellular slime molds. In this exercise we will only focus on the gymnamoebas and mention the entamoebas because of their importance as pathogens.

- **Gymnamoebas**

  These amoebozoa are primarily free living, occurring in aquatic habitats and in the soil. They achieve motility by amoeboid movement in which pseudopodia are extended and the cell cytoplasm streams into the tip of the pseudopodium. Microfilaments associated with the cytoplasmic membrane aid in the overall process. Pseudopodia are also utilized by the amoebas for entrapping and surrounding food by phagocytosis. They can range in size from 15μ to above 700μ. An example is shown in figure 5.1, illustration 5.

- **Entamoebas**

  These amoebas are strictly parasites of invertebrates and vertebrates. An important pathogen in humans is *Entamoeba histolytica,* which causes amoebic dysentery. The organism is transmitted in a cyst form by fecal contamination of water and food. In the intestine it causes ulceration, resulting in a bloody diarrhea.

Algae The algae are a diverse group of organisms that obtain their carbon requirements from oxygenic photosynthesis in which carbon dioxide is fixed into cellular materials and water is split to evolve oxygen. They are typically smaller and less complex in their structure than land plants, but they are similar to plants because they possess photosynthetic pigments such as chlorophyll and carotenoids that harvest light energy from the sun. They range in size from microscopic unicellular forms to the seaweeds that form giant kelp beds in the oceans. The microscopic algae form filaments or colonies that are comprised of several individuals loosely held together in an organized fashion. Many of the unicellular algae are motile by means of a flagellum. Reproduction can occur by both asexual and sexual mechanisms.

The algae are diverse in their ecology. They occur primarily in aquatic habitats such as freshwater lakes and streams and in the oceans, where they are important members of the phytoplankton. Algae also occur in terrestrial habitats. Some grow on snow, imparting a pink color. They enter into unique symbiotic relationships with fungi called lichens, which are found on trees and rocks. Although unique morphologically, the algae are related genetically to other protists.

Red and Green Algae The red and green algae belong to the *rhodophytes* and *chlorophytes*, respectively. They may be unicellular (figure 5.3, illustrations 1–5), colonial

(figure 5.3, illustrations 6, 7, 8, 12, 13), or filamentous (figure 5.4). Many seaweeds are included in the red algae. Agar used in microbiological media is extracted from a seaweed, *Gelidium*, a member of the red algae.

- **Unicellular Red Algae**

  The unicellular red algae occur primarily in marine habitats, with a few freshwater varieties. Cells may have one or more flagella. They contain chloroplasts, which carry out photosynthesis. Their primary photosynthetic pigment is chorophyll *a*; however, they are unique because they lack chorophyll *b*. In addition, they also have phycobiliproteins, which assist in harvesting light. Their red coloration is in fact due to the presence of *phycoerythrin*, one of the phycobiliproteins. These pigments are also found in the cyanobacteria.

- **Unicellular Green Algae**

  The majority of algae observed in fresh water such as ponds and lakes are green algae and hence, these will be the ones that you will most likely encounter in this exercise. Like green plants, they contain chlorophyll *a* and *b*. They store starch granules as an energy reserve. *Chlamydomonas* is an excellent example of a unicellular green alga (figure 5.3, illustration 1). This organism has been the object of extensive studies, especially involving photosynthesis. Colonial forms of green algae include *Pandorina, Eudorina, Gonium,* and *Volvox* (figure 5.3, illustrations 7, 8, 12, 13). *Vaucheria* and *Tribonema* (figure 5.4, illustrations 5 and 6) are not green algae but have been recently reclassified as Tribophyceae based on genetic analysis.

  A unique group of green algae are the "desmids," whose cells are composed of two halves, called semi-cells (figure 5.4, illustration 12, and figure 5.5, illustrations 16–20). The two halves of the cell are separated by a constriction called the isthmus. Some of the organisms such as *Spirogyra* form a filament, whereas others such as *Desmidium* form a pseudofilament.

Prokaryotes

Cyanobacteria

The cyanobacteria first appeared on the earth some 2.7 billion years ago. They were critical to the evolution of life on the earth because they were the first to carry out oxygenic photosynthesis in which oxygen was released into the atmosphere. The accumulation of oxygen over time paved the way for the development of aerobic bacteria and eukaryotic organisms that required oxygen for respiratory metabolism.

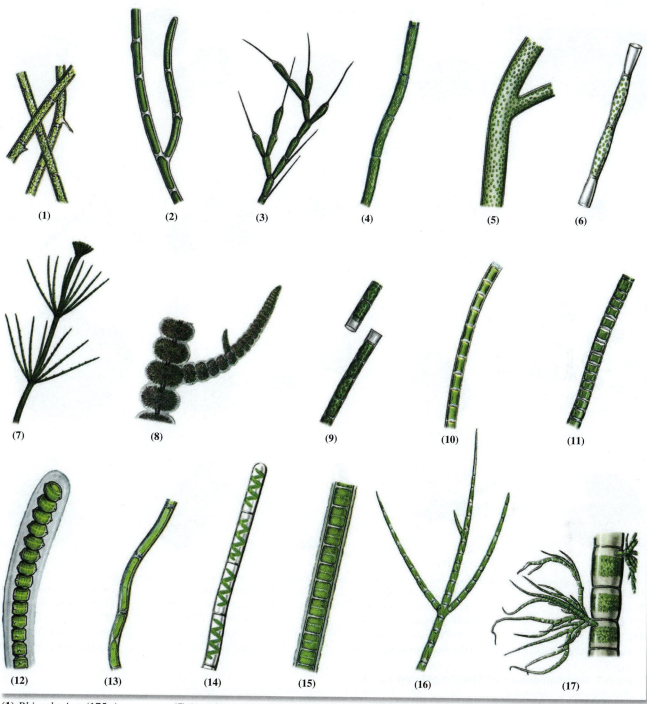

Figure 5.4 Filamentous algae.
(Courtesy of the U.S. Environmental Protection Agency, Office of Research & Development, Cincinnati, OH 45268)

(1) *Rhizoclonium* (175×)
(2) *Cladophora* (100×)
(3) *Bulbochaete* (100×)
(4) *Oedogonium* (350×)

(5) *Vaucheria* (100×)
(6) *Tribonema* (300×)
(7) *Chara* (3×)
(8) *Batrachospermum* (2×)

(9) *Microspora* (175×)
(10) *Ulothrix* (175×)
(11) *Ulothrix* (175×)
(12) *Desmidium* (175×)

(13) *Mougeotia* (175×)
(14) *Spirogyra* (175×)
(15) *Zygnema* (175×)
(16) *Stigeoclonium* (300×)
(17) *Draparnaldia* (100×)

Prior to the cyanobacteria's appearance, the earth was anaerobic, and only organisms capable of anaerobic metabolism could survive. Evidence also suggests that these bacteria may have invaded a primitive cell and established an endosymbiotic relationship, thus giving rise to chloroplasts in algae and plants.

Cyanobacteria are extremely diverse and occur in a variety of environments, from the tropics to the poles. They primarily occur in aquatic habitats such as the oceans, lakes, and streams. The cyanobacteria play a critical role in the global carbon cycle, accounting for 20–30% of the carbon fixed on the earth by

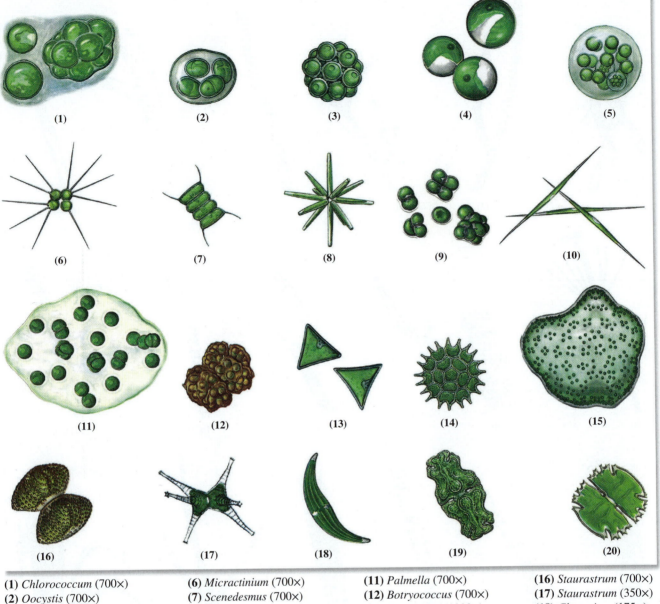

(1) *Chlorococcum* (700×)
(2) *Oocystis* (700×)
(3) *Coelastrum* (350×)
(4) *Chlorella* (350×)
(5) *Sphaerocystis* (350×)

(6) *Micractinium* (700×)
(7) *Scenedesmus* (700×)
(8) *Actinastrum* (700×)
(9) *Phytoconis* (700×)
(10) *Ankistrodesmus* (700×)

(11) *Palmella* (700×)
(12) *Botryococcus* (700×)
(13) *Tetraedron* (1000×)
(14) *Pediastrum* (100×)
(15) *Tetraspora* (100×)

(16) *Staurastrum* (700×)
(17) *Staurastrum* (350×)
(18) *Closterium* (175×)
(19) *Euastrum* (350×)
(20) *Micrasterias* (175×)

Figure 5.5 Nonfilamentous and nonflagellated algae.
(Courtesy of the U.S. Environmental Protection Agency, Office of Research & Development, Cincinnati, OH 45268)

photosynthesis. This, coupled with their ability to fix nitrogen, makes them primary producers in the environment. They comprise part of the phytoplankton in the oceans and are therefore extremely important in food chains in these habitats.

Their taxonomy is not completely settled at this time. They can be divided into five morphological groups: (1) unicellular, dividing by binary fission; (2) unicellular, dividing by multiple fission; (3) filamentous, containing nitrogen-fixing heterocysts; (4) filamentous, containing no heterocysts; and (5) branching filamentous species. Over 1000 species

have been described. They were once classified as "blue-green algae" primarily because they contain chlorophyll *a* and release oxygen from photosynthesis. However, they are prokaryotes because they lack an organized nucleus or chloroplast and do not contain other organelles. They have 70S ribosomes and contain peptidoglycan in their cell walls. They differ from the green sulfur and purple sulfur bacteria because the latter contain bacteriochlorophyll and carry out anoxygenic photosynthesis in which oxygen is not produced. Figure 5.7 illustrates only a random few that are frequently seen.

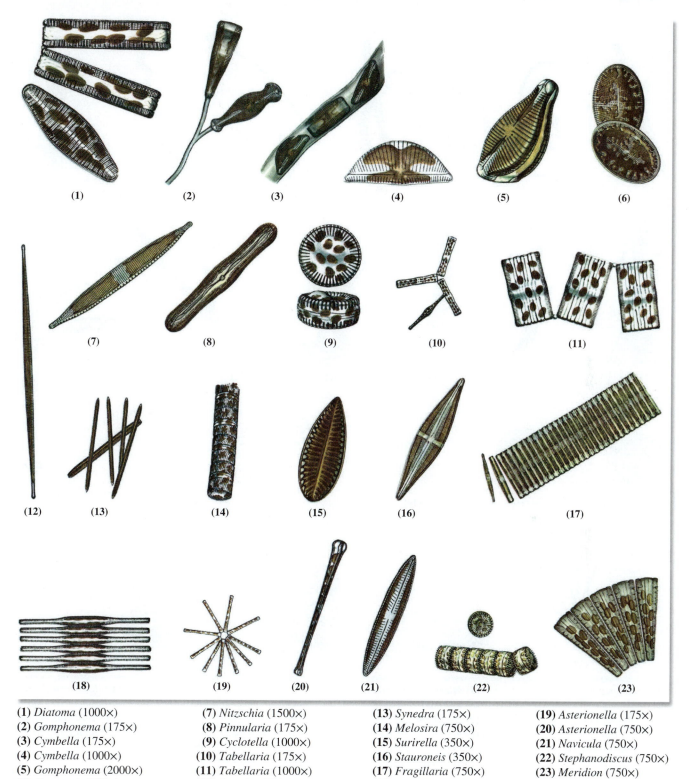

(1) *Diatoma* (1000×)
(2) *Gomphonema* (175×)
(3) *Cymbella* (175×)
(4) *Cymbella* (1000×)
(5) *Gomphonema* (2000×)
(6) *Cocconeis* (750×)

(7) *Nitzschia* (1500×)
(8) *Pinnularia* (175×)
(9) *Cyclotella* (1000×)
(10) *Tabellaria* (175×)
(11) *Tabellaria* (1000×)
(12) *Synedra* (350×)

(13) *Synedra* (175×)
(14) *Melosira* (750×)
(15) *Surirella* (350×)
(16) *Stauroneis* (350×)
(17) *Fragillaria* (750×)
(18) *Fragillaria* (750×)

(19) *Asterionella* (175×)
(20) *Asterionella* (750×)
(21) *Navicula* (750×)
(22) *Stephanodiscus* (750×)
(23) *Meridion* (750×)

Figure 5.6 Diatoms.
(Courtesy of the U.S. Environmental Protection Agency, Office of Research & Development, Cincinnati, OH 45268)

The previous designation of these organisms as "blue-green" was somewhat misleading as many are black, purple, red, and various shades of green, including blue-green. The varying colors are due to the differing proportions of photosynthetic pigments present in the cells, which include chlorophyll *a*, carotenoids, and phycobiliproteins. The latter pigments consist of allophycocyanin, phycocyanin, and phycoerythrin that combine with protein molecules to form **phycobiliproteins,** which serve as light-harvesting molecules

41

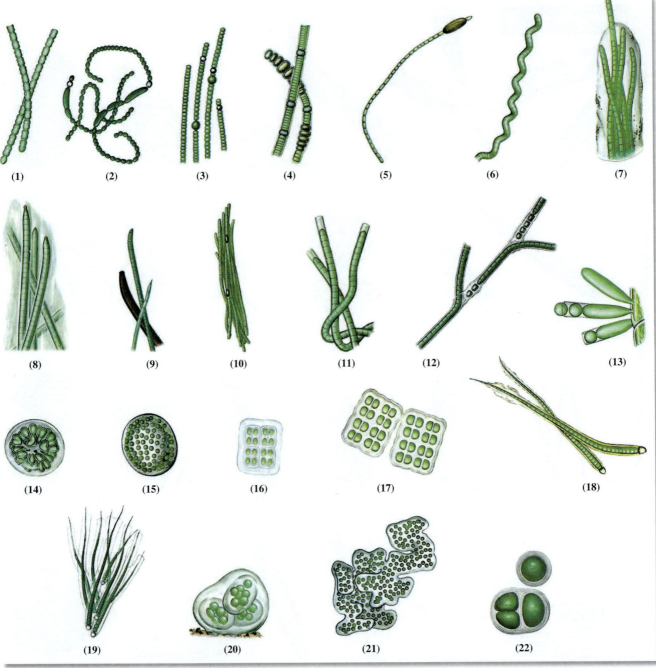

(1) (2) (3) (4) (5) (6) (7)

(8) (9) (10) (11) (12) (13)

(14) (15) (16) (17) (18)

(19) (20) (21) (22)

(1) *Anabaena* (350×)
(2) *Anabaena* (350×)
(3) *Anabaena* (175×)
(4) *Nodularia* (350×)
(5) *Cylindrospermum* (175×)
(6) *Arthrospira* (700×)

(7) *Microcoleus* (350×)
(8) *Phormidium* (350×)
(9) *Oscillatoria* (175×)
(10) *Aphanizomenon* (175×)
(11) *Lyngbya* (700×)
(12) *Tolypothrix* (350×)

(13) *Entophysalis* (1000×)
(14) *Gomphosphaeria* (1000×)
(15) *Gomphosphaeria* (350×)
(16) *Agmenellum* (700×)
(17) *Agmenellum* (175×)
(18) *Calothrix* (350×)

(19) *Rivularia* (175×)
(20) *Anacystis* (700×)
(21) *Anacystis* (175×)
(22) *Anacystis* (700×)

Figure 5.7 Cyanobacteria.
(Courtesy of the U.S. Environmental Protection Agency, Office of Research & Development, Cincinnati, OH 45268)

for the photosystems. Interestingly, these structures are also found in the red algae. Because an organized chloroplast is not present in these bacteria, their pigments and photosystems are organized in **thylakoids,** which are parallel arrays of stacked membranes.

Laboratory Report

Answer the questions in Laboratory Report 5.

5 Microbiology of Pond Water—Protists, Algae, and Cyanobacteria

A. Results

In this study of freshwater microorganisms, record your observations in the following tables. The number of organisms to be identified will depend on the availability of time and materials. Your instructor will indicate the number of each type that should be recorded.

Record the genus of each identifiable organism. Also, indicate the group to which the organism belongs. Microorganisms that you cannot identify should be sketched in the space provided. It is not necessary to draw those that are identified.

1.   **Protozoa**

GENUS	GROUP	BOTTLE NO.	SKETCHES OF UNIDENTIFIED

2.   **Algae**

GENUS	GROUP	BOTTLE NO.	SKETCHES OF UNIDENTIFIED

3.   **Cyanobacteria**

GENUS	BOTTLE NO.	SKETCHES OF UNIDENTIFIED

B. Short-Answer Questions

1. In which domains are algae, protozoa, and cyanobacteria classified?

   _____

   _____

2. (a) Name one similarity between algae and plants. (b) Name one difference.

   _____

   _____

3. Compare and contrast the three mechanisms of motility displayed by protozoa.

   _____

   _____

4. (a) What organisms were formerly known as "blue-green algae"? (b) Why are these organisms not algae?

   _____

   _____

5. What makes "red tides" red?

   _____

   _____

6. What is the genus of the causative agent of malaria? In what group does it belong?

   _____

   _____

7. What are mitosomes? In what organisms do they occur?

   _____

   _____

8. What are thylakoids? What function do they have in photosynthesis?

   _____

   _____

9. In the alveolates, what structure may control osmotic balance?

   _____

   _____

10. What function do the micronuclei play in the ciliates?

   _____

   _____

11. What is a frustule, and what unique compound comprises this structure?

   _____

   _____

12. Besides chlorophyll, what additional light-harvesting pigments are found in the cyanobacteria?

   _____

   _____

C. Fill-in-the-Blanks Question

1. For each type of organism, place a check mark in the box to indicate whether the cellular characteristic or function is present.

CHARACTERISTIC OR FUNCTION	PROTOZOA	ALGAE	CYANOBACTERIA
Nucleus			
Flagella			
Pseudopodia			
Cilia			
Photosynthetic pigment(s)			
Chloroplasts			
Cell wall			

MICROBIOLOGY

Ubiquity of Bacteria

Figure 6.1 Nutrient agar plate exposed to the environment.
© Andreas Reh/Getty Images

Bacteria are the most widely distributed organisms in the biosphere. The major portion of biomass on the earth is made up from bacteria and other microbial life. Bacteria occur as part of the normal flora of humans and animals, where they occupy sites on surfaces such as the skin and intestinal tract. In humans it is estimated that the normal flora outnumber our own cells by approximately tenfold. In plants they are found on leaf surfaces and in nodules on roots, where they form a symbiotic partnership with the plant to fix nitrogen from the atmosphere. They are also responsible for a variety of diseases in both plants and animals. We also depend on bacteria to break down our sewage and waste. They are important in the soil, where they are active in the nitrogen and carbon cycles, thus contributing to soil fertility. Bacteria are abundant in the oceans, where they contribute to cycles and mineralization. In the oceans, some have unique metabolic capabilities that allow them to grow next to superheated hydrothermal vents and utilize hydrogen sulfide and other gases emitted from volcanic activity associated with these vents. Bacteria have been isolated from core samples taken from deep within the earth's crust. Archaea form colorful blooms in sulfur hot springs found in places such as Yellowstone National Park. In essence, Bacteria and Archaea have been found in almost every place humans have searched for them. Thus, they are almost ubiquitous in their distribution on the earth.

When Robert Koch first studied bacteria in his laboratory, one of the first challenges was to devise a method to grow bacteria in culture so that populations could be separated into individual species. Initial approaches used pieces of vegetables such as potatoes and carrots, or the addition of gelatin to meat broths. It was found that bacterial cells would grow as visible, discrete **colonies** on a solid medium. A **colony** is a visible mass of cells usually resulting from the division of a single cell (figure 6.1), and the number of cells in a single colony can exceed one

billion (10^9). However, a colony can arise from more than one cell, for example when a chain of cells such as streptococci grow on a nutrient medium. The cultivation of bacterial cells was vastly improved in Koch's laboratory when Frau Hesse, the wife of an early coworker of Koch, suggested the use of agar-agar as a solidifying agent. A major advantage was that agar media were easily manipulated, and agar, unlike gelatin, was not degraded by the pathogenic bacteria Koch was studying. As a solidifying agent, agar could be added to rich broths to form a solid medium on which isolated colonies would develop after inoculation. The purity of cultures was further improved when R. J. Petri, a worker in Koch's lab, introduced a covered dish that protected the nutrient surface of media from contamination by bacteria in the environment, especially the air. These methods were extremely important in the accomplishments of Koch's lab because they provided a means for separating and culturing the pathogens that they were studying, including anthrax caused by *Bacillus anthracis*. Microbiologists still use agar and the petri dish in the cultivation of bacteria today.

No single medium exists that will support the growth of all bacteria, owing to the diverse metabolic capabilities and requirements of bacteria. However, many bacteria will grow on extracts of meat or vegetables that have been solidified with agar, an example being nutrient agar. During this laboratory period, you will be provided with sterile bacteriological media that you will expose to the environment in various ways. Any bacterial cells that occur in the environment and are deposited on the agar medium or in the broth will subsequently undergo cell division to produce colonies or cause the broth to become turbid. The idea is to appreciate that bacteria are ubiquitous in the environment and can be spread by convection currents in the air. It is important to also understand that the morphology of individual bacterial colonies that develop on an agar medium usually differ from one species to another. This may involve different aspects of the colony such as the regularity of its edge, pigmentation, the configuration and texture of its surface, and its elevation. Examples of these characteristics are given in Exercise 35, figure 35.4.

You will be provided with three kinds of sterile bacteriological media that you will expose to the environment in various ways. To ensure that your exposures cover as wide a spectrum as possible, specific assignments will be made for each student. This may involve the use of a swab to remove bacteria from an object or surface; in other instances a petri plate containing a nutrient medium will be exposed to the air in different environments. A number will designate your assignment, as detailed in the chart below.

Materials

per student:
- 1 tube of nutrient broth
- 1 petri plate of trypticase soy agar (TSA)
- 1 sterile cotton swab
- Sharpie marking pen

per two or more students:
- 1 petri plate of blood agar

1. Scrub down your desktop with a disinfectant (see Exercise 8, Aseptic Technique).
2. Expose your TSA plate according to your assignment in the chart below. *Label the bottom* of your plate with your initials, your assignment number, and the date.

3. Moisten a sterile swab by immersing it into a tube of nutrient broth and expressing most of the broth out of it by pressing the swab against the inside wall of the tube.
4. Rub the moistened swab over a part of your body such as a finger or ear, or some object such as a doorknob or telephone mouthpiece, and return the swab to the tube of broth. It may be necessary to break off the stick end of the swab so that you can replace the cap on the tube.
5. Label the tube with your initials and the source of the bacteria.
6. Expose the blood agar plate by coughing onto it. Label the bottom of the plate with the initials of the individuals who cough onto it. Be sure to date the plate.
7. Incubate the plates and tube at 37°C for 48 hours.

Evaluation

After 48 hours of incubation, examine the tube of nutrient broth and two plates. Shake the tube vigorously without wetting the cap. Is it cloudy or clear? Compare it with an uncontaminated tube of broth. What is the significance of cloudiness? Do you see any colonies growing on the blood agar plate? Are the colonies all the same size and color? If not, what does this indicate? Group together a set of TSA plates representing all nine types of exposure. Record your results in the Laboratory Report.

Your instructor will indicate whether these tubes and plates are to be used for making slides in Exercise 11 (Simple Staining). If the plates and tubes are to be saved, containers will be provided for their storage in the refrigerator. Place the plates and tubes in the designated containers.

Laboratory Report

Record your results in Laboratory Report 6.

EXPOSURE METHOD FOR TSA PLATE	STUDENT NUMBER
1. To the air in laboratory for 30 minutes	1, 10, 19, 28
2. To the air in room other than laboratory for 30 minutes	2, 11, 20, 29
3. To the air outside of building for 30 minutes	3, 12, 21, 30
4. Blow dust onto exposed medium	4, 13, 22, 31
5. Moist lips pressed against medium	5, 14, 23, 32
6. Fingertips pressed lightly on medium	6, 15, 24, 33
7. Several coins pressed temporarily on medium	7, 16, 25, 34
8. Hair combed over exposed medium (10 strokes)	8, 17, 26, 35
9. Optional: Any method not listed above	9, 18, 27, 36

6   Ubiquity of Bacteria

A. Results

1. After examining your TSA and blood agar plates, record your results in the following table and on a similar table that your instructor has drawn on the chalkboard. With respect to the plates, we are concerned with a quantitative evaluation of the degree of contamination and differentiation as to whether the organisms are bacteria or molds. Quantify your recording as follows:

0	no growth		
+	1 to 10 colonies	+ + +	51 to 100 colonies
+ +	11 to 50 colonies	+ + + +	over 100 colonies

    After shaking the tube of broth to disperse the organisms, look for cloudiness (turbidity). If the broth is clear, no bacterial growth occurred. Record no growth as 0. If the tube is turbid, record + in the last column.

STUDENT INITIALS	PLATE EXPOSURE METHOD		COLONY COUNTS		BROTH	
	TSA	Blood Agar	Bacteria	Mold	Source	Result

Use the class results to answer the following questions.

2. Using the number of colonies as an indicator, which habitat sampled by the class appears to contain the most bacteria? _____

3. Why do you suppose this habitat contains such a high microbial count? _____

_____

4. a. Were any plates completely lacking in colonies?

   _____

   b. Do you think that the habitat sampled was really sterile?

   _____

   c. If your answer to *b* is *no,* then how can you account for the lack of growth on the plate?

   _____

   d. If your answer to *b* is *yes,* defend it.

   _____

B. Short-Answer Questions

1. In what ways do the macroscopic features of bacterial colonies differ from those of molds?

   _____

2. Why is the level of contamination measured as number of colonies rather than size of colonies?

   _____

3. Should one be concerned to find bacteria on the skin? How about molds? Explain.

   _____

4. How can microbial levels be controlled on the skin? On surfaces in the environment? In the air?

   _____

5. Compare the following features of bacteria to those of eukaryotic microorganisms:
   a. size.

   _____

   b. organization of genetic material.

   _____

   c. ribosomes.

   _____

   d. cell wall.

   _____

   e. respiration and photosynthesis.

   _____

   f. motility mechanisms.

   _____

The Fungi:
Molds and Yeasts

Learning Outcomes

After completing this exercise, you should be able to

1. Differentiate between yeast and fungi.

2. Identify some common fungi based on their microscopic morphology and colony characteristics.

3. Identify the fruiting structures of yeasts and molds based on their microscopic morphology.

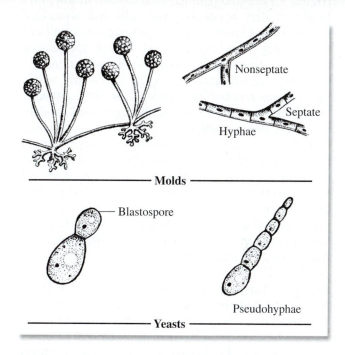

Figure 7.1 **Structural differences between molds and yeasts.**

The fungi comprise a large and diverse group of saprophytic, eukaryotic organisms. Saprophytes are organisms that obtain their nutritional needs from degrading organic materials in the environment. This group includes the molds, mushrooms, and yeasts. Unlike plants, which have cell walls composed of cellulose, fungal cell walls are composed of chitin, a polymer of the carbohydrate N-acetyl-glucosamine. They mostly occur in terrestrial environments, although some are found exclusively in aquatic habitats. They secrete **exoenzymes** that break down polysaccharides and proteins into their monomeric components of sugars, peptides, and amino acids, which are utilized for nutrition. They are responsible for the decay of dead plant and animal material, and thus they are essential for the mineralization and recycling of organic and biological material in the environment. For example, the wood-rotting fungi play an almost exclusive role in the degradation and turnover of dead trees. Wood has a complex structure composed of cellulose and lignin, a polymer of phenolic compounds. Fungi are unique in their capacity to degrade the compounds that comprise wood. Fungi can grow under widely varying environmental conditions, with the ability to withstand low pH values and temperatures up to 62°C. Because of their unique metabolic capabilities and diverse growth capacities, they are often nuisance organisms. They are responsible for significant economic losses of foods such as fruits and vegetables. They especially thrive in moist habitats such as bathrooms, where they are responsible for the formation of mildew on surfaces.

Molds are fungi that form colonies composed of microscopic, rounded, intertwining filaments called **hyphae** (hypha, singular, figure 7.1). There are two different structural types of hyphae in fungi. One type

is characterized by individual cells that are separated by the formation of *septa* or cross-walls, with each cell having a typical eukaryotic structure. In this type of hyphae, pores in the septa allow streaming of cytoplasm between cells. A second type of hyphae is *coenocytic*, in which cross-walls are not present to form individual cells. Repeated nuclear divisions in coenocytic hyphae result in nuclei that are transported by cytoplasmic streaming throughout the fungal colony. Growth of a mold colony occurs at the ends of hyphae, called apical tips. Hyphae cover the surface of a substrate to form a branching, filamentous network called a **mycelium.** A visible mold colony is a mat formed by the hyphae that make up this substrate mycelium. The oldest part of the colony is at the center of the colony. Apical tips are primarily found at the edge of the colony where colony growth is occurring. A mold colony often has a characteristic pigmentation that is useful in identification. Structures that produce spores such as sporangiophores and conidiophores arise when branches of the substrate mycelium form and undergo cellular differentiation to form these specialized reproductive structures.

Yeasts are fungi that do not ordinarily form hyphae. The primary mode of reproduction for yeasts such as *Saccharomyces cerevisiae* is by the process of budding, whereby a bud forms and eventually separates from the main cell to form a new yeast cell. However, sometimes buds may not separate but instead form a chain of cells called **pseudohyphae** (figure 7.1). Some pathogenic fungi are **dimorphic.** In tissue, they occur as yeast cells and reproduce by budding. However, when cultured onto nutrient media, they form mycelia and typical sporulation structures. An example is *Histoplasma capsulatum.*

Fungi are important pathogens in humans, where they can cause minor infections of hair, skin, and nails or serious deep mycoses in internal organs that result in death. The **dermatophytes** such as *Trichophyton* infect hair, skin, and nails in humans to cause diseases such as athlete's foot and ringworm. A number of different fungi cause subcutaneous mycoses, which are characterized by lesions that develop at the site of inoculation. Two examples are *Sporothrix*, which causes sporotrichosis, and *Madurella*, which usually infects the feet to cause edema and swelling in a disease called mycetoma. The true pathogenic fungi cause mycoses that may involve all internal organs of the body. Many are acquired by inhaling infective spores into the respiratory tract. These include the following: *Blastomyces*, North American blastomycosis; *Coccidioides*, coccidioidomycosis; *Histoplasma*, histoplasmosis; and *Paracoccidioidomycosis*, South American blastomycosis. Progressive infections by *Histoplasma* can involve the liver, spleen, and lymph nodes. Fungi that cause deep mycoses are usually dimorphic because they produce mycelia when grown at 25°C on culture media but produce budding yeast forms when growing in tissue at 37°C.

Yeasts are also responsible for infections in humans. *Candida albicans* causes infections in various body sites. This organism can infect the mouth and tongue, especially in newborns, to cause **thrush,** causing the tissue to have a chalky appearance. The latter results when *Candida* is introduced from the mother's vaginal tract during birth, and because newborns initially lack a normal flora in the mouth, *Candida* can establish a presence and thrive. This organism also causes yeast infections in the vaginal tract as well as the intestinal tract. It can occasionally invade the heart and central nervous system to cause endocarditis and meningitis, respectively. Another species of *Candida, C. glabrata,* is a primary cause of infections in patients with compromised immune systems, such as AIDS patients. *Cryptococcus neoformans* is responsible for various infections involving the skin, skeleton, and the central nervous system. It occurs in approximately 8% of AIDS patients.

Fungi are also important pathogens of plants, where they cause significant economic losses each year. They infect crops such as corn, grains, grasses, and peanuts. Carcinogenic (cancer-causing) aflatoxins are produced by fungi growing in infected peanuts and grains. These toxins are monitored by the U.S. government, which has set allowable limits for human consumption of peanut products. The fungus *Claviceps* infects grain to produce ergot alkaloids that can cause hallucinations and even death. Some have theorized that the Salem witch hunts and trials in colonial Massachusetts involved ergot poisoning from infected grain used to make bread.

Fungi are an important part of our diet as well as producers of certain foods. Fermentation of grapes and grains by yeasts produces beverages such as wine and beer. Roquefort and blue cheese are produced by the growth of *Penicillium* in cheeses made from the milk of cattle, sheep, and goats. Mushrooms such as *Agaricus* and morels, members of the Basidiomycetes, are used in a variety of foods consumed by humans. However, some mushrooms such as *Amanita* are extremely poisonous because they produce very potent toxins that cause death when ingested.

Fungi can form important and beneficial symbiotic relationships with organisms such as plants and cyanobacteria. **Mycorrhizae** are close symbiotic associations between fungi and plant roots. The fungus facilitates the uptake of minerals such as phosphate and water by the plant, and in return, the fungus is supplied with carbohydrates from the plants that are used for nutrition.

Fungi also form symbiotic associations with algae and cyanobacteria called **lichens.** Approximately 85% of these associations involve filamentous or unicellular eukaryotic algae, 10% involve cyanobacteria, and 4% consist of both an alga and a cyanobacterial partner. It is estimated that 20% of all fungi form lichens, presumably because these associations are a reliable way for fungi to obtain fixed carbon from the photosynthetic partner and also obtain nitrogen from a cyanobacterial partner. Ascomycetes account for about 98% of fungal partners. Lichens are prevalent in the environment and can be found on the surfaces of trees, rocks, and even the roofs of houses. The fungus anchors the lichen colony to the surface and protects the photosynthetic partner from erosion by rain and winds and from drying. The fungal partner also synthesizes lichen acids that dissolve and chelate inorganic nutrients in the environment that are needed by the photosynthetic partner. The fungus also can facilitate the uptake of water needed by the photosynthetic partner. Lichens are sensitive to many toxic chemicals in the environment. Owing to their sensitivity to compounds such as sulfur dioxide and metals, they are important indicators of pollution such as acid rain.

Many diverse organisms have been traditionally categorized as fungi, including water molds, puff balls, bracket fungi, yeasts, and molds. These organisms do not have a uniform genetic background, and they are believed to have evolved from at least

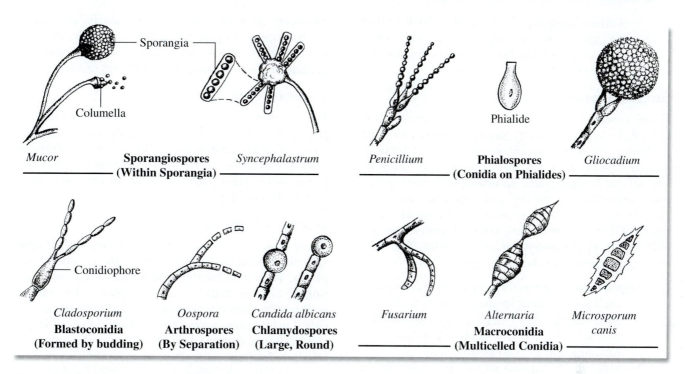

Figure 7.2 **Types of asexual spores seen in fungi.**

two ancestral lines. Classification of fungi is a complex and dynamic process. Traditional classification schemes relied primarily on morphology and reproductive mechanisms to classify fungal groups. However, modern schemes employ genetic analysis to ascertain relatedness between groups and species. Recent genetic analyses of fungi indicate that fungal classification based on morphological characteristics does not reflect evolutionary or phylogenetic relationships between organisms. However, for the beginning laboratory, identification and characterization of fungi is still most easily performed using morphological characteristics. In this exercise, you will examine prepared slides and those made from living cultures of molds and yeasts. The spores of fungi are abundant in the air and can be easily cultured on appropriate media for both macroscopic and microscopic study. Additionally, you will examine various fungi microscopically and attempt to identify them based on their morphological characteristics. Before attempting to identify various molds or yeasts, familiarize yourself with the figures that depict the characteristics of the groups. Note that yeasts differ from molds in that yeasts are unicellular and do not form a true mycelium (figure 7.1).

Fungal Spores: Asexual and Sexual

Fungal spores are essential in the reproduction of fungi and are produced by either asexual or sexual means. Traditional taxonomy relied greatly on the

morphological characteristics of fungal spores and spore-related structures. Classification was based on morphological characteristics such as variation in spore size, color, the appearance of spores, and the type of fruiting body. These two types of spores are described next.

Asexual Spores

Asexual spores are produced by mitotic division and differentiation of specialized hyphae that extend above the colony. Types of asexual spores are shown in figure 7.2.

Sporangiospores: Sporangiospores form within a thick-walled sac called a sporangium. They can be either motile or nonmotile.

Conidia: Conidia are nonmotile asexual spores that form on specialized hyphae called conidiophores. They include the following:

Phialospores: These spores are produced on a vase- or flask-shaped cell called a *phialide.* They are found in *Penicillium* and *Gliocladium.*

Blastoconidia: These are present in some filamentous fungi and occur by budding in yeast cells. *Cladosporium* and *Candida* produce blastoconidia.

Arthrospores: Arthrospores form by the fragmentation and formation of crosswalls in preexisting hyphae. They occur in *Geotrichum* and *Galactomyces.*

Chlamydospores: Asexual, thick-walled spores that are round or irregular. They occur in most fungi and may function in survival.

Sexual Spores

Fungi produce sexual spores by the union of two unicellular and genetically distinct gametes to form a diploid cell which undergoes meiosis and mitosis. Sexual reproduction can also occur by the fusion of specialized hyphae called *gametangia*. Sexual spores are resistant to heat, drying, and some chemical compounds but are not as resistant as bacterial endospores. Germination of both kinds of spores, asexual and sexual, produces hyphae and mycelia. The various sexual spores are shown in figure 7.3.

Zygospores are formed by the fusion and genetic exchange between hyphae which have formed gametangia. The hyphae are genetically distinct (+, −). The common bread mold *Rhizopus* forms zygospores.

Ascospores are haploid sexual spores formed in the interior of an oval or elongated structure called an ascus. An example is the fungus *Chaetomium*.

Basidiospores are sexual haploid spores produced externally on a club-shaped *basidium*. Basidospores are produced by mushrooms such as *Agaricus campestris*.

Subdivision of Fungi

The fungi are divided into five groups based on genetic analysis of ribosomal RNA:

Chytridiomycetes

These fungi are primarily aquatic. They may be unicellular or form hyphae. They produce flagellated zoospores, an adaptation of their aquatic habitat. Some are pathogens on reptiles such as *Batrachochytrium dendrobatidis*, which causes an infection in frogs that inhibits respiration across the epidermis of the animal.

Zygomycetes

The zygomycetes produce coenocytic (multinucleate) hyphae. They form sexual zygospores or nonmotile asexual sporangiospores. Some are important in food spoilage, such as *Rhizopus stolonifer*, a common bread mold. They prefer to grow in an atmosphere with high humidity.

Glomeromycetes

These fungi are a small group of about 160 species that almost exclusively form ectomycorrhizae on plants. They may have played a role in the colonization of land habitats by vascular plants. The fungal hyphae penetrate the plant cell wall to form structures called arbuscules, which are swollen vesicles. These structures are thought to aid the plant in the uptake of minerals from the soil.

Ascomycetes

The ascomycetes are a large and diverse group that are represented by single-celled organisms such as *Saccharomyces cerevisiae* but also by filamentous organisms such as *Neurospora crassa*, another type of bread mold. Sexual spores called *ascospores* are produced in saclike structures called asci (figure 7.3). Some asci occur in structures called ascocarps. Asexual reproduction involves the formation of conidiospores. Some ascomycetes are the fungal partners in the ectomycorrhizae on plant roots and in lichens with cyanobacteria.

Basidiomycetes

The basidiomycetes contain more than 30,000 species described to date. Mushrooms and toadstools are found

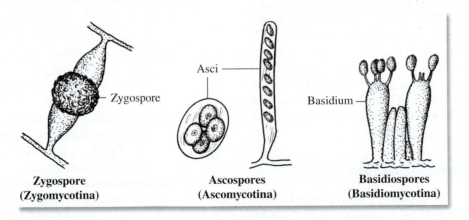

Figure 7.3 **Types of sexual spores seen in the fungi.**

in this of group of fungi. The differentiation between mushrooms and toadstools is that mushrooms, such as *Agaricus campestris*, are edible, whereas toadstools, such as *Amanita*, are considered poisonous. Normally a basidiomycete grows as haploid mycelium in the soil. However, distinct mating strains of mycelium occur which can fuse to form dikaryotic mycelia containing two nuclei per cell. When growth conditions are favorable, for instance, when moisture is abundant, the mycelium can differentiate into a *basidiocarp* which occurs above ground and is the mushroom structure that is normally seen. Gills, which are flat plates beneath the basidiocarp, contain the nuclei that fuse to form a basidium ("small pedestal") with diploid nuclei (figure 7.3). Two meiotic divisions of the diploid nuclei finally yield four haploid nuclei in the basidium. Each nucleus develops into a basidiospore, which can be distributed by winds and other natural means to habitats where new basidiomycete colonies are established to repeat the cycle of growth. One mushroom, *Coprinus*, distributes its basidiospores by producing chitinase that degrades its basidiocarp.

Not all of the fungi described here will be available for study in this exercise. Most will probably belong to zygomycetes and ascomycetes. Follow the directions of your laboratory instructor.

Laboratory Procedures

Several options are provided here for the study of fungi. The procedures to be followed will be outlined by your instructor.

Yeast Study

The organism *Saccharomyces cerevisiae,* which is used in bread making and alcohol fermentation, will be used for this study (figure 7.4). Either prepared slides or living organisms may be used.

Materials

- prepared slides of *Saccharomyces cerevisiae*
- broth cultures of *Saccharomyces cerevisiae*
- methylene blue stain
- microscope slides and coverslips

Prepared Slides If prepared slides are used, they may be examined under high-dry or oil immersion. Look for typical **blastospores** and **ascospores**. Space is provided on the Laboratory Report for drawing the organisms.

Living Materials If broth cultures of *Saccharomyces cerevisiae* are available, they should be examined on a wet mount slide with phase-contrast or brightfield

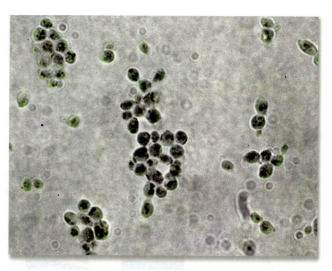

Figure 7.4 *Saccharomyces cerevisiae*, **methylene blue stain.**
© McGraw-Hill Education/Lisa Burgess, photographer.

optics. Place two or three loopfuls of the organisms on the slide with a drop of methylene blue stain. Oil immersion will reveal the greatest amount of detail. Look for the **nucleus** and **vacuole.** The nucleus is the smaller body. Draw a few cells on the Laboratory Report.

Mold Study

Examine a petri plate of Sabouraud's agar that has been exposed to the air for about an hour and incubated at room temperature for 3–5 days. This medium has a low pH, which makes it selective for fungi. A good plate will have many different-colored colonies. Note the characteristic "cottony" nature of the colonies. Also, look at the bottom of the plate and observe how the colonies differ in color here. Colony surface color, underside pigmentation, hyphal structure, and the types of spores produced are important phenotypic characteristics used in the identification of fungi.

Figure 7.5 reveals how some common molds appear when grown on Sabouraud's agar. Keep in mind that the appearance of a fungal colony can change appreciably as it gets older. The photographs in figure 7.5 are of colonies that are 10–21 days old.

Conclusive identification cannot be made unless a microscope slide is made to determine the type of hyphae and spores that are present. Figure 7.6 reveals, diagrammatically, the microscopic differences to look for when identifying fungal genera.

Two Options In making slides from fungal colonies, one can make either wet mounts directly from the colonies by the procedure outlined here or make cultured slides as outlined in Exercise 20. The following steps should be used for making stained slides directly from the colonies. Your instructor will indicate the number of identifications that are to be made.

Mold	Top	Reverse

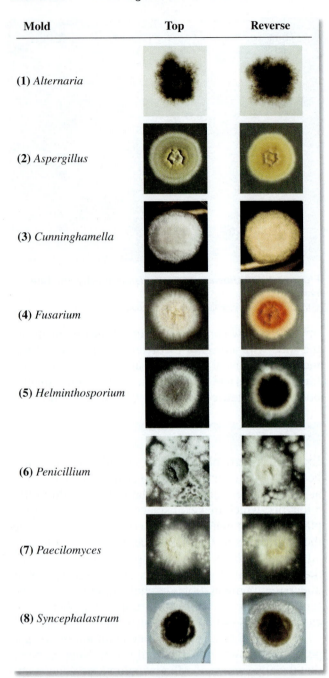

(1) *Alternaria*

(2) *Aspergillus*

(3) *Cunninghamella*

(4) *Fusarium*

(5) *Helminthosporium*

(6) *Penicillium*

(7) *Paecilomyces*

(8) *Syncephalastrum*

Figure 7.5 Colony characteristics of some of the more common molds.
© Harold Benson

Materials

- fungal cultures on Sabouraud's agar
- microscope slides and coverslips
- lactophenol cotton blue stain
- sharp-pointed scalpels or dissecting needles
- prepared slide of a fungal culture

1. Place an uncovered plate on a dissecting microscope and examine the colony. Look for hyphal structures and spore arrangement. Increase the magnification if necessary to more clearly see spores. Ignore white colonies as they are usually young and have not begun the sporulation process.
2. Consult figures 7.5 and 7.6 to make a preliminary identification based on colony characteristics and low-power magnification of hyphae and spores.
3. Make a wet mount slide by transferring a small amount of mycelium with a scalpel, dissecting needle, or toothpick to a drop of lactophenol cotton blue. Gently tease apart the mycelium with the dissecting needles. Cover the specimen with a coverslip and examine with the low-power objective. Look for hyphae that have spore structures. Go to the high-dry objective to discern more detail about the spores. Compare your specimen to figure 7.6 and see if you can identify the culture based on microscopic morphology.
4. Repeat the procedure for each colony.

Note: An alternative procedure that preserves the fruiting structures is the **cellophane tape method.** Place 1–2 drops of lactophenol cotton blue on a microscope slide. Using a piece of *clear* cellophane tape slightly smaller than the length of the slide, gently touch the surface of a fungal colony with the sticky side of the tape. Transfer the tape containing the material from the fungal colony to the lactophenol cotton blue stain and press the tape onto the slide, making sure that the culture material is in the stain. Observe with the low-power and high-dry lens.

Prepared Slides   If prepared slides are available, first examine them with high-dry objective and then switch to the oil immersion for greater detail. Look for typical structures such as sporangiospores, phialospores, and conidiospores. Space is provided on the Laboratory Report for drawing the organisms.

Laboratory Report

After recording your results on the Laboratory Report, answer all the questions.

(1)   (2)   (3)   (4)   (5)

(6)   (7)   (8)   (9)   (10)

(11)   (12)   (13)   (14)   (15)   (16)

(17)   (18)   (19)   (20)   (21)

(1) *Penicillium*–bluish-green; brush arrangement of phialospores.

(2) *Aspergillus*–bluish-green with sulfur-yellow areas on the surface. *Aspergillus niger* is black.

(3) *Verticillium*–pinkish-brown, elliptical microconidia.

(4) *Trichoderma*–green, resemble *Penicillium* macroscopically.

(5) *Gliocadium*–dark-green; conidia (phialospores) borne on phialides, similar to *Penicillium*; grows faster than *Penicillium*.

(6) *Cladosporium (Hormodendrum)*–light green to grayish surface; gray to black back surface; blastoconidia.

(7) *Pleospora*–tan to green surface with brown to black back; ascospores shown are produced in sacs borne within brown, flask-shaped fruiting bodies called pseudothecia.

(8) *Scopulariopsis*–light-brown; rough-walled microconidia.

(9) *Paecilomyces*–yellowish-brown, elliptical microconidia.

(10) *Alternaria*–dark greenish-black surface with gray periphery; black on reverse side; chains of macroconidia.

(11) *Bipolaris*–black surface with grayish periphery; macroconidia shown.

(12) *Pullularia*–black, shiny, leathery surface; thick-walled; budding spores.

(13) *Diplosporium*–buff-colored wooly surface; reverse side has red center surrounded by brown.

(14) *Oospora (Geotrichum)*–buff-colored surface; hyphae break up into thin-walled rectangular arthrospores.

(15) *Fusarium*–variants, of yellow, orange, red, and purple colonies; sickle-shaped macroconidia.

(16) *Trichothecium*–white to pink surface; two-celled conidia.

(17) *Mucor*–a zygomycete; sporangia with a slimy texture; spores with dark pigment.

(18) *Rhizopus*–a zygomycete; spores with dark pigment.

(19) *Syncephalastrum*–a zygomycete; sporangiophores bear rod-shaped sporangioles, each containing a row of spherical spores.

(20) *Nigrospora*–conidia black, globose, one-celled, borne on a flattened, colorless vesicle at the end of a conidiophore.

(21) *Montospora*–dark gray center with light gray periphery; yellow-brown conidia.

Figure 7.6 Microscopic appearance of some of the more common molds.

7 The Fungi: Molds and Yeasts

A. Results

1. **Yeast Study**

   Draw a few representative cells of *Saccharomyces cerevisiae* in the appropriate circles below. Blastospores (buds) and ascospores, if seen, should be shown and labeled.

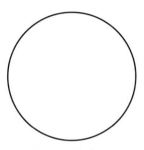

Prepared Slide

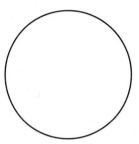

Living Cells

2. **Mold Study**

   In the following table, list the genera of molds identified in this exercise. Under colony description, give the approximate diameter of the colony, its topside color, and backside (bottom) color. For microscopic appearance, make a sketch of the organism as it appears on the slide preparation.

GENUS	COLONY DESCRIPTION	MICROSCOPIC APPEARANCE (DRAWING)

B. Short-Answer Questions

1. What does the term "coenocytic" mean?

   _____

   _____

2. What criteria are the basis for traditional classification schemes? What modern approach to classification has shown that traditional schemes do not apply?

   _____

   _____

3. What unique compound is found in the cell walls of fungi but is absent in plant cell walls?

   _____

   _____

4. What is one of the fungi that is responsible for infections of skin and nails?

   _____

   _____

5. What does the term "dimorphic" refer to? Give an example of an organism that is dimorphic and what disease it causes.

   _____

   _____

6. How do zygospores differ from conidiospores?

   _____

   _____

7. What foods are produced by fungi?

   _____

   _____

8. What is considered to be the difference between mushrooms and toadstools?

   _____

   _____

9. Why might fungi have been theorized to be involved in the Salem witch trials?

   _____

10. What are mycotoxins? In what popular food might they be found?

    _____

11. What are the ectomycorrhizae?

    _____

12. What components of wood must be degraded for its turnover by the wood-rotting fungi?

    _____

Manipulation of Microorganisms

One of the most critical practices that any beginning microbiology student must learn is **aseptic technique**. This technique ensures that an aseptic environment is maintained when handling microorganisms. This means two things:

1. no contaminating microorganisms are introduced into cultures or culture materials and
2. the microbiologist is not contaminated by cultures that are being manipulated.

Aseptic technique is crucial in characterizing an unknown organism. Multiple transfers must often be made from a stock culture to various test media. It is imperative that only the desired organism is transferred each time and that no foreign bacteria are introduced during the transfer. Aseptic technique is also important when isolating and purifying bacteria from a mixed source of organisms. The

© Adam Gault/age fotostock

streak plate and pour plate techniques are methods that can be used to isolate an individual species from a mixed culture. Once an organism is in pure culture and stored as a stock culture, aseptic technique insures that the culture remains pure when it is necessary to obtain a sample of the organism.

Individuals who work with pathogenic bacteria must be sure that any pathogen that is being handled is not accidently released, causing harm to themselves or to coworkers. Failure to observe aseptic technique can obviously pose a serious threat to many.

In the following exercises, you will learn the techniques that allow you to handle and manipulate cultures of microorganisms. Once you have mastered these procedures, you will be able to make transfers of microorganisms from one kind of medium to another with confidence. You will also be able to isolate an organism from a mixed culture to obtain a pure isolate. It is imperative that you have a good grasp of these procedures, as they will be required over and over in the exercises in this manual.

Aseptic Technique

Learning Outcomes

After completing this exercise, you should be able to

1. Aseptically transfer a bacterial culture from one broth tube to a sterile broth tube.
2. Aseptically transfer a bacterial culture from an agar slant to a sterile agar slant.
3. Aseptically transfer a bacterial colony from an agar plate to a sterile agar slant.

The use of aseptic technique ensures that no contaminating organisms are introduced into culture materials during handling or inoculation. It also ensures that the organisms that are being handled do not contaminate the handler or others who may be present. Finally, its use means that no contamination remains after you have worked with cultures.

Please note that the use of aseptic technique does not mean that you are working in a sterile environment, as that would require the absence of all microbes, including endospores and other resistant forms. However, because the methods used remove or kill a large number of microbes, aseptic technique does reduce the likelihood of contamination.

As you work with these procedures over time, they will become routine and second nature to you. Any time you work with microbial cultures, you should use the set of procedures outlined below. This may involve the transfer of a broth culture to a plate for streaking, or inoculating an isolated colony from a plate onto a slant culture to prepare a stock culture. It may also involve inoculating many tubes of media and agar plates from a stock culture in order to characterize and identify an unknown bacterium. Ensuring that only the desired organism is transferred in each inoculation is of paramount importance in the identification process. The general procedure for aseptic technique follows.

Work Area Disinfection The work area is first treated with a disinfectant to kill any microorganisms that may be present. This process destroys vegetative cells and viruses but may not destroy endospores.

Loops and Needles The transfer of cultures will be achieved using inoculating loops and needles. These tools

must be sterilized before transferring any culture. A loop or needle is sterilized by inserting it into a Bunsen burner or incinerator until it is red-hot. This will incinerate any contaminating organisms that may be present. Allow the loop to cool completely before picking up inoculum. This will ensure that viable cells are transferred.

Working with Culture Tubes Prior to inserting a cooled loop or needle into a culture tube, the cap is removed and the mouth of the tube may be flamed. The cap should always be kept in your hand, rather than placed on the lab bench. If the tube is a broth tube, the loop is inserted into the tube and twisted several times to ensure that the organisms on the loop are delivered to the liquid. If the tube is an agar slant, the surface of the slant is inoculated by drawing the loop up the surface of the slant from the bottom of the slant to its top in a zigzag pattern. For stab cultures, a needle is inserted into the agar medium by stabbing it into the agar. After the culture is inoculated, the mouth of the tube may be reflamed and the tube is recapped.

Working with Agar Plates Loops are used to inoculate or streak agar plates. The plate cover is raised and held diagonally over the plate to protect the surface from any contamination in the air. The loop containing the inoculum is then streaked gently over the surface of the agar. It is important not to gouge or disturb the surface of the agar with the loop. The cover is replaced and the loop is flamed.

Final Flaming of the Loop or Needle After the inoculation is complete, the loop or needle is flamed to destroy any organisms that remain on these implements. The loop or needle is then returned to its receptacle for storage. It should never be placed on the desk surface.

Final Disinfection of the Work Area When all work for the day is complete, the work area is treated with disinfectant to ensure that any organism that might have been deposited during any of the procedures is killed.

To gain some practice in aseptic transfer of bacterial cultures, three simple transfers will be performed in this exercise:

1. broth culture to broth tube
2. agar slant culture to an agar slant and
3. agar plate to an agar slant.

Transfer from Broth Culture to Another Broth

Complete a broth tube to broth tube inoculation using the following technique. Figure 8.1 illustrates the procedure for removing organisms from a broth culture, and figure 8.2 shows how to inoculate a tube of sterile broth.

Materials

- broth culture of *Escherichia coli*
- tubes of sterile nutrient broth
- inoculating loop

- Bunsen burner or incinerator
- disinfectant for desktop and paper towels
- Sharpie marking pen

1. Prepare your desktop by swabbing down its surface with a disinfectant. Use a sponge or paper towels.
2. With a marking pen, label a tube of sterile nutrient broth with your initials and *E. coli*.
3. Sterilize your inoculating loop by flaming it *until it becomes bright red*. The entire wire must be heated. See illustration 1, figure 8.1.
4. Using your free hand, gently shake the tube to disperse the culture (illustration 2, figure 8.1).

(1) Inoculating loop is heated until it is red-hot.

(2) Organisms in culture are dispersed by shaking tube.

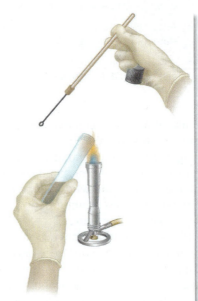

(3) Tube enclosure is removed and mouth of tube is flamed.

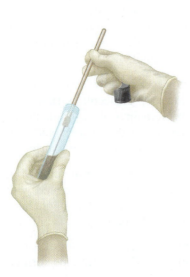

(4) A loopful of organisms is removed from tube.

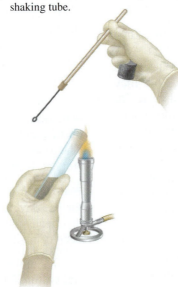

(5) Loop is removed from culture and tube mouth is flamed.

(6) Tube enclosure is returned to tube.

Figure 8.1 **Procedure for removing organisms from a broth culture with inoculating loop.**

5. Grasp the tube cap with the little finger of your hand holding the inoculating loop and remove it from the tube. Flame the mouth of the tube, as shown in illustration 3, figure 8.1. Note: If an incinerator is used, the tube is not flamed.

6. Insert the inoculating loop into the culture (illustration 4, figure 8.1).

7. Remove the loop containing the culture, flame the mouth of the tube again (illustration 5, figure 8.1), and recap the tube (illustration 6). Place the culture tube back on the test-tube rack.

8. Grasp a tube of sterile nutrient broth with your free hand, carefully remove the cap with your little finger, and flame the mouth of this tube (illustration 1, figure 8.2).

9. Without flaming the loop, insert it into the sterile broth, inoculating it (illustration 2, figure 8.2). To disperse the organisms into the medium, move the loop back and forth in the tube.

10. Remove the loop from the tube and flame the mouth (illustration 3, figure 8.2). Replace the cap on the tube (illustration 4, figure 8.2).

11. Sterilize the loop by flaming it (illustration 5, figure 8.2). Return the loop to its container.

12. Incubate the culture you just inoculated at 37°C for 24–48 hours.

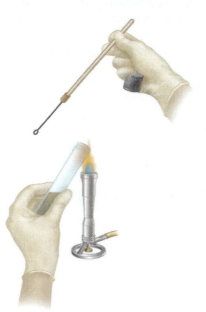

(1) Cap is removed from sterile broth and tube mouth is flamed.

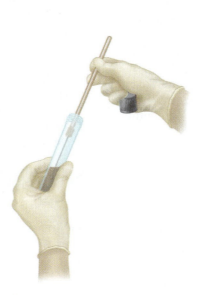

(2) Sterile, cooled loop is inserted into tube of sterile broth.

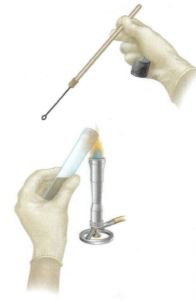

(3) Loop is removed from broth and tube mouth is flamed.

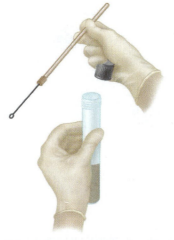

(4) Tube enclosure is returned to tube.

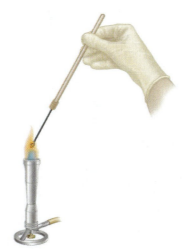

(5) Loop is flamed and returned to receptacle.

Figure 8.2 **Procedure for inoculating a nutrient broth.**

Transfer of Bacteria from a Slant

To inoculate a sterile nutrient agar slant from an agar slant culture, use the following procedure. Figure 8.3 illustrates the entire process.

Materials

- agar slant culture of *E. coli*
- sterile nutrient agar slant
- inoculating loop
- Bunsen burner or incinerator
- Sharpie marking pen

1. If you have not already done so, prepare your desktop by swabbing down its surface with a disinfectant.
2. With a marking pen, label a nutrient agar slant with your initials and *E. coli*.
3. Sterilize your inoculating loop by holding it over the flame of a Bunsen burner *until it becomes bright red* (illustration 1, figure 8.3). The entire wire must be heated. Allow the loop to cool completely.
4. Using your free hand, pick up the slant culture of *E. coli* and remove the cap using the little finger of the hand that is holding the loop (illustration 2, figure 8.3).
5. Flame the mouth of the tube and insert the cooled loop into the tube. Pick up some of the culture on

the loop (illustration 3, figure 8.3) and remove the loop from the tube.
6. Flame the mouth of the tube (illustrations 4 and 5, figure 8.3) and replace the cap, being careful not to burn your hand. Return tube to rack.
7. Pick up a sterile nutrient agar slant with your free hand, remove the cap with your little finger as before, and flame the mouth of the tube (illustration 6, figure 8.3).
8. Without flaming the loop containing the culture, insert the loop into the tube and gently inoculate the surface of the slant by moving the loop back and forth over the agar surface, while moving up the surface of the slant (illustration 7, figure 8.3). This should involve a type of serpentine or zigzag motion.
9. Remove the loop, flame the mouth of the tube, and recap the tube (illustration 8, figure 8.3). Replace the tube in the rack.
10. Flame the loop, heating the entire wire to red-hot (illustration 9, figure 8.3), allow to cool, and place the loop in its container.
11. Incubate the inoculated agar slant at 37°C for 24–48 hours.

Working with Agar Plates

(Inoculating a slant from an agar plate)

The transfer of organisms from colonies on agar plates to slants or broth tubes is very similar to the

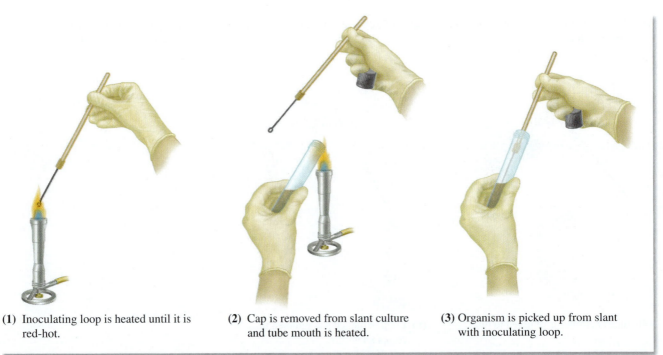

(1) Inoculating loop is heated until it is red-hot.

(2) Cap is removed from slant culture and tube mouth is heated.

(3) Organism is picked up from slant with inoculating loop.

continued

Figure 8.3 **Procedure for inoculating a nutrient agar slant from a slant culture.**

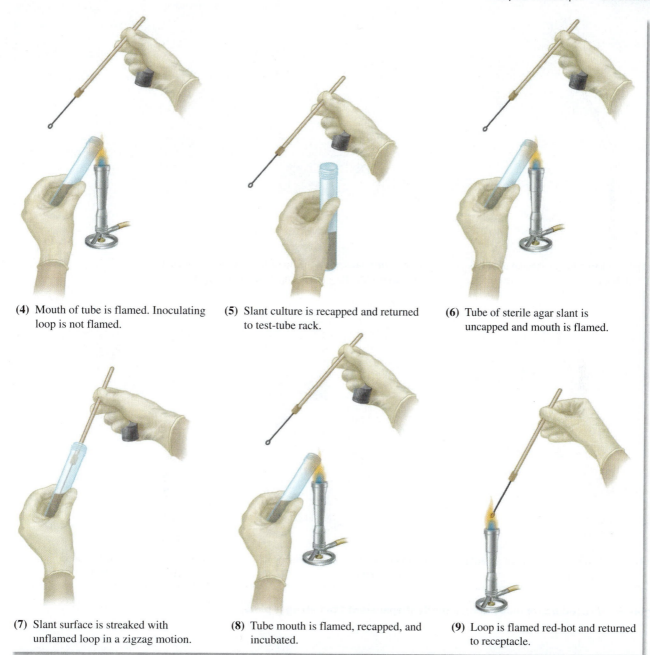

(4) Mouth of tube is flamed. Inoculating loop is not flamed.

(5) Slant culture is recapped and returned to test-tube rack.

(6) Tube of sterile agar slant is uncapped and mouth is flamed.

(7) Slant surface is streaked with unflamed loop in a zigzag motion.

(8) Tube mouth is flamed, recapped, and incubated.

(9) Loop is flamed red-hot and returned to receptacle.

Figure 8.3 *(continued)*

procedures used in the last two transfers (broth to broth and slant to slant). The following rules should be observed:

Loops and Needles   Loops are routinely used when streaking agar plates and slants. When used properly, a loop will not gouge or tear the agar surface. Needles are used in transfers involving stab cultures.

Plate Handling   Media in plates must always be protected against contamination. To prevent exposure to air contamination, covers should always be left closed. When organisms are removed from a plate

culture, the cover should be only partially opened as shown in illustration 2, figure 8.4.

Flaming Procedures   Inoculating loops or needles must be flamed in the same manner that you used when working with previous tubes. One difference when working with plates is that plates are never flamed!

Plate Labeling and Incubation   Agar plates should be labeled on the bottom of the plate (the part containing the agar). Inoculated agar plates are almost always incubated upside down. This prevents moisture from condensing on the agar surface and spreading the inoculated organisms.

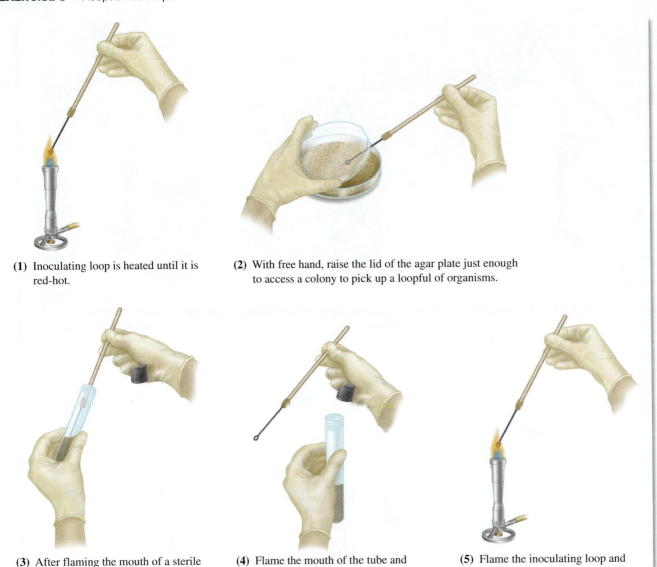

(1) Inoculating loop is heated until it is red-hot.

(2) With free hand, raise the lid of the agar plate just enough to access a colony to pick up a loopful of organisms.

(3) After flaming the mouth of a sterile slant, streak its surface.

(4) Flame the mouth of the tube and recap the tube.

(5) Flame the inoculating loop and return it to receptacle.

Figure 8.4 **Procedure for inoculating a nutrient agar slant from an agar plate.**

To transfer organisms from a petri plate to an agar slant, use the following procedure. Figure 8.4 illustrates the entire process.

Materials

- nutrient agar plate with bacterial colonies
- sterile nutrient agar slant
- inoculating loop
- Sharpie marking pen

1. If you have not done so, clean your work area with disinfectant. Allow area to dry.
2. Label a sterile nutrient agar slant with your name and organism to be transferred.
3. Flame an inoculating loop until it is red-hot (illustration 1, figure 8.4). Allow the loop to cool.

4. As shown in illustration 2, figure 8.4, raise the lid of a petri plate sufficiently to access a colony with your sterile loop.

Do not gouge the agar with your loop as you pick up organisms. Simply allow the loop to gently glide over the gelatin-like surface of the agar. Do not completely remove the lid while inoculating or removing organisms from the agar plate. This will expose the agar surface to air and potential contamination. Always close the lid once you have removed organisms from the plate.
5. With your free hand, pick up the sterile nutrient agar slant tube. Remove the cap by grasping the cap with the little finger of the hand that is holding the loop.

6. Flame the mouth of the tube and insert the loop into the tube to inoculate the surface of the slant, using a serpentine motion (illustration 3, figure 8.4). Avoid disrupting the agar surface with the loop.

7. Remove the loop from the tube and flame the mouth of the tube. Replace the cap on the tube (illustration 4, figure 8.4).

8. Flame the loop (illustration 5, figure 8.4) and place it in its container.

9. Incubate the nutrient agar slant at 37°C for 24–48 hours.

Results

Examine all three tubes and record your results in Laboratory Report 8.

8 Aseptic Technique

A. Results

1. Were all your transfers successful? _____

2. How do you know your transfer to a broth was successful? How do you know your transfers to agar slants were successful? _____
    _____
    _____

3. If any of your transfers were unsuccessful, suggest possible errors that may have been made in the transfer process. _____
    _____

B. Short-Answer Questions

1. Provide three reasons why the use of aseptic technique is essential when handling microbial cultures in the laboratory.
    _____
    _____

2. Provide two examples of how heat is used during inoculation of a tube culture.
    _____
    _____

3. How is air contamination prevented when an inoculating loop is used to introduce or take a bacterial sample to/from an agar plate?
    _____
    _____

4. Where should a label be written on an agar plate?
    _____

5. How should agar plates be incubated? Why?
    _____
    _____

6. Disinfectants are effective against which types of organisms? Which types of organisms may remain on the lab bench even after disinfection? What disinfectant(s) is used in your laboratory?
    _____
    _____

7. Compare and contrast the growth of bacteria in different physical types of media (broths, slants, and agar plates). What might be the advantages and disadvantages of using each type?

_____

_____

_____

C. Multiple Choice

Select the answer that best completes the following statements.

1. A disinfectant is used on your work surface
   a. before the beginning of laboratory procedures.
   b. after all work is complete.
   c. after any spill of live microorganisms.
   d. Both (b) and (c) are correct.
   e. All of the above are correct.

2. To retrieve a sample from a culture tube with an inoculating loop, the cap of the tube is
   a. removed and held in one's teeth.
   b. removed and held with the fingers of the loop hand.
   c. removed with the fingers of the loop hand and placed in the fingers of the tube hand.
   d. removed with the fingers of the loop hand and placed on the laboratory bench.
   e. Any of these methods can be used.

3. An inoculating loop or needle is sterilized using heat
   a. by one brief passage.
   b. for exactly 5 minutes.
   c. until the entire wire is bright red.
   d. until the handle is bright red.
   e. until the tip is bright red.

4. Which of the following would be a correctly labeled agar plate?
   a. *Staph* on the bottom
   b. *S. aureus* on the bottom
   c. *S. aureus* on the lid
   d. *Staphylococcus aureus* on the lid

5. Noah wanted to transfer *Staphylococcus aureus* from a broth to an agar plate. He picked up the broth culture, removed the cap, and flamed the mouth of the tube. He inserted an inoculating loop to obtain a bacterial sample. Then, he flamed the mouth of the tube and replaced the cap. Noah opened the lid of a labeled agar plate diagonally and used the loop to streak the surface of the agar. After closing the lid, he flamed the loop in an incinerator and put it back in its container. The plate was incubated upside down for 24–48 hours. What did Noah do wrong in this transfer?
   a. He did not use the transfer tool correctly.
   b. He did not handle the culture tube correctly.
   c. He did not handle the agar plate correctly.
   d. He used the wrong tool in transfer.

ANSWERS

Multiple Choice

1. _____

2. _____

3. _____

4. _____

5. _____

Pure Culture Techniques

Learning Outcomes

After completing this exercise, you should be able to

1. Obtain isolated colonies of a mixed culture using the streak plate method.

2. Obtain isolated colonies from a bacterial culture using the loop dilution pour plate method.

3. Differentiate between surface and subsurface colonies in a pour plate.

4. Evaluate the purity of your isolated colonies by transferring a single colony to an agar slant and obtaining the growth of a single type of organism.

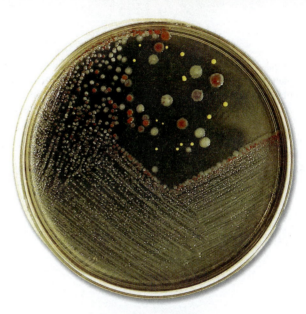

Figure 9.1 A streak plate demonstrating well-isolated colonies of three different bacteria.
© Harold Benson

When we try to study the microbiota of the body, soil, water, or just about any environment, we realize quickly that bacteria exist in natural environments as mixed populations. It is only in very rare instances that they occur as a single species. Robert Koch, the father of medical microbiology, was one of the first to recognize that if he was going to prove that a particular bacterium causes a specific disease, it would be necessary to isolate the agent from all other bacteria and characterize the pathogen. From his studies on pathogenic bacteria, his laboratory contributed many techniques to the science of microbiology, including isolation methods for obtaining **pure cultures** of bacteria. A pure culture contains only a single kind of an organism, whereas a mixed culture contains more than one kind of organism. A contaminated culture contains a desired organism but also unwanted organisms. With a pure culture, we can study the cultural, morphological, and physiological characteristics of an individual organism.

Several methods for obtaining pure cultures are available to the microbiologist. Two commonly used isolation procedures are the **streak plate** and the **pour plate.** Both procedures involve diluting the bacterial cells in a sample to an end point where single bacterial cells are spread out across the surface or within the agar of a plate so that when each cell divides, it gives rise to an isolated pure **colony.** The colony is therefore assumed to be the identical progeny of the original cell and can be picked and used for further study of the bacterium.

In this exercise, you will use both the streak plate and pour plate methods to separate a mixed culture of bacteria. The bacteria may be differentiated by the characteristics of the colony, such as color, shape, and other colony characteristics. Isolated colonies can then be subcultured and stained microscopic slides prepared to check for purity.

Streak Plate Method

The streak plate method is the procedure most often used by microbiologists to obtain pure cultures. It is quick and requires very few materials. However, it requires a certain level of skill which is only obtained through practice. Your instructor may want you to try more than one method or only concentrate on one of the streak patterns illustrated in figure 9.2. Figure 9.1 illustrates how colonies of a mixed bacterial culture should be spread out and separated on a properly made quadrant streak plate using method B shown in figure 9.2. Good spacing between colonies on the plate is critical so that a single pure colony can be aseptically isolated from quadrant 4 and used for further testing and study. This will ensure that you are not working with a mixed or contaminated culture.

Quadrant Streak
(Method A)

(1) Streak one loopful of organisms over Area 1 near edge of the plate. Apply the loop lightly. Don't gouge into the medium.

(2) Flame the loop, cool 5 seconds, and make 5 or 6 streaks from Area 1 through Area 2. Momentarily touching the loop to a sterile area of the medium before streaking insures a cool loop.

(3) Flame the loop again, cool it, and make 6 or 7 streaks from Area 2 through Area 3.

(4) Flame the loop again, and make as many streaks as possible from Area 3 through Area 4, using up the remainder of the plate surface.

(5) Flame the loop before putting it aside.

Quadrant Streak
(Method B)

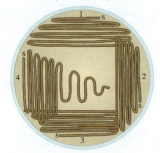

(1) Streak one loopful of organisms back and forth over Area 1, starting at point designated by "s." Apply loop lightly. Don't gouge into the medium.

(2) Flame the loop, cool 5 seconds, and touch the medium in a sterile area momentarily to insure coolness.

(3) Rotate dish 90 degrees while keeping the dish closed. Streak Area 2 with several back and forth strokes, hitting the original streak a few times.

(4) Flame the loop again. Rotate the dish and streak Area 3 several times, hitting the last area several times.

(5) Flame the loop, cool it, and rotate the dish 90 degrees again. Streak Area 4, contacting Area 3 several times and drag out the culture as illustrated.

(6) Flame the loop before putting it aside.

Radiant Streak

(1) Spread a loopful of organisms in a small area near the edge of the plate in Area 1. Apply the loop lightly. Don't gouge into the medium.

(2) Flame the loop and allow it to cool for 5 seconds. Touching a sterile area will insure coolness.

(3) **From the edge** of Area 1 make 7 or 8 straight streaks to the opposite side of the plate.

(4) Flame the loop again, cool it sufficiently, and cross streak over the last streaks, **starting near Area 1.**

(5) Flame the loop before putting it in its receptacle.

Figure 9.2 **Three different streak patterns that can be used to obtain isolated colonies.**

Figure 9.2 shows three procedures for producing a streak plate that will yield isolated colonies. By far the most popular and most utilized procedure is the **quadrant streak plate,** shown in methods A and B. All of the methods depend upon the physical dilution of cells over the plate surface until a single cell is deposited in an area and grows to produce an isolated bacterial colony. It is important for beginning students to master the streak plate as success in future exercises will depend on using this technique to obtain isolated cultures. This is especially true for the exercises in Part 8 involving the identification of an unknown bacterium.

Materials

- electric hot plate and beaker of water
- Bunsen burner or incinerator
- inoculating loop, thermometer, and Sharpie marking pen
- 20 ml nutrient agar pour and 1 sterile petri plate
- mixed culture of *Serratia marcescens* and *Micrococcus luteus* (or a mixed culture of *Escherichia coli* and *Chromobacterium violaceum*)

1. Prepare your tabletop by disinfecting its surface with the disinfectant provided in the laboratory. Use paper towels to scrub it clean.
2. Using the marking pen, label the bottom surface of a sterile petri plate with your name and date.
3. Liquefy a tube of nutrient agar, cool to 50°C, and pour the medium into the bottom of the plate, following the procedure illustrated in figure 9.3. Be sure to flame the mouth of the tube prior to pouring to destroy any bacteria around the end of the tube.

   After pouring the medium into the plate, gently rotate the plate so that it becomes evenly distributed, but do not splash any medium up over the sides.

   Agar, the solidifying agent in this medium, becomes liquid when boiled and resolidifies at around 42°C. Failure to cool it prior to pouring into the plate will result in condensation of moisture on the cover. Any moisture on the cover is undesirable because it can become deposited on the agar surface, causing the organisms to spread randomly over the surface and thereby defeating the purpose of the isolation procedure. Note: This step can be skipped if you are working with pre-poured agar plates.
4. Streak your plate using one of the methods shown in figure 9.2. Method B is the most commonly used procedure.

Caution

Be sure to follow the routine in figure 9.4 for obtaining the organism from culture using correct aseptic technique.

---

5. Incubate the plate in an inverted position for 24 to 48 hours. If the plate is not incubated in an inverted position, condensate from the dish lid will be deposited on the agar surface, dispersing the organisms, disrupting the desired growth pattern, and preventing the formation of individual colonies.

(1) Liquefy a nutrient agar pour by placing it in a beaker of boiling water for 5 minutes.

(2) Cool down the nutrient agar pour to 50°C. Hold at 50°C for 5 minutes.

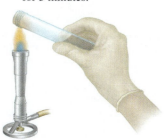

(3) Remove the cap from the tube and flame the open end of the tube.

(4) Pour the contents of the tube into the bottom of the agar plate and allow it to solidify.

Figure 9.3 **Procedure for pouring an agar plate for streaking.**

(1) Shake the culture tube from side to side to suspend organisms. Do not moisten cap on tube.

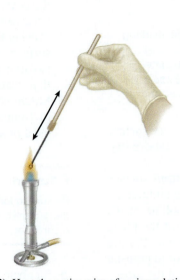

(2) Heat the entire wire of an inoculating loop until it is red-hot.

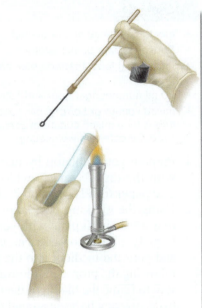

(3) Remove the cap and flame the mouth of the tube. Do not place the cap down on the table.

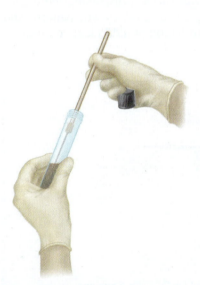

(4) After allowing the loop to cool for at least 5 seconds, remove a loopful of organisms. Avoid touching the side of the tube.

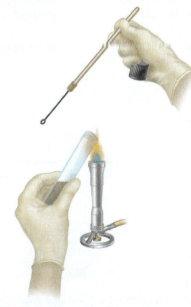

(5) Flame the mouth of the culture tube again.

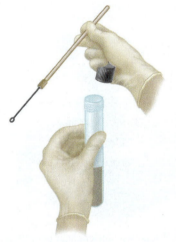

(6) Return the cap to the tube and place the tube in a test-tube rack.

continued

Figure 9.4 Routine for obtaining an inoculum and streaking an agar plate.

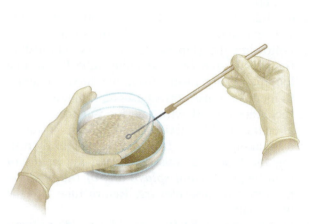

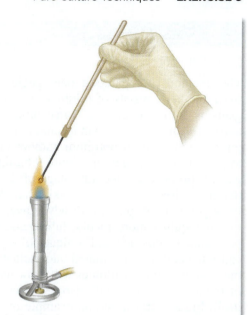

(7) Streak the plate using one of the patterns shown in figure 10.2, holding the lid at a diagonal as shown. Do not gouge into the medium with the loop.

(8) Flame the loop before placing it in its container.

Figure 9.4 *(continued)*

Results for isolating colonies using the three different streaking patterns are shown in figure 9.5. The quadrant streak methods shown in (a) and (b) are the most commonly used and usually give consistent results.

(a)

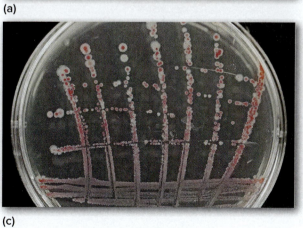

(b)

(c)

Figure 9.5 Streak plates using the three inoculation patterns for obtaining isolated colonies as shown in figure 9.2: (a) quadrant streak method A; (b) quadrant streak method B; (c) radiant streak.

© McGraw-Hill Education/Lisa Burgess, photographer

Pour Plate Method

(Loop Dilution)

This method of separating one species of bacteria from another consists of diluting one loopful of organisms across a series of three tubes of liquefied nutrient agar in such a manner that one of the plates poured will have an optimum number of organisms to provide good isolation. Figure 9.6a illustrates the general procedure. One advantage of this method is that it requires somewhat less skill than that required for a good streak plate; a disadvantage, however, is that it requires more media, tubes, and plates. The pour plate technique yields unique colony growth because the cells are inoculated into melted agar before the plate is poured. Colonies will grow on the surface of the agar, but they will also grow within the agar itself. Figure 9.6b shows an example of a pour plate and the difference between surface and subsurface colonies.

Proceed as follows to make three loop dilution pour plates, using the same mixed culture you used for your streak plate.

Materials

- mixed culture of bacteria
- 3 nutrient agar pours
- 3 sterile agar plates
- electric hot plate
- beaker of water
- thermometer
- inoculating loop and Sharpie marking pen

1. Label the three nutrient agar pours **I, II,** and **III** with a marking pen and place them in a beaker of water on an electric hot plate to be liquefied.

To save time, start with hot tap water if it is available.
2. While the tubes of media are being heated, label the bottoms of the three empty agar plates **I, II,** and **III.**
3. Cool down the tubes of media to 50°C, using the same method that was used for the streak plate.
4. Following the step-by-step process outlined in figure 9.7, start by inoculating tube I with one loopful of organisms from the mixed culture. Remember to use correct aseptic technique.
5. Inoculate tube II with one loopful from tube I after thoroughly mixing the organisms in tube I by shaking the tube from side to side or by rolling the tube vigorously between the palms of both hands. ***Do not splash any of the medium up onto the tube closure.*** Return tube I to the water bath.
6. Gently shake tube II from side to side to completely disperse the organisms and inoculate tube III with one loopful from tube II. Return tube II to the water bath.
7. Gently shake tube III, flame its mouth, and pour its contents into plate III.
8. Flame the mouths of tubes I and II and pour their contents into their respective plates.
9. After the medium has completely solidified, incubate the *inverted* plates at 22°C (room temperature) for 24 to 48 hours.

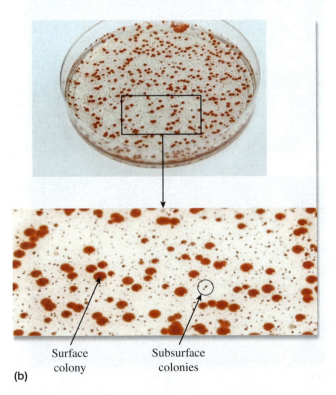

(a)

(b)

Figure 9.6 (a) General steps of the loop dilution pour plate technique. (b) Pour plate results illustrating both surface and subsurface growth.

© McGraw-Hill Education/ Gabriel Guzman, photographer

Evaluation of the Two Isolation Methods

After 24 to 48 hours of incubation, examine your streak plate and your three pour plates. Look for colonies that are well isolated from the others. Note how crowded the colonies appear on pour plate I as compared with pour plates II and III. It is likely that plate II or III will have the most favorable isolation of colonies. Can you pick out well-isolated colonies of the two different bacterial species that were present in the mixed culture?

Draw the appearance of your streak plate and pour plates on the Laboratory Report.

Subculturing Techniques

The next step in the development of a pure culture is the transfer of an isolated colony from a streak or pour plate to a tube of nutrient broth or a slant of nutrient agar. In this exercise, you will use your loop to carefully pick up an isolated colony from one of your plates and aseptically transfer that colony to a sterile nutrient agar slant. You will do this for each of the organisms that you isolated from the mixed culture, putting each species into its own agar slant. Ideally, this subculturing process should yield a pure culture of each organism.

(1) Liquefy three nutrient agar pours, cool to 50°C, and let stand for 10 minutes.

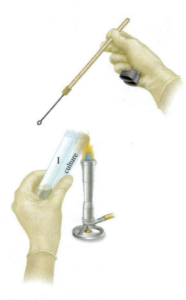

(2) After shaking the mixed culture to disperse the organisms, flame the loop and mouths of the tubes.

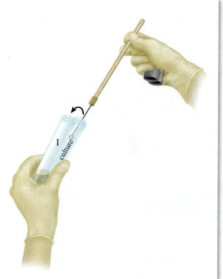

(3) Transfer one loopful of the mixed culture to tube I.

(4) Flame the loop and the mouths of both tubes.

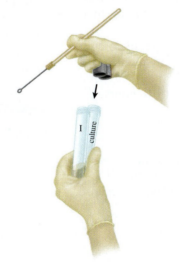

(5) Replace the caps on the tubes and return mixed culture to the test-tube rack.

(6) Disperse the organisms in tube I by gently shaking the tube or rolling it between the palms.

continued

Figure 9.7 Detailed procedure for inoculating and pouring plates using the loop dilution pour plate technique.

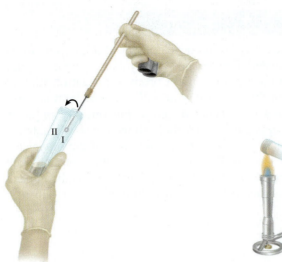
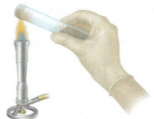
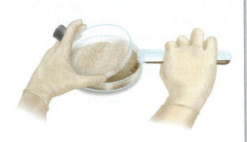

(7) Transfer one loopful from tube I to tube II. Remember to flame the loop and the mouths of both tubes before and after the transfer. Return tube I to the water bath.

(8) After shaking tube II and transferring one loopful to tube III using aseptic technique, flame the mouth of each tube.

(9) Pour the inoculated pours into their respective agar plates.

Figure 9.7 (continued)

Materials

- two nutrient agar slants
- inoculating loop
- Bunsen burner

1. Label one sterile agar slant *Micrococcus luteus* and a second *Serratia marcescens.*
2. Select a well-isolated red colony on either your streak plate or one of your pour plates, and use your inoculating loop to aseptically transfer it to the tube labeled *S. marcescens.*
3. Repeat this procedure for a yellow/cream-colored colony and transfer the colony to the tube labeled *M. luteus.*
4. Incubate the tubes at 22°C for 24 to 48 hours.

Evaluation of Slants

After incubation, examine the slants. Is the *S. marcescens* culture red? If the culture was incubated at a temperature higher than 22°C, it may not be red because higher temperatures inhibit production of the organism's red pigment. Draw the appearance of the slant in the Laboratory Report. What color is the *M. luteus* culture? Draw the slant.

You cannot be sure that your cultures are pure until you have made a microscopic examination of the respective cultures. It is entirely possible that the *S. marcescens* culture harbors some contaminating *M. luteus* and vice versa. If time allows, you can prepare Gram-stained smears (see Exercise 14) to verify the purity of your cultures, and record those results in the Laboratory Report. *S. marcescens* is a gram-negative short rod, whereas *M. luteus* is a gram-positive coccus.

Laboratory Report

Complete Laboratory Report 9 for this exercise.

9 Pure Culture Techniques

A. Results

1. **Evaluation of Streak Plate**

   Show within the circle the distribution of the colonies on your streak plate. To identify the colonies, use red for *Serratia marcescens* and yellow for *Micrococcus luteus*. (Use purple for *Chromobacterium violaceum* and yellow for *E. coli* if these species were included in your mixed culture.) If time permits, your instructor may inspect your plate and enter a grade where indicated.

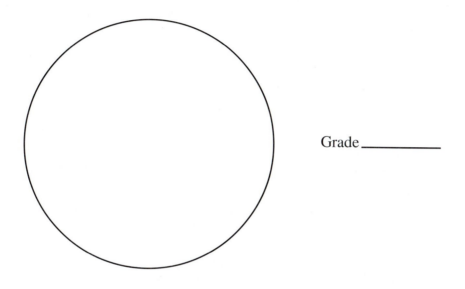

Grade _____

2. **Evaluation of Pour Plates**

   Show the distribution of colonies on plates II and III, using only the quadrant section for plate II. If plate III has too many colonies, follow the same procedure. Use colors to illustrate the different bacterial species.

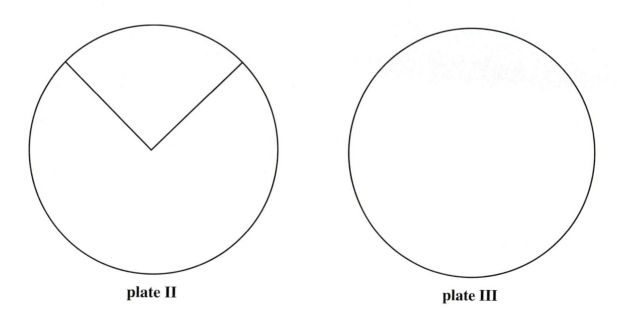

plate II   **plate III**

3. **Subculture Evaluation**

   With colored pencils, sketch the appearance of the growth on the slant diagrams below. If you prepared and observed Gram-stained slides of the subcultured bacteria, draw a few cells of each organism as revealed by Gram staining in the adjacent circle.

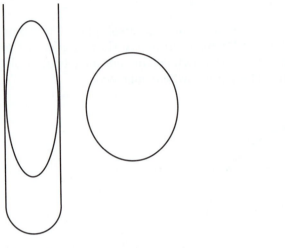

Serratia marcescens
(or *Escherichia coli*)

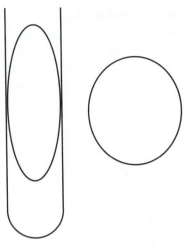

Micrococcus luteus
(or *Chromobacterium violaceum*)

4. Compare the results of your streak and pour plates. Which method achieved the best separation of species?

   _____

   _____

5. Do your slants contain pure cultures? How would you confirm their purities?

   _____

   _____

B. Short-Answer Questions

1. Define the term "colony" as it relates to bacterial growth on solid media.

   _____

   _____

2. What colony characteristics can be used for differentiation of bacterial species? As an example, compare the properties of colonies of *Serratia marcescens* and *Micrococcus luteus* on your streak plate.

   _____

   _____

3. Why is dilution a necessary part of pure culture preparation?

   _____

   _____

4. What advantage(s) does the streak plate method have over the pour plate method?

   _____

   _____

5. What advantage(s) does the pour plate method have over the streak plate method?

   _____

   _____

6. Why is the loop flamed before it is placed in a culture tube? Why is it flamed after completing the inoculation?

   _____

   _____

7. Before inoculating and pouring molten nutrient agar into a plate, why must the agar first be cooled to 50°C?

   _____

   _____

8. Explain why plates should be inverted during incubation.

   _____

   _____

9. Describe the difference between the appearance of surface and subsurface colonies in a pour plate. If this is the same bacterial species, why do these differences in colonial growth occur?

   _____

   _____

   _____

Staining and Observation of Microorganisms

The eight exercises in this unit include the procedures for 10 slide techniques that one might use to study the morphology of bacteria. A culture method in Exercise 17 is also included as a substitute for slide techniques when pathogens are encountered.

These exercises are intended to serve two equally important functions: (1) to help you to develop the necessary skills in making slides and (2) to introduce you to the morphology of bacteria. Although the title of each exercise pertains to a specific technique, the organisms chosen for each method have been carefully selected so that you can learn to recognize certain morphological features. For example, in the exercise on simple staining (Exercise 11), a single stain applied to the selected organism can be used to demonstrate cell morphology, cell arrangement, and internal storage materials such as metachromatic granules. In Exercise 14 (Gram Staining), you will learn how to perform an important differential stain that uses

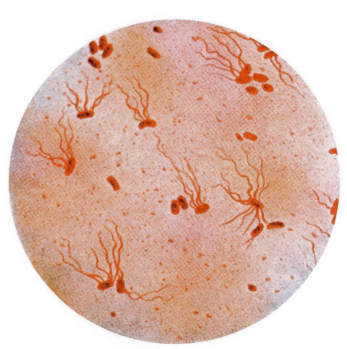

Source: Centers for Disease Control

more than one stain. This procedure allows you to taxonomically differentiate between two different kinds of bacteria as well as distinguish their cell morphology—cocci or rods.

The importance of mastering these techniques cannot be overemphasized. Although one is seldom able to make species identification on the basis of morphological characteristics alone, it is a very significant starting point. This fact will become increasingly clear with subsequent experiments.

Although the steps in the various staining procedures may seem relatively simple, student success is often quite unpredictable. Unless your instructor suggests a variation in the procedure, try to follow the procedures exactly as stated, without improvisation. Photomicrographs in color have been provided for many of the techniques; use them as a guide to evaluate the slides you have prepared. If your results do not turn out as expected, consider the possible sources of error. If time allows, repeat the procedure until you are successful.

Smear Preparation

Learning Outcomes

After completing this exercise, you should be able to

1. Prepare a thin smear of bacteria on a microscope slide by transferring bacteria from both liquid and solid cultures.

2. Fix the bacterial cells using gentle heat to create a smear that can be stained.

The success of virtually all staining procedures depends upon the preparation of a good **smear.** The procedure outlined in this exercise should be followed for most stains unless you are instructed otherwise. Good smears are critical for discerning: (1) the morphology of cells, such as rods, cocci, and other bacterial shapes; (2) the arrangement of cells, such as single cells, chains, or clusters; and (3) internal structures, such as endospores and cell inclusions. Learning early how to prepare good smears will ensure success in the staining exercises and later exercises for the identification of an unknown bacterium.

There are several goals in preparing a smear. The first goal is to adhere the cells to the microscope slide so that they are not washed off during subsequent staining and washing procedures. Second, it is important to ensure that shrinkage of cells does not occur during staining; otherwise, distortion and artifacts can result. A third goal is to prepare thin smears because the thickness of the smear will determine if you can visualize individual cells, their arrangement, or details regarding microstructures associated with cells. The arrangement of cells such as streptococci in chains or staphylococci in clusters is diagnostic for these groups of organisms. Also, internal structures such as polyphosphate granules (volutin or metachromatic granules) are important for identifying organisms such as *Corynebacterium diphtheriae.* Thick smears of cells with large clumps can obscure details about arrangement and the presence of internal structures. Furthermore, stain can become entrapped in the clumps of cells, preventing its removal by destaining and washing and leading to erroneous results for staining reactions. The procedure for making a smear is illustrated in figure 10.1.

The first step in preparing a bacteriological smear differs according to the source of the organisms. If the bacteria are growing in a liquid medium (broths, milk, saliva, urine, etc.), one starts by placing two or more loopfuls of the liquid medium directly on the slide.

From solid media such as nutrient agar, blood agar, or some part of the body, one starts by placing one or two loopfuls of water on the slide and then using an inoculating loop to disperse the organisms in the water. Bacteria growing on solid media tend to cling to each other and must be dispersed sufficiently by dilution in water; unless this is done, the smear will be too thick. *The most difficult concept for students to understand about making slides from solid media is that it takes only a very small amount of material to make a good smear.* When your instructor demonstrates this step, pay very careful attention to the amount of material that is placed on the slide.

The slides made in this exercise may be used in subsequent staining exercises. Your instructor will indicate exactly what smears you should make.

From Broth Cultures

(Broths, saliva, milk, etc.)

If you are preparing a bacterial smear from liquid media, follow this routine, which is depicted on the left side of figure 10.1.

Materials

- clean microscope slides
- broth cultures of *Staphylococcus, Streptococcus,* and *Bacillus*
- Bunsen burner or incinerator
- inoculating loop
- Sharpie marking pen
- slide holder (clothespin)

1. Obtain a new or clean slide, and be sure to handle the slide by its edges throughout the process.
2. Write the initials of the organism or organisms on the left-hand side of the slide with a Sharpie marking pen.
3. To provide a target on which to place the organisms, make a $\frac{1}{2}''$ circle on the *bottom* side of the slide, centrally located, with a Sharpie marking

pen. Later on, when you become more skilled, you may wish to omit the use of this "target circle."

4. Shake the culture vigorously and transfer two loopfuls of organisms to the center of the slide over the target circle. Follow the routine for obtaining the organism sample as shown in figure 10.2. *Be sure to flame the loop after it has touched the slide.*

Caution

Be sure to cool the loop completely before inserting it into a medium. A loop that is too hot will spatter the medium and move bacteria into the air.

5. Spread the organisms over the area of the target circle.

6. Allow the slide to dry by normal evaporation of the water. Don't apply heat.

7. After the smear has become completely air dried, place the slide in a clothespin and pass the slide several times through the Bunsen burner flame. If you are using an incinerator, place the slide onto its slide holder for a few minutes or hold the bottom of the slide to the opening of the incinerator for about 10 seconds.

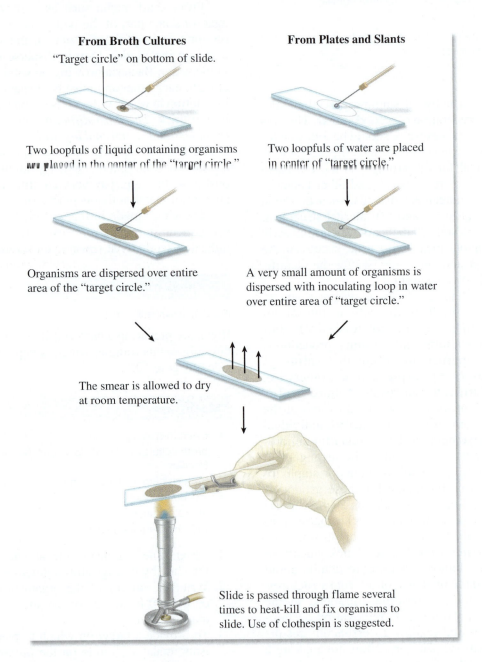

From Broth Cultures

"Target circle" on bottom of slide.

Two loopfuls of liquid containing organisms are placed in the center of the "target circle."

Organisms are dispersed over entire area of the "target circle."

From Plates and Slants

Two loopfuls of water are placed in center of "target circle."

A very small amount of organisms is dispersed with inoculating loop in water over entire area of "target circle."

The smear is allowed to dry at room temperature.

Slide is passed through flame several times to heat-kill and fix organisms to slide. Use of clothespin is suggested.

Figure 10.1 Procedure for making a bacterial smear.

Caution

Avoid prolonged heating of the slide as this can result in the slide shattering and injuring you. The underside of the slide should feel warm to the touch.

Note that in this step one has the option of using a clothespin to hold the slide. *Use the option preferred by your instructor.*

From Plates and Slants

When preparing a bacterial smear from solid media, such as nutrient agar or a part of the body, follow the routine depicted on the right side of figure 10.1.

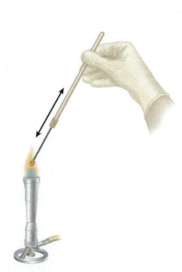
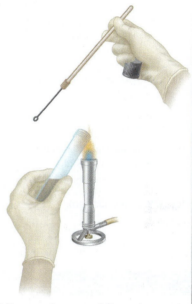

(1) Shake the culture tube from side to side to suspend organisms. Do not moisten the cap on the tube during shaking.

(2) Heat loop and wire to red-hot. Flame the handle slightly also.

(3) Remove the cap and flame the mouth of the tube. Do not place the cap down on the table.

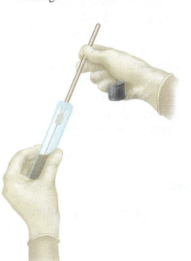
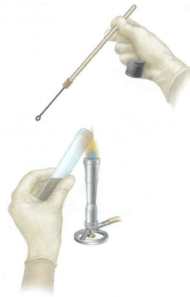
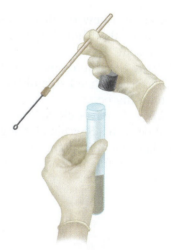

(4) After allowing the loop to cool for at least 5 seconds, aseptically remove a loopful of organisms from the tube. Avoid touching the sides of the tube.

(5) Flame the mouth of the culture tube again.

(6) Return the cap to the tube and place the tube in a test-tube rack.

continued

Figure 10.2 **Aseptic procedure for organism removal.**

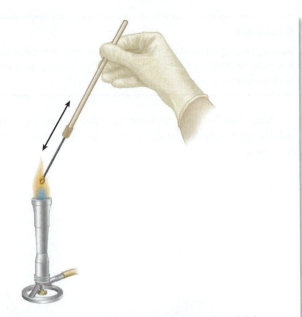

(7) Place the loopful of organisms in the center of the target circle on the slide.

(8) Flame the loop again before removing another loopful from the culture or setting the inoculating loop aside.

Figure 10.2 (continued)

Materials

- clean microscope slides
- inoculating needle and loop
- Sharpie marking pen
- slide holder (clothespin)
- Bunsen burner or incinerator

1. Obtain a new or clean slide, and remember to handle the slide by its edges throughout the process.
2. Write the initials of the organism or organisms on the left-hand side of the slide with a Sharpie marking pen.
3. Mark a "target circle" on the bottom side of the slide with a Sharpie marking pen.
4. Flame an inoculating loop, let it cool, and transfer two loopfuls of water to the center of the target circle.

5. Flame an inoculating needle and then let it cool. Pick up *a very small amount of the organisms,* and mix it into the water on the slide. Disperse the mixture over the area of the target circle. Be certain that the organisms have been well emulsified in the liquid. *Be sure to flame the inoculating needle before placing it in its holder.*
6. Allow the slide to dry by normal evaporation of the water. Don't apply heat.
7. After the slide has become completely dry, place it in a clothespin and pass it several times through the flame of a Bunsen burner or use an incinerator to heat-fix the slide. Avoid prolonged heating of the slide as it can shatter from excessive exposure to heat.

Laboratory Report

Complete the Laboratory Report for this exercise.

10   Smear Preparation

A. Short-Answer Questions

1. How does smear preparation of cells from a liquid medium differ from preparation of cells from a solid medium?

   _____

   _____

2. Why is it important to limit the quantity of cells used to prepare a smear?

   _____

   _____

3. Describe the potential consequences of making a smear that is too thick.

   _____

   _____

4. For preparation of a smear on a slide, what is the purpose of heat fixation? What problems can arise when the slide is heated in a flame?

   _____

   _____

Simple Staining

Learning Outcomes

After completing this exercise, you should be able to

1. Prepare a thin smear of bacterial cells and stain them with a simple stain.

2. Understand why staining is necessary to observe bacteria with a brightfield microscope.

3. Observe the different morphologies of bacterial cells.

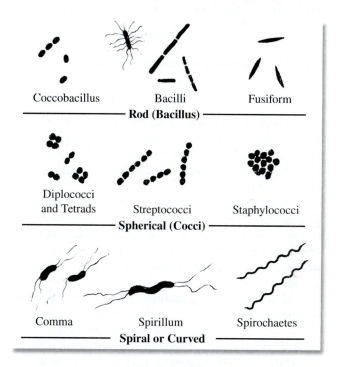

Figure 11.1 Bacterial morphology.

Because bacterial cells are composed of approximately 80% water, there is very little contrast between the cell and the surrounding aqueous environment in which most cells occur. This lack of contrast makes it extremely difficult to visualize cells or their internal details in an aqueous suspension using a brightfield microscope. To enhance the contrast of bacterial cells so they can be visualized with a brightfield microscope, a smear of cells is prepared on a microscope slide and usually heat-fixed to adhere the cells to the slide and, importantly, to preserve their structural integrity. Smears are then stained using various dyes to enhance cell features and structures.

The use of a single stain to color a bacterial cell is referred to as **simple staining.** Commonly used dyes for performing simple staining are methylene blue, basic fuchsin, and crystal violet. These are referred to as **basic dyes** because they have color-bearing ionic groups (*chromophores*) that are positively charged (cationic). They work well with bacterial cells that have chemical groups on their surfaces that confer a net negative charge to the cell. Therefore, there is a pronounced electrostatic attraction between the cell and the cationic chromophore of the stain, and the result is stained cells. In contrast to cationic chromophores, dyes that have anionic chromophores (negatively charged) are called **acidic dyes.** These stains are repelled by bacterial cells and stain the background of the slide instead.

Simple stains can be used to determine the morphology of bacterial cells. Figure 11.1 illustrates most of the common shapes of bacterial cells. They can be grouped into three morphological types: **bacilli** (rods), **cocci** (spherical), and **spirals** (corkscrew-shaped rods). Rods or bacilli can have rounded, flat,

or tapered ends. The fusiform bacteria are rods with tapered ends and are prevalent in the human mouth. Cocci may occur singly, in chains, in tetrads (packets of four), or in irregular clusters. Most streptococci occur in chains, whereas the staphylococci occur in clusters that resemble grapes. The spiral-shaped bacteria can exist as spirochaetes, as spirilli, or as comma-shaped, curved rods. *Treponema pallidum,* the causative agent of syphilis, is a spirochaete that is too thin to be observed by brightfield microscopy. *Vibrio cholerae,* the bacterium responsible for cholera, is a comma-shaped bacterium.

Cultures (or smears already prepared in Exercise 10) of *Staphylococcus aureus, Streptococcus lactis, Bacillus megaterium,* and *Corynebacterium xerosis* will be used in this simple staining exercise. *C. xerosis* is related to *Corynebacterium diphtheriae,* the bacterium responsible for diphtheria in humans. The basic dye methylene blue (methylene$^+$ chloride$^-$) will be used to stain these cells, allowing you to visualize their different morphologies and cell arrangements. It is informative to do the simple stain

(1) A bacterial smear is stained with methylene blue for 1 minute.

(2) Stain is briefly washed off slide with water.

(3) Water drops are carefully blotted off slide with bibulous paper.

Figure 11.2 **Procedure for simple staining.**

with corynebacteria because the stain demonstrates unique characteristics of these bacteria that are used in their identification. These characteristics are: pleomorphism, metachromatic granules, and palisade arrangement of cells.

Pleomorphism pertains to irregularity of form: that is, demonstrating several different shapes. While *C. diphtheriae* is basically rod-shaped, it also appears club-shaped, spermlike, or needle-shaped. *Bergey's Manual* uses the terms "pleomorphic" and "irregular" interchangeably. (See figure 11.3.)

Metachromatic granules are distinct reddish-purple granules within cells that show up when the organisms are stained with methylene blue. These granules are masses of *volutin*, a polymetaphosphate.

Palisade arrangement pertains to parallel arrangement of rod-shaped cells. This characteristic, also called "picket fence" arrangement, is common to many corynebacteria.

Materials

- slant cultures (or heat-fixed smears) of *S. aureus, S. lactis, B. megaterium,* and *C. xerosis*
- methylene blue (Loeffler's)
- wash bottle
- bibulous paper

Procedure

1. If you have not done so already, prepare heat-fixed smears of *S. aureus, S. lactis, B. megaterium,* and *C. xerosis* using the procedure in Exercise 10.
2. Use the steps in figure 11.2 to stain each of your heat-fixed smears.
3. Examine the slides under the microscope, and compare them with the photomicrographs in figures 11.3 and 11.4.
4. Record your observations in Laboratory Report 11.

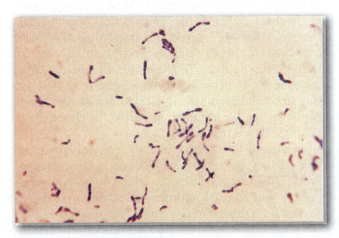

Figure 11.3 **Methylene blue stain of *Corynebacterium* showing club-shaped cells.**
Source: Centers for Disease Control

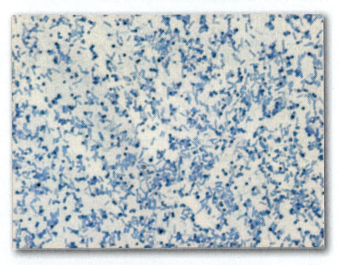

Figure 11.4 **Simple stain of *Bacillus subtilis* and *Staphylococcus aureus*.**
© McGraw-Hill Education. Auburn University Research Instrumentation Facility. Michael Miller, photographer.

11 Simple Staining

A. Results

1. What three noteworthy physical characteristics of *Corynebacterium xerosis* are visible after performing a simple stain? Draw cells from your slide to demonstrate these characteristics.

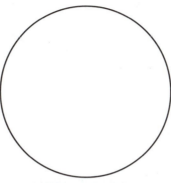

Corynebacterium xerosis

2. Describe the arrangement and cell morphology of the streptococci. Draw cells from your slide to demonstrate these characteristics.

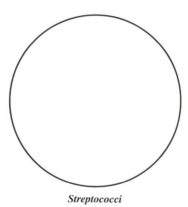

Streptococci

3. Describe the arrangement and cell morphology of staphylococci. Draw cells from your slide to demonstrate these characteristics.

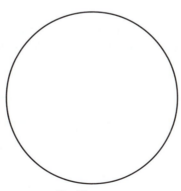

Staphlyococcus

4. Describe the arrangement and cell morphology of members of the genus *Bacillus*. Draw cells from your slide to demonstrate these characteristics.

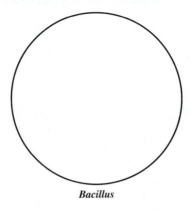

Bacillus

B. Short-Answer Questions

1. What are chromophores?

   _____

2. What is the difference between basic and acidic dyes?

   _____

3. Why do acidic dyes not stain bacterial cells?

   _____

4. Crystal violet is an example of what type of stain?

   _____

5. What is meant by palisade arrangement of cells?

   _____

6. What shape does *Vibrio cholerae* have?

   _____

7. Where are the fusiform bacteria usually found in humans?

   _____

8. If you were working with an unlabeled simple stained smear, would you be able to identify the bacterial species by observing the slide under the microscope? Why or why not?

   _____

   _____

   _____

Negative Staining

Learning Outcomes

After completing this exercise, you should be able to

1. Prepare a negative stain of bacterial cells using the slide-spreading or loop-spreading techniques.

2. Use the negative stain to visualize cells from your teeth and mouth.

3. Discern different morphological types of bacterial cells in a negative stain.

Negative stains can be useful in studying the morphology of bacterial cells and characterizing some of the external structures, such as capsules, that are associated with bacterial cells. Negative stains are acidic and thus have a negatively charged chromophore that does not penetrate the cell but rather is repelled by the similarly charged bacterial cell. The background surrounding the cell is colored by a negative stain, resulting in a negative or indirect staining of the cell. Usually cells appear as transparent objects against a dark background (see figure 12.1). Examples of negative stains are India ink and nigrosin. The negative stain procedure consists of mixing the organism with a small amount of stain and spreading a very thin film over the surface of a microscope slide. For negative stains, cells are *not usually heat-fixed* prior to the

application of the negative stain. Sometimes negative staining can be combined with positive staining to better demonstrate structures such as capsules. In this case, the capsule can be seen as a halo surrounding a positively stained cell against a dark background.

Negative staining can also be useful for accurately determining cell dimensions. Because heat fixation is not performed, no shrinkage of cells occurs and size determinations are more accurate than those determined on fixed material. Avoiding heat fixation is also important if the capsule surrounding the cell is to be observed because heat fixation will severely shrink this structure. The negative stain is also useful for observing spirochaetes, which tend to be very thin cells that do not readily stain with positive stains.

Two Methods

Negative staining can be done by two methods. Figure 12.2 illustrates the more commonly used method in which the organisms are mixed in a drop of nigrosin and spread over the slide with another slide. The goal is to produce a smear that is thick at one end and feather-thin at the other end. Somewhere between the too thick and too thin areas will be an ideal spot to study the organisms.

Figure 12.3 illustrates a second method, in which organisms are mixed in only a loopful of nigrosin instead of a full drop. In this method, the organisms are spread over a smaller area in the center of the slide with an inoculating needle. No spreader slide is used in this method.

Note in the procedure below that slides may be prepared with organisms from between your teeth or from specific bacterial cultures. Your instructor will indicate which method or methods you should use and demonstrate some basic aseptic techniques. Various options are provided here to ensure success.

Materials

- microscope slides (with polished edges)
- nigrosin solution or India ink
- slant cultures of *S. epidermidis* and *B. megaterium*
- inoculating needle and loop
- sterile toothpicks
- Bunsen burner or incinerator
- Sharpie marking pen

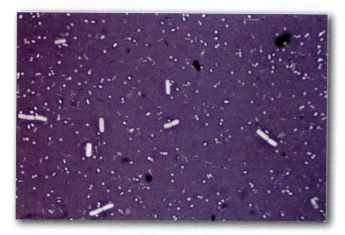

Figure 12.1 **Negative stain of *Bacillus* and *Staphylococcus* using nigrosin.**

© McGraw-Hill Education/Lisa Burgess, photographer

(1) Organisms are dispersed into a small drop of nigrosin or India ink. Drop should not exceed ⅛″ diameter and should be near the end of the slide.

(2) Spreader slide is moved toward drop of suspension until it contacts the drop, causing the liquid to be spread along its spreading edge.

(3) Once spreader slide contacts the drop on the bottom slide, the suspension will spread out along the spreading edge as shown.

(4) Spreader slide is pushed to the left, dragging the suspension over the bottom slide. After the slide has air-dried, it may be examined under oil immersion. *Remember: This slide should not be heat-fixed.*

Figure 12.2 Negative staining technique using a spreader slide.

(1) A loopful of nigrosin or India ink is placed in the center of a clean microscope slide.

(2) A sterile inoculating needle is used to transfer the organisms to the liquid and mix the organisms into the stain.

(3) Suspension of bacteria is spread evenly over an area of one or two centimeters with the needle.

(4) Once the preparation has completely air-dried, it can be examined under oil immersion. No heat should be used to hasten drying.

Figure 12.3 Negative staining technique using an inoculating needle.

1. Clean your tabletop with disinfectant in preparation for making slides.
2. Obtain two or three new or clean microscope slides.
3. Referring to figures 12.2 or 12.3, place the proper amount of stain on the slide.
4. **Oral Organisms:** Remove a small amount of material from between your teeth with a sterile straight toothpick and mix it into the stain on the slide. Be sure to break up any clumps of organisms with the toothpick or a sterile inoculating loop. When using a loop, *be sure to flame it first to make it sterile.*

5. **From Cultures:** With a *sterile* inoculating needle, transfer a very small amount of bacteria from the slant to the center of the stain on the slide.
6. Spread the mixture over the slide according to the procedure used in figures 12.2 or 12.3.
7. Allow the slide to air-dry and examine under oil immersion.

Laboratory Report

Draw a few representative types of the organisms you observe on your slides in Laboratory Report 12. If the slide is of oral organisms, look for yeasts and hyphae as well as bacteria. Spirochaetes may also be present.

Caution

If you use a toothpick, discard it into a beaker of disinfectant.

Laboratory Report

12

Student: _____

Date: _____ Section: _____

12 Negative Staining

A. Results

1. Draw the different types of microorganisms that were found in the negative stain of the oral sample. How would you differentiate between oral streptococci, yeasts, and spirochaetes in your sample?

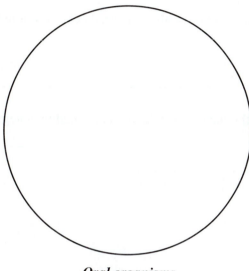

Oral organisms
(nigrosin stain)

2. Draw your slides made from bacterial cultures. Label the species under your drawing, and indicate the morphology of the bacterium.

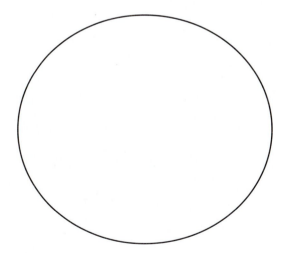 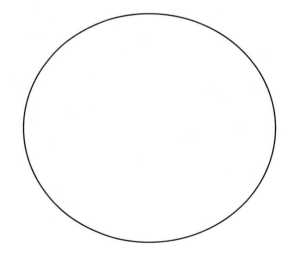

Bacterium: _____   Bacterium: _____

Morphology: _____   Morphology: _____

B. Short-Answer Questions

1. What type of chromophore is associated with a negative stain?

   _____

   _____

2. What is an example of a negative stain?

   _____

   _____

3. What step normally associated with staining bacterial cells is omitted when the dimensions of cells are determined? Why?

   _____

   _____

4. What external bacterial cell structures can be demonstrated by a negative stain?

   _____

   _____

Capsular Staining

Learning Outcomes

After completing this exercise, you should be able to

1. Prepare and stain a smear of an encapsulated bacterium using the Anthony capsule staining method.

2. Visualize the capsule and differentiate it from the cell body.

Many bacterial cells are surrounded by an extracellular gel-like layer that occurs outside of the cell wall. If the layer is distinct and gelatinous, it is referred to as a **capsule.** If the layer is diffuse and irregular, it is called a **slime layer.** The capsule or slime layer can vary in its chemical composition. If it is made up of polysaccharides, it is known as a **glycocalyx,** which literally means "sugar shell." However, the capsule found in *Bacillus anthracis* is composed of poly-D-glutamic acid, which forms a proteinaceous matrix.

Capsules or slime layers perform very important functions for a cell. In pathogens such as *Streptococcus pneumoniae,* they are protective structures because they prevent phagocytic white blood cells from engulfing and destroying the pathogen, enabling the organism to invade the lungs and cause pneumonia. Another function for capsules or slime layers is attachment of the bacterial cell to solid surfaces in the environment. For example, *Streptococcus mutans* produces a capsule that facilitates the attachment of the organism to the tooth surface, resulting in the formation of dental plaque. If not removed, plaque will contribute to the formation of dental caries. Evidence supports the view that probably all bacterial cells have some amount of slime layer, but in most cases the amount is not enough to be readily discerned.

Staining of the bacterial capsule cannot be accomplished by ordinary staining procedures. If smears are heat-fixed prior to staining, the capsule shrinks or is destroyed and therefore cannot be seen in stains. In the Anthony method (figure 13.1), smears are prepared

(1) Two loopfuls of the organism are used to prepare a thin smear.

(2) The smear is allowed to only air-dry; do not heat-fix.

(3) Stain the smear with 1% crystal violet for 2 minutes.

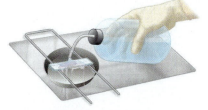

(4) The crystal violet is washed off the slide with 20% copper sulfate into a proper waste container. Do not wash copper sulfate into the sink.

(5) Blot dry with bibulous paper or allow it to fully air-dry.

Figure 13.1 **Steps of the Anthony capsule staining method.**

from cultures grown in skim milk broth. These smears are air-dried but not heat-fixed. Then, they are stained with crystal violet for 2 minutes. The crystal violet not only stains the cells, but it also binds to the milk proteins from the culture medium, causing the background of the slide to stain purple. The stain is then washed off with an aqueous solution of 20% copper sulfate. This reagent functions as a decolorizer, removing the crystal violet from the capsules. It also serves as a counterstain, staining the capsules a very light blue. Under oil immersion, the capsules will appear as clear or light blue halos around the cells and the cells will be dark purple (figure 13.2). You will use this procedure to stain the capsules of *Klebsiella pneumoniae*.

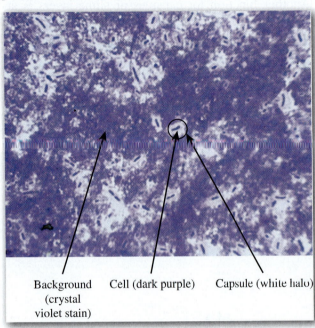

Background   Cell (dark purple)   Capsule (white halo)
(crystal
violet stain)

Figure 13.2 Anthony capsule stain of *Klebsiella pneumoniae*.
© McGraw-Hill Education/Lisa Burgess, photographer

Materials

- 36–48 hour skim milk culture of *Klebsiella pneumoniae*
- 1% (wt/vol) crystal violet
- 20% (wt/vol) aqueous copper sulfate
- waste containers for the copper sulfate
- disinfectant solutions for used slides

Procedure

1. Prepare a thin smear of *Klebsiella pneumoniae* on a microscope slide.
2. Allow the smear to air-dry only. **Do not heat-fix as this will cause the capsule to shrink or be destroyed.**
3. Apply 1% crystal violet and allow it to remain on the slide for 2 minutes.
4. With the slide over the proper waste container provided, gently wash off the crystal violet with 20% copper sulfate. **Caution: Do not wash the copper sulfate and stain directly into the sink.**
5. Gently blot the slide dry with bibulous paper or allow it to fully air-dry.
6. Observe with the oil immersion lens and compare your stain with figure 13.2.
7. Be sure to dispose of your used slide in the disinfectant container when you are finished.

Note: The capsule stain may also be done using the procedure for the negative stain in Exercise 12. Either nigrosin or India ink can be used as the stain. Cells will have a halo around them against a dark background.

Laboratory Report

Record your results in Laboratory Report 13.

13   Capsular Staining

A. Results

1. Draw cells that display a capsule from your stained slide of *Klebsiella pneumoniae*. Explain how the capsule is visualized without the use of dyes that adhere to a capsule.

_____

_____

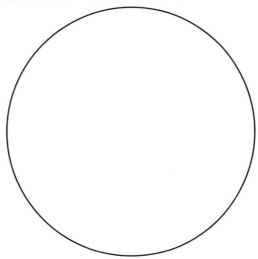

Klebsiella pneumoniae
(capsular stain)

B. Short-Answer Questions

1. What are two functions of the capsule or slime layer in bacterial cells?

_____

_____

2. What biological molecules can make up the bacterial capsule or slime layer?

_____

_____

3. What function does the capsule have for *Streptococcus mutans*?

_____

_____

4. Look up the term *virulence factor* if you do not already know its definition. Explain why the bacterial capsule is considered a virulence factor.

   _____

   _____

5. A student heat-fixes his smear intended for capsule staining. What result might the student expect?

   _____

   _____

Gram Staining

Learning Outcomes

After completing this exercise, you should be able to

1. Explain the importance of the Gram stain in microbiology.

2. Summarize the differences between gram-positive and gram-negative cell walls.

3. Explain how each step of the Gram stain procedure works to differentially stain the two types of bacteria.

4. Prepare thin smears of gram-positive and gram-negative cells, and stain them using the Gram stain method.

5. Differentiate between gram-positive and gram-negative cells under the microscope.

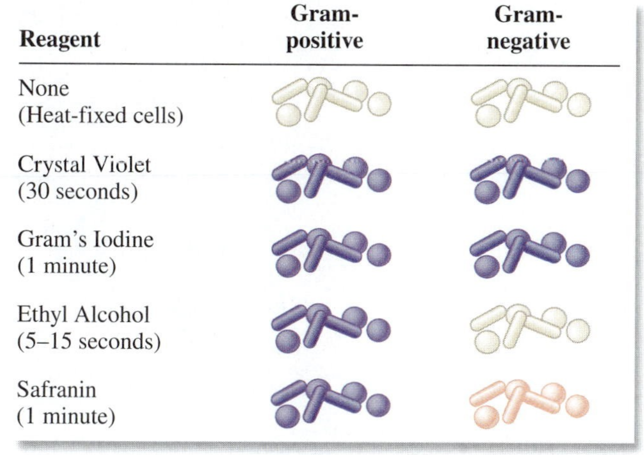

Reagent	Gram-positive	Gram-negative
None (Heat-fixed cells)		
Crystal Violet (30 seconds)		
Gram's Iodine (1 minute)		
Ethyl Alcohol (5–15 seconds)		
Safranin (1 minute)		

Figure 14.1 Color changes that occur at each step in the Gram staining process.

In 1884, the Danish physician Hans Christian Gram was trying to develop a staining technique that would differentiate bacterial cells from eukaryotic nuclei in diseased lung tissue. He discovered that certain stains were retained by some types of bacterial cells but removed from others during the staining process. His published work served as the foundation of what would become the most important stain in bacteriology, the Gram stain.

Gram staining is a valuable diagnostic tool used in the clinical and research setting. Although newer molecular techniques have been developed, the Gram stain is still a widely used method for the identification of unknown bacteria. It is often the first test conducted on an unknown species in the laboratory, and in some cases, it can provide presumptive identification of the organism. For medical professionals, a Gram stain of a clinical specimen may be used to determine an appropriate treatment for a bacterial infection.

The Gram stain is an example of a differential stain. Differential staining reactions take advantage of the fact that cells or structures within cells display dissimilar staining reactions that can be distinguished by the use of different dyes. In the Gram stain, two kinds of cells, gram-positive and gram-negative, are differentiated based on their cell wall structure and composition. These types of cells can be identified by their respective colors, purple and pink or red, after performing the staining method. The procedure is

based on the fact that gram-positive bacteria retain a purple dye complex, whereas gram-negative bacteria are decolorized and must be counterstained with a red dye in order to be visualized by microscopy.

Figure 14.1 illustrates the appearance of cells after each step in the Gram stain procedure. Initially, both gram-positive and gram-negative cells are stained by the **primary stain,** crystal violet. In the second step of the procedure, Gram's iodine is added to the smear. Iodine is a **mordant** that combines with the crystal violet and forms an insoluble complex in gram-positive cells. At this point, both types of cells will still appear as purple. During **decolorization** with alcohol and/or acetone, gram-positive cells retain the crystal violet-iodine complex, and therefore these cells will appear purple under the microscope. Alternatively, the dye-mordant complex is removed from gram-negative cells, leaving them colorless. Safranin is applied as a counterstain, coloring the gram-negative cells pink or red. The safranin also sticks to the gram-positive cells, but their appearance is unchanged because the crystal violet is a much more intense stain than safranin.

The mechanism for how the Gram stain works is not completely understood, but it is known to be related to structural and chemical differences in the cell walls of gram-positive and gram-negative bacteria (figure 14.2). When viewed by electron microscopy, gram-positive cells have a thick layer of **peptidoglycan** that comprises the cell wall of these organisms.

Figure 14.2 **Comparison of gram-positive and gram-negative cell walls.**

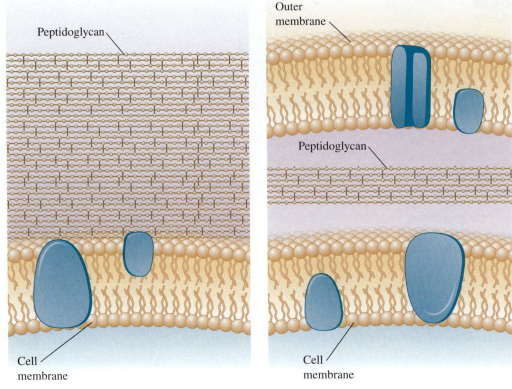

Gram-positive cell   **Gram-negative cell**

In contrast, the cell wall in gram-negative cells consists of an outer membrane that covers a much thinner layer of peptidoglycan. It is believed that the thick, tightly linked peptidoglycan molecules of gram-positive cells trap the crystal violet–iodine complexes, preventing their removal when the smear is correctly decolorized. In contrast, the decolorizer dissolves the lipids in the outer membrane of gram-negative bacteria, allowing the dye-mordant complexes to escape through the thin peptidoglycan layer.

Some bacteria are considered gram-variable because some cells will retain the crystal violet stain, while others will not and appear red from the counterstain. Other bacteria, called acid-fast bacteria, have a unique cell wall made of "waxy" lipids. Mycobacteria, the causative agents of tuberculosis and Hansen's disease (leprosy), are acid-fast bacteria. These cells may appear as either nonreactive or gram-positive after the Gram stain technique, but a special acid-fast staining technique (see Exercise 16) can be used to identify bacteria with this type of cell wall.

Although the Gram stain technique may seem quite simple, performing it with a high degree of reliability requires some practice and experience. Several factors can affect the outcome of the procedure:

1. It is important to use cultures that are 16–18 hours old. Gram-positive cultures older than this can convert to gram-variable or gram-negative and give erroneous results. (It is important to note that gram-negative bacteria never convert to gram-positive.)

2. It is critical to prepare thin smears. Thin smears allow the observation of individual cells and any arrangement in which the cells occur. However, thick smears can entrap the primary stain, preventing decolorization. Cells that occur in the entrapped stain may falsely appear gram-positive.

3. Decolorization is the most critical step in the Gram stain procedure. If the decolorizer is overapplied, the dye-mordant complex may be removed from gram-positive cells, causing them to incorrectly appear as gram-negative cells.

During this laboratory period, you will be provided an opportunity to stain several different kinds of bacteria to see if you can achieve the degree of success that is required. Remember, if you don't master this technique now, you may have difficulty with future lab exercises, so repeat stains that yield incorrect results as much as time allows.

Materials

- broth cultures of *Staphylococcus aureus, Pseudomonas aeruginosa, Escherichia coli,* and *Moraxella catarrhalis* (Note: *Staphylococcus epidermidis* may be used instead of *Staphylococcus aureus.*)
- nutrient agar slant culture of *Bacillus megaterium*
- Gram staining kit and wash bottle
- bibulous paper

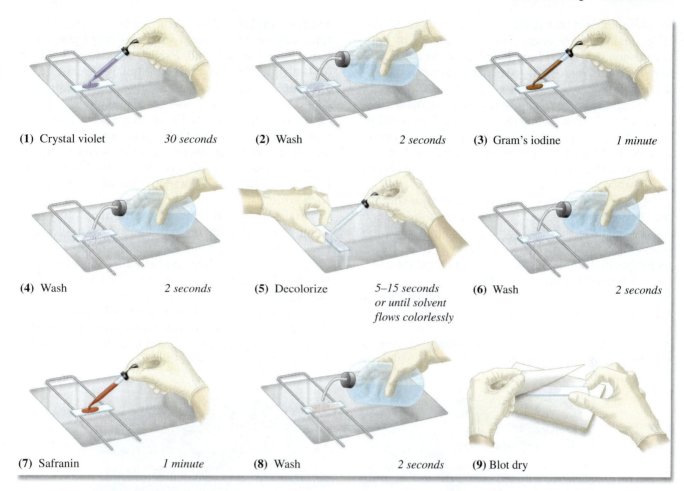

(1) Crystal violet   *30 seconds*   **(2)** Wash   *2 seconds*   **(3)** Gram's iodine   *1 minute*

(4) Wash   *2 seconds*   **(5)** Decolorize   *5–15 seconds or until solvent flows colorlessly*   **(6)** Wash   *2 seconds*

(7) Safranin   *1 minute*   **(8)** Wash   *2 seconds*   **(9)** Blot dry

Figure 14.3 **The Gram staining procedure.**

Gram Stain Procedure

1. Cover a heat-fixed smear with **crystal violet** and let stand for *30 seconds* (see figure 14.3).
2. Briefly wash off the stain, using a wash bottle of distilled water. Drain off excess water.
3. Cover the smear with **Gram's iodine** solution and let it stand for *1 minute*. (Your instructor may prefer only 30 seconds for this step.) Wash off the Gram's iodine.
4. Hold the slide at a 45-degree angle and apply the **decolorizer,** allowing it to flow down the surface of the slide. Do this until the decolorizer is colorless as it flows from the smear down the surface of the slide. *This should take no more than 15 seconds for properly prepared smears.* **Note:** Thick smears can take longer for decolorization.
5. Stop decolorization by washing the slide with a gentle stream of water.
6. Cover the smear with **safranin** for 1 minute.
7. Wash gently for a few seconds and blot dry with bibulous paper.
8. Examine the slide under oil immersion.

Gram Staining Exercises

The organisms that will be used here for Gram staining represent a diversity of morphology and staining characteristics. Some of the rods and cocci are gram-positive; others are gram-negative. Once you practice the technique on a known gram-positive and a known gram-negative organism, you will use these bacteria as controls for determining the Gram reaction of other bacterial species. You will also observe the Gram stain reaction of a spore-forming bacterium.

Gram Stain Practice Slides   Prepare two slides with three smears on each slide. On the left portion of each slide, make a thin smear of *Staphylococcus aureus*. On the right portion of each slide, make a thin smear

107

of *Pseudomonas aeruginosa*. In the middle of the slide, make a thin smear that is a mixture of both organisms, using two loopfuls of each organism. ***Be sure to flame the loop sufficiently to avoid contaminating cultures.***

Using the Gram stain procedure shown in figure 14.3, stain one slide first, saving the other one if you need to repeat the technique. If done properly, *Staphylococcus aureus* should appear as purple cocci, and *Pseudomonas aeruginosa* should appear as pink or red rods, as shown in figure 14.4a.

Call your instructor over to evaluate your slide. If the slide is improperly stained, the instructor may be able to tell what went wrong by examining all three smears. He or she will inform you how to correct your technique and if time allows, you can repeat the exercise with the other practice slide.

Record your results in Laboratory Report 14 by drawing a few cells in the appropriate circle.

Slides to Determine Gram Reaction Using a similar procedure to the previous exercise, you will

again be making slides with triple smears. As before, you should make a smear of *Staphylococcus aureus* on the left and *Pseudomonas aeruginosa* on the right. In this exercise, make one slide with a mixture of *Bacillus megaterium* and *Escherichia coli* in the center. On another slide, make a smear of *Moraxella catarrhalis* in the center. Because you already know the Gram reaction of *S. aureus* and *P. aeruginosa*, these bacterial smears can be used as controls to determine if the Gram stain procedure was done correctly for the bacteria on each slide.

Stain both of these slides using the Gram stain technique, and then observe under oil immersion. Draw a few cells in the appropriate circles on your Laboratory Report sheet, and identify the Gram reaction for all three types of bacteria. As you examine the slide with *Bacillus megaterium*, look for clear areas in the rods which represent endospores. Since endospores are impermeable to the dyes used in this technique, they will appear as transparent holes in the cells as seen in figure 14.4d.

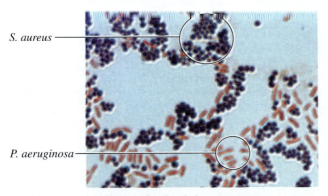

(a) Gram stain: *Staphylococcus aureus* and *Pseudomonas aeruginosa*

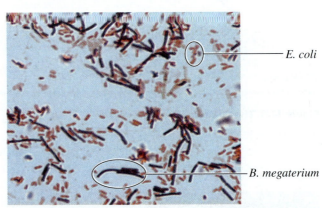

(b) Gram stain: *Bacillus megaterium* and *Escherichia coli*

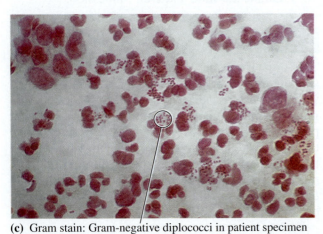

(c) Gram stain: Gram-negative diplococci in patient specimen

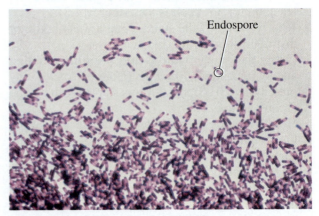

(d) Gram stain: Endospore-forming rods

Figure 14.4 Photomicrographs of Gram-stained bacteria.
(a), (b) © McGraw-Hill Education. Auburn University Photographic Services; (c) Source: Dr. Gilda Jones/Centers for Disease Control; (d) Source: Dr. Norman Jacobs/Centers for Disease Control

14   Gram Staining

A. Results

Draw cells from the Gram-stained slides, and label each bacterial species. Based on your results, note the Gram reaction, cell shape, and cell arrangement of each bacterium in the given table. Verify these results to determine if you are correct.

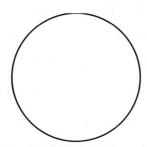

P. aeruginosa and *S. aureus*

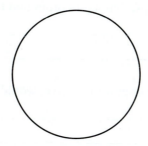

B. megaterium and *E. coli*

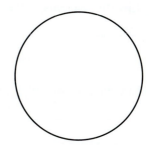

M. catarrhalis

BACTERIAL SPECIES	GRAM REACTION	CELL SHAPE	CELL ARRANGEMENT
Staphylococcus aureus			
Pseudomonas aeruginosa			
Bacillus megaterium			
Escherichia coli			
Moraxella catarrhalis			

B. Short-Answer Questions

1. Why is the Gram stain considered a differential stain?

   _____

   _____

2. How do gram-positive and gram-negative bacteria differ in cellular structure, and how does this contribute to their differential staining properties?

   _____

   _____

3. How does the age of a culture affect the Gram stain reaction? What is an optimum culture age for a valid Gram reaction?

   _____

   _____

4. Which step in the Gram stain procedure is most prone to error? If done incorrectly, how might that step affect the end result?

_____

_____

5. What is the function of a mordant, and which reagent serves this purpose in the Gram stain procedure?

_____

_____

6. List the reagents of the Gram stain technique in order and their general role in the staining process.

_____

_____

7. In what type of cell, gram-positive or gram-negative, would you find lipopolysaccharide in its cell wall?

_____

_____

C. Matching Questions

Match the expected result (purple, red, or colorless) to the following descriptions of Gram-stained cells. Consult your chart at the beginning of this lab report if you need help remembering the correct Gram reaction for each species. Choices may be used more than once.

1. _____ *Staphylococcus aureus* before the primary stain

2. _____ *Pseudomonas aeruginosa* after the primary stain

3. _____ *Bacillus megaterium* after the addition of the mordant

4. _____ *Staphylococcus aureus* after decolorization

5. _____ *Moraxella catarrhalis* after decolorization

6. _____ *Pseudomonas aeruginosa* after decolorization

7. _____ *Bacillus megaterium* after adding the counterstain

8. _____ *Escherichia coli* under the microscope if you forgot to apply safranin

9. _____ *Escherichia coli* under the microscope if you forgot to apply decolorizer

10. _____ *Bacillus megaterium* under the microscope if you forgot to apply iodine

| A. Purple |
| B. Red |
| C. Colorless |

Spore Staining:
Two Methods

Learning Outcomes

After completing this exercise, you should be able to

1. Prepare an endospore stain of bacterial cells and observe endospores in the stained preparation.

2. Differentiate between vegetative cells and endospores.

When species of bacteria belonging to the genera *Bacillus* and *Clostridia* exhaust essential nutrients, they undergo a complex developmental cycle that produces dormant structures called **endospores.** Endospores survive environmental conditions that are not favorable for normal bacterial growth. If nutrients once again become available, the endospore can go through the process of germination to form a new vegetative cell, and growth will resume. Endospores are very dehydrated structures that are not actively metabolizing. Furthermore, they are resistant to heat, radiation, acids, and many chemicals, such as disinfectants, that normally harm or kill vegetative cells. Their chemical resistance is due in part to the fact that they have a protein coat that forms a protective barrier around the spore. Heat resistance is associated with the water content of endospores. The higher the water content of an endospore, the less heat resistant the endospore will be. During sporulation, the water content of the endospore is reduced to 10–30% of the vegetative cell. During endospore formation, calcium dipicolinate and spore-specific proteins form a cytoplasmic gel that reduces the protoplasmic volume of the endospore to a minimum. In addition, a thick cortex forms around the endospore, and contraction of the cortex results in a smaller dehydrated structure. Calcium dipicolinate is not present in vegetative cells. The gel formed by this chemical and the spore-specific proteins controls the amount of water that can enter the endospore, thus maintaining its dehydrated state.

Since endospores are not easily destroyed by heat or chemicals, they define the conditions necessary to establish sterility. For example, to destroy endospores by heating, they must be exposed for 15–20 minutes to steam under pressure, which generates temperatures of 121°C. Such conditions are produced in an **autoclave**.

The resistant properties of endospores also mean that they are not easily penetrated by stains. For example in Exercise 14, you observed that endospores did not readily Gram stain. If endospore-containing cells are stained by basic stains such as crystal violet, the spores appear as unstained areas in the vegetative cell. However, if heat is applied while staining with malachite green, the stain more readily penetrates and becomes entrapped in the endospore. The malachite green is not removed by subsequent washing with decolorizing agents or water. In this instance, heat is acting as a mordant to facilitate the uptake of the stain.

Schaeffer-Fulton Method

The Schaeffer-Fulton method utilizes malachite green to stain the endospore and safranin to stain the vegetative portion of the cell. Utilizing this technique, a properly stained spore-former will have a green endospore contained in a red sporangium. Figure 15.1 reveals what *Bacillus* and *Clostridium* should look like when correctly stained with this staining method.

Materials

- 24–36 hour nutrient agar slant culture of *Bacillus megaterium*
- electric hot plate and small beaker (25 ml)
- spore-staining kit consisting of a bottle each of 5% malachite green and safranin
- wash bottle
- bibulous paper

Prepare a smear of *Bacillus megaterium* and allow the smear to air-dry. Heat-fix the dried smear and follow the steps for staining outlined in figure 15.2.

Note: A variation of the Schaeffer-Fulton method that does not require steaming may be used on 36-hour cultures of *Bacillus megaterium*. In this variation, a heat-fixed smear is immediately flooded with malachite green and allowed to stand for at least 10 minutes. After washing the smear with water, it is then stained with safranin for 1 minute. Results will be green spores and red vegetative cells, but may be lighter than with the original method that uses steam.

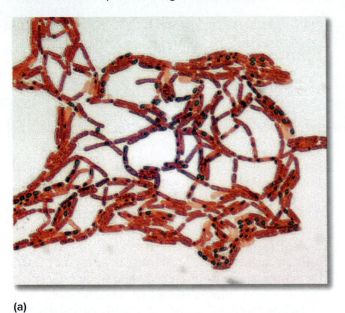

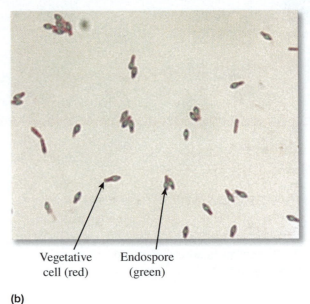

Vegetative   Endospore
cell (red)   (green)

(a)   (b)

Figure 15.1 **Spore stain of (a) *Bacillus* and (b) *Clostridium*.**
Source: Larry Stauffer, Oregon State Public Health Laboratory. Centers for Disease Control.

(1) Cover smear with small piece of paper toweling and saturate it with malachite green. Steam over boiling water for *5 minutes*. The paper toweling should be small enough that it does not hang over the edges of the slide, and additional stain should be added as needed to keep the toweling saturated.

(2) After the slide has cooled sufficiently, remove the paper toweling and rinse with water for *30 seconds*.

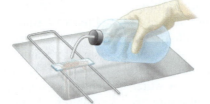

(3) Counterstain with safranin for about *30 seconds*.

(4) Rinse briefly with water to remove safranin.

(5) Blot dry with bibulous paper, and examine slide under oil immersion.

Figure 15.2 **The Schaeffer-Fulton spore stain method.**

Dorner Method

The Dorner method for staining endospores produces a red spore within a colorless sporangium. Nigrosin stains the background of the slide for contrast. The six steps involved in this technique are shown in figure 15.3. Although both the sporangium and endospore are stained during boiling in step 3, the sporangium is decolorized by the diffusion of carbolfuchsin molecules into the nigrosin.

Materials

- carbolfuchsin
- nigrosin

- electric hot plate and small beaker (25 ml)
- small test tube (10 × 75 mm size)
- test-tube holder
- 24–36 hour nutrient agar slant culture of *Bacillus megaterium*

Prepare a slide of *Bacillus megaterium* that utilizes the Dorner method. Follow the steps in figure 15.3.

Laboratory Report

After examining the organisms under oil immersion, draw a few cells in the appropriate circles in Laboratory Report 15.

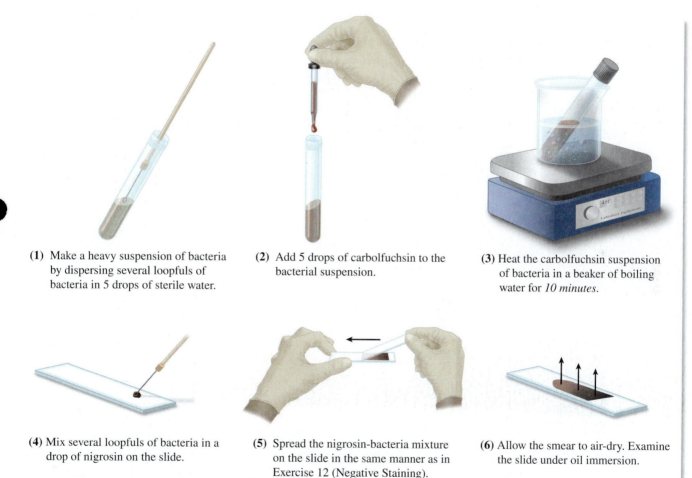

(1) Make a heavy suspension of bacteria by dispersing several loopfuls of bacteria in 5 drops of sterile water.

(2) Add 5 drops of carbolfuchsin to the bacterial suspension.

(3) Heat the carbolfuchsin suspension of bacteria in a beaker of boiling water for *10 minutes*.

(4) Mix several loopfuls of bacteria in a drop of nigrosin on the slide.

(5) Spread the nigrosin-bacteria mixture on the slide in the same manner as in Exercise 12 (Negative Staining).

(6) Allow the smear to air-dry. Examine the slide under oil immersion.

Figure 15.3 **The Dorner spore stain method.**

15 Spore Staining: Two Methods

A. Results

Draw cells from the spore slides. Label your drawings to differentiate endospores from vegetative cells.

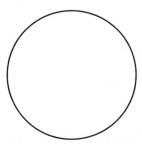

B. megaterium
(Schaeffer-Fulton method)

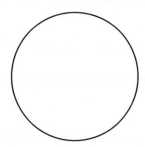

B. megaterium
(Dorner method)

B. Short-Answer Questions

1. What are the functions of endospores in bacteria?

   _____

   _____

2. What external structure on the endospore acts as a protective barrier? What is its composition?

   _____

   _____

3. Compared to a vegetative cell, how much less water is present in an endospore?

   _____

   _____

4. What is the mordant in the spore stain?

   _____

   _____

5. What is the stimulus for endospore production in bacteria?

   _____

   _____

6. What conditions are necessary to destroy endospores? In what device are these conditions achieved?

    _____

    _____

7. What is the color of endospores after Gram staining? After spore staining?

    _____

    _____

8. What is the secondary stain in the spore stain?

    _____

    _____

9. What is the color of the vegetative cell after the spore stain?

    _____

10. Of these three genera of bacteria, which does not produce endospores: *Clostridium, Mycobacterium,* or *Bacillus*? _____

11. Are bacterial endospores reproductive structures? Explain why or why not.

    _____

    _____

12. Give three examples of diseases caused by an endospore-forming bacterium and the name of the specific bacterial agent involved.

    _____

    _____

Acid-Fast Staining:
Kinyoun Method

Learning Outcomes

After completing this exercise, you should be able to

1. Prepare an acid-fast stain of bacterial cells.

2. Differentiate between acid-fast and non-acid-fast cells in a mixed stain.

3. Explain the basis for the stain and why the stain is important in clinical microbiology.

Bacteria such as *Mycobacterium* and some *Nocardia* have cell walls with a high lipid content. One of the cell wall lipids is a waxy material called **mycolic acid.** This material is a complex lipid that is composed of fatty acids and fatty alcohols that have hydrocarbon chains up to 80 carbons in length. It significantly affects the staining properties of these bacteria and prevents them from being stained by many of the stains routinely used in microbiology. The acid-fast stain is an important diagnostic tool in the identification of *Mycobacterium tuberculosis,* the causative agent of tuberculosis, and *Mycobacterium leprae,* the bacterium that causes Hansen's disease (leprosy) in humans.

To facilitate the staining of acid-fast bacteria, it is necessary to use methods that make the cells more permeable to stains because the mycolic acid in their cell walls prevents the penetration of most stains. In the Ziehl-Neelsen staining method, the primary stain, carbolfuchsin, contains phenol, and the cells are heated for 5 minutes during the staining procedure. Phenol and heat facilitate the penetration of the carbolfuchsin into the cell. Heat is acting as a mordant to make the mycolic acid and cell wall lipids more permeable to the stain. Subsequent treatment of the cells with acid-alcohol, a decolorizer, does not remove the entrapped stain from the cells. Hence, these bacteria are termed **acid-fast.** In order for non-acid-fast bacteria to be visualized in the acid-fast procedure, they must be counterstained with methylene blue, as the primary stain is removed from these bacteria by the acid-alcohol. In the Ziehl-Neelsen method, the application of heat to cells during staining with carbolfuchsin and phenol is not without safety concerns. Phenol can vaporize when heated, giving rise to noxious fumes that are toxic to the eyes and mucous

membranes. The **Kinyoun acid-fast method** is a modification in which the concentrations of primary stain, basic fuchsin (substituted for carbolfuchsin), and phenol are increased, making it unnecessary to heat the cells during the staining procedure. The increased concentrations of basic fuchsin and phenol are sufficient to allow the penetration of the stain into acid-fast cells, and the basic fuchsin is not removed during destaining with acid-alcohol. This procedure is safer because phenol fumes are not generated during staining of the cells.

In the acid-fast staining method, acid-fast bacteria such as *Mycobacterium* are not decolorized by acid-alcohol and are therefore stained pink to red by the basic fuchsin. Because non-acid-fast bacteria such as *Staphylococcus* are decolorized by the acid-alcohol, a secondary stain, methylene blue, must be applied to visualize these cells in stained preparations. These appear blue after staining is completed (figure 16.1).

In the following exercise, you will prepare an acid-fast stain of a mixture of *Mycobacterium smegmatis* and *Staphylococcus epidermidis* using the Kinyoun method for acid-fast staining. *M. smegmatis* is a non-pathogenic, acid-fast rod that occurs in soil and on the external genitalia of humans. *S. epidermidis* is a

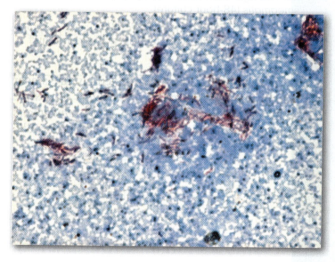

Figure 16.1 Acid-fast stain of *Mycobacterium smegmatis* (red) and *Staphylococcus aureus* (blue).
© McGraw-Hill Education/Auburn University Research Instrumentation Facility/Michael Miller, photographer

1. Measure the correct amount of water needed to make up your batch. The following volumes required per tube must be taken into consideration:

```
pours . . . . . . . . . . . . . . . . . . . . . . . . . . . 12 ml
deeps . . . . . . . . . . . . . . . . . . . . . . . . . . . 6 ml
slants . . . . . . . . . . . . . . . . . . . . . . . . . . . 4 ml
broths . . . . . . . . . . . . . . . . . . . . . . . . . . . 5 ml
broths with fermentation tubes . . . . . . . 5–7 ml
```

2. Consult the label on the bottle to determine how much powder is needed for 1000 ml and then determine by proportionate methods how much you need for the amount of water you are using. Weigh this amount on a balance and add it to the beaker of water (figure 18.2). If the medium does not contain agar, the mixture usually goes into solution without heating (figure 18.3).

3. **If the medium contains agar,** heat the mixture on a stirring hot plate (figure 18.4) or on an electric hot plate until it comes to a boil. To safeguard against water loss, *before heating, mark the level of the top of the medium on the side of the beaker with a Sharpie.* As soon as

it starts to boil, turn off the heat. If an electric hot plate is used, the medium must be removed from the hot plate or it will boil over the sides of the container.

Caution

Be sure to keep stirring the medium so that it does not char on the bottom of the beaker.

4. Check the level of the medium with the mark on the beaker to note if any water has been lost. Add sufficient distilled water as indicated. Keep the temperature of the medium at about 60°C to avoid solidification. The medium will solidify at around 40°C.

Adjusting the pH

Although dehydrated media contain buffering agents to keep the pH of the medium in a desired range, the pH of a batch of medium may differ from that stated on the label of the bottle. Before the medium is tubed, therefore, one should check the pH and make any necessary adjustments.

If a pH meter (figure 18.5) is available and already standardized, use it to check the pH of your medium. If the medium needs adjustment, use the bottles of HCl and NaOH to correct the pH. If no meter is available, pH papers will work about as well. Make pH adjustment as follows:

Materials

- beaker of medium
- acid and base kits (dropping bottles of 1N and 0.1N HCl and NaOH)
- glass stirring rod
- pH papers
- pH meter (optional)

1. Dip a piece of pH test paper into the medium to determine the pH of the medium.
2. **If the pH is too high,** add a drop or two of HCl to lower the pH. For large batches use 1N HCl. If the pH difference is slight, use the 0.1N HCl. Use a glass stirring rod to mix the solution as the drops are added.
3. **If the pH is too low,** add NaOH, one drop at a time, to raise the pH. For slight pH differences, use 0.1N NaOH; for large differences use 1N NaOH. Use a glass stirring rod to mix the solution as the drops are added.

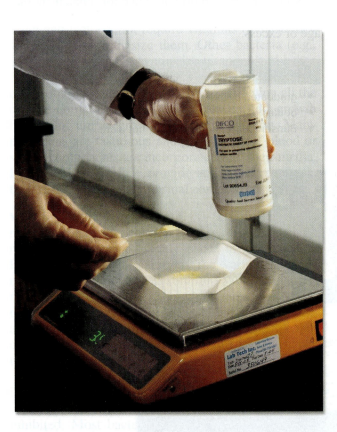

Figure 18.2 Correct amount of dehydrated medium is carefully weighed on a balance.
© McGraw-Hill Education/Auburn University Photographics Services

3. If
bea
tair
pla   n the
ring

4. If th
a D
tub
ope
diu
may
sub

Cappi

The las
closure
are suit
the tube
of the t
to escap
are usin
should
instead,
about a

If no
it may b
A prope
tube so t

Figure 18.3 **Dehydrated medium is dissolved in a measured amount of water.**
© McGraw-Hill Education/Auburn University Photographics Services

Figure 18.4 **If the medium contains agar, it must be heated to dissolve the agar.**
© McGraw-Hill Education/Auburn University Photographics Services

Figure 18.5 **The pH of the medium is adjusted by adding acid or base as per recommendations.**
© McGraw-Hill Education/Auburn University Photographics Services

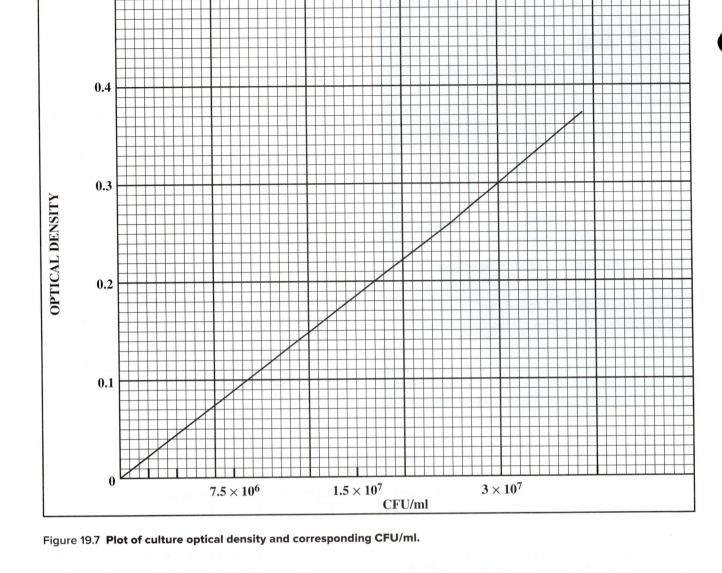

Figure 19.7 Plot of culture optical density and corresponding CFU/ml.

Filling

Once the
dispensed
chine is to
to be set u
can be adj
desired sp
filled, the

Materi

• autom

1. Follov
   tor fo
   will ir
   dium
   bly ot
   pipett
   up me
   amou
   pipett
2. Place
   ceed t
   of del
   will h

of cells in a suspension by only measuring the absorbance of the culture. This is achieved by determining the increase in bacterial cell numbers in a culture over time using the standard plate count method and also determining the absorbance values for each of the samples that were plated. The absorbance values are then graphed versus the plate counts (CFU/ml) to generate a standard curve. For the bacterium growing under the same defined conditions, new absorbance values are determined and used to ascertain the corresponding number of cells in the sample from the standard curve (figure 19.7). For example, in figure 19.7, an absorbance of 0.3 is equal to 3×10^7 CFU/ml. It is important to understand that this method is only valid when the medium and growth conditions are defined for a particular organism. One cannot use numbers generated for an *Escherichia coli* culture to determine the number of cells in a *Staphylococcus aureus*

culture grown under different conditions. Also, the wavelength used to determine absorbance will vary for each type of bacterium.

In the following exercise, you will demonstrate the relationship between absorbance and cell turbidity by measuring the absorbance values for various dilutions of a culture. There should be a proportional relationship between the concentration of bacterial cells and the absorbance or optical density of a culture. To demonstrate this relationship, you will measure the absorbance values of various dilutions provided to you. These values will be plotted on a graph as a function of culture dilution. For lower values of absorbance there may be a linear relationship between the concentration of cells and absorbance. At higher values, however, the relationship may not be linear. That is, for a doubling in cell concentration, there may be less than a doubling of absorbance.

(a) Turn on the instrument with the on/off switch (left front knob). Allow the spectrophotometer to warm up for 15 minutes. Set the wavelength to 550 nm and position the filter to correspond to this wavelength. Select the Transmittance mode and make sure the cuvette holder cover is closed. Adjust the digital readout to zero % transmittance using the left front (on/off) knob.

(b) Set the mode to Absorbance. Insert a cuvette containing sterile nutrient broth into the sample holder and close the cover. Adjust the Absorbance to 0 (zero) using the right front knob. This may require several turns of the knob.

(c) Remove the blank and insert a cuvette with one of the bacterial cell samples. Close the cover and read the absorbance/O.D.

(d) Occasionally blank the instrument to zero with the sterile nutrient broth during the course of reading the cell samples.

Figure 19.8 Calibration procedure for the B & L Spectronic 200 digital spectrophotometer.
© McGraw-Hill Education/Auburn University Photographics Services

Materials

- broth culture of *E. coli* (same one as used for plate count)
- spectrophotometer cuvettes (2 per student)
- 4 small test tubes and test-tube rack
- 5 ml pipettes
- bottle of sterile nutrient broth (20 ml per student)

1. Calibrate the spectrophotometer using the procedure described in figure 19.8. These instructions apply specifically to the Bausch and Lomb Spectronic 200 digital spectrophotometer. It is important to blank the spectrophotometer by adjusting the instrument to an absorbance of 0 (zero) using uninoculated nutrient broth. The medium contains components that cause it to be slightly colored and, hence, it will absorb some light, adding to the light absorbance of the bacterial culture tubes. Blanking the instrument using the uninoculated medium will subtract the absorbance resulting from the medium. In handling the cuvettes, keep the following in mind:

a. Rinse the cuvette with distilled or deionized water to clean it before using.

b. Keep the lower part of the cuvette free of liquids, smudges, and fingerprints by carefully wiping the surface only with Kim wipes or

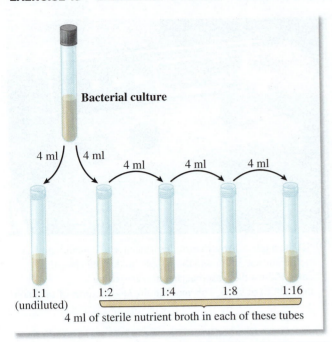

Bacterial culture

4 ml / 4 ml / 4 ml / 4 ml / 4 ml

1:1 (undiluted)   1:2   1:4   1:8   1:16

4 ml of sterile nutrient broth in each of these tubes

Figure 19.9 Dilution procedure for cuvettes.

lint-free tissue provided. Do not use paper towels or handkerchiefs for this purpose. If smudges, liquids, or fingerprints occur on the cuvette surface, they can contribute to light absorbance and erroneous readings.

c. Insert the cuvette into the sample holder with the index line aligned with the index line on the cuvette holder. Properly seat the cuvette by exactly aligning the lines on the cuvette and holder.

d. Handle cuvettes with care as they are of optical quality and expensive.

2. Label a cuvette 1:1 (near top of tube) and four test tubes 1:2, 1:4, 1:8, and 1:16. These tubes will be used for the serial dilutions shown in figure 19.9.

3. With a 5 ml pipette, dispense 4 ml of sterile nutrient broth into tubes 1:2, 1:4, 1:8, and 1:16.

4. Shake the culture of *E. coli* vigorously to suspend the organisms, and with the same 5 ml pipette, transfer 4 ml to the 1:1 cuvette and 4 ml to the 1:2 test tube.

5. Mix the contents in the 1:2 tube by drawing the mixture up into the pipette and discharging it into the tube three times.

6. Transfer 4 ml from the 1:2 tube to the 1:4 tube, mix three times, and go on to the other tubes in a similar manner. Tube 1:16 will have 8 ml of diluted organisms.

7. Measure the optical density of each of the five tubes, starting with the 1:16 tube first. The contents of each of the test tubes must be transferred to a cuvette for measurement. Be sure to close the lid on the sample holder when making measurements. A single cuvette can be used for all the measurements.

Laboratory Report

Record the O.D. values in the table of Laboratory Report 19 and then plot the values on the provided graph.

19 Enumeration of Bacteria: The Standard Plate Count

A. Results

1. Quantitative Plating Method

    a. Record your plate counts in this table:

DILUTION BOTTLE	ml PLATED	DILUTION	DILUTION FACTOR	NUMBER OF COLONIES
b (1:10,000)	1.0	1:10,000	10^4	
b (1:10,000)	0.1	1:100,000	10^5	
c (1:1,000,000)	1.0	1:1,000,000	10^6	
c (1:1,000,000)	0.1	1:10,000,000	10^7	

    b. How many cells per milliliter were in the undiluted culture?_____

2. Optical Density Determination

    a. Record the optical density values for your dilutions in the following table.

DILUTION	OPTICAL DENSITY
1:1	
1:2	
1:4	
1:8	
1:16	

b. Plot the optical densities versus the concentration of organisms. Complete the graph by drawing a line between plot points.

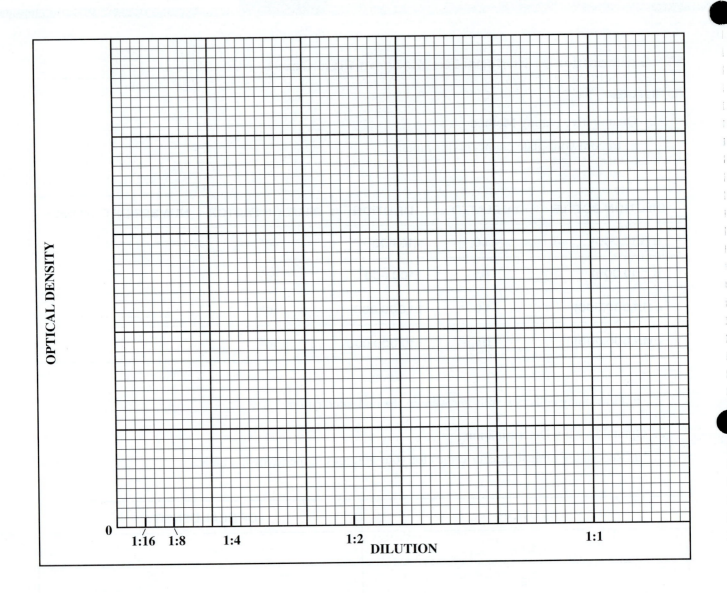

c. What is the maximum O.D. that is within the linear portion of the curve? _____

d. What is the corrected or true O.D. of the undiluted culture? (Hint: If the O.D. for the 1:2 dilution but not the 1:1 dilution is within the linear portion of the curve, then the O.D. of the 1:1 dilution should not be considered correct. The correct or true O.D. of the undiluted culture in this example could be estimated by multiplying the O.D. of the 1:2 dilution by 2.) _____

e. What is the correlation between corrected O.D. and cell number for your culture? _____

_____

_____

B. Short-Answer Questions

1. Why is CFU more applicable to a culture of *Streptococcus* than to a culture of *E. coli*?

   _____

   _____

2. How would you inoculate a plate to get a 1:10 dilution? A 1:100 dilution?

   _____

   _____

3. Why is it necessary to perform a plate count in conjunction with the turbidimetry procedure?

   _____

   _____

4. For the following methods of bacterial enumeration, does the method determine total count or viable count?
   a. MPN_____
   b. Microscopic count_____
   c. Standard plate count_____
   d. Turbidity_____

Slide Culture:
Fungi

The isolation, culture, and microscopic examination of fungi require the use of suitable selective media and special microscopic slide techniques. Simple wet mounts prepared from fungal cultures usually do not reveal the arrangement of spores on fruiting bodies because the manipulation of the culture disrupts the fruiting structures and the hyphae of the culture. The type of fruiting structure and spore arrangement and morphology are important in the identification and taxonomy of these microorganisms. One way to preserve the integrity of the fruiting structure is to prepare a slide culture that can then be stained. This allows the observation of the fruiting structure *in situ* and does not disrupt the arrangement of the spores. In this exercise, a slide culture method will be used to prepare stained slides of molds. The method is superior to wet mounts in that the hyphae, sporangiophores, and spores remain more or less intact when stained.

When fungi are collected from the environment, as in Exercise 7, Sabouraud's agar is most frequently used. It is a simple medium consisting of 1% peptone, 4% glucose, and 2% agar-agar. The pH of the medium is adjusted to 5.6, which favors the growth of fungi but inhibits most bacterial growth.

Unfortunately, for some fungi the pH of Sabouraud's agar is too low and the glucose content is too high. A better medium for these organisms is one suggested by C. W. Emmons that contains only 2% glucose, with 1% neopeptone, and an adjusted pH of 6.8–7.0. To inhibit bacterial growth, 40 mg of chloramphenicol is added to 1 liter of the medium.

In addition to the above two media, cornmeal agar, Czapek solution agar, and others are available for special applications in culturing molds.

Figure 20.1 illustrates the procedure that will be used to produce a fungal culture that can be stained directly on the slide. Note that a sterile cube of Sabouraud's agar is inoculated on two sides with spores from a mold colony. Figure 20.2 illustrates how the cube is held with a scalpel blade as inoculation takes place. The cube is placed in the center of a microscope slide with one of the inoculated surfaces placed against the slide. On the other inoculated surface of the cube is placed a cover glass. The assembled slide is incubated at room temperature for 48 hours in a moist chamber (petri dish with a small amount of water). After incubation, the cube of medium is carefully separated from the slide and discarded.

During incubation the fungal culture will grow over the glass surfaces of the slide and cover glass. By adding a little stain to the slide, a semipermanent slide can be made by placing a cover glass over it. The cover glass can also be used to make another slide by placing it on another clean slide with a drop of stain on it. Before the stain (lactophenol cotton blue) is used, it is desirable to add to the hyphae a drop of alcohol, which acts as a wetting agent.

🕐 First Period

(Slide Culture Preparation)

Proceed as follows to make slide cultures of one or more mold colonies:

Materials

- petri dishes, glass, sterile
- filter paper (9 cm dia, sterile)
- glass U-shaped rods
- fungal culture plate (mixture)
- 1 petri plate of Sabouraud's agar or Emmons' medium per 4 students
- scalpels
- inoculating loop
- sterile water
- microscope slides and cover glasses (sterile)
- forceps

1. Aseptically, with a pair of forceps, place a sheet of sterile filter paper in a petri dish.
2. Place a sterile U-shaped glass rod on the filter paper. (Rod can be sterilized by flaming, if held by forceps.)
3. Pour enough sterile water (about 4 ml) on filter paper to completely moisten it.

Figure 20.1 **Procedure for making two stained slides from slide culture.**

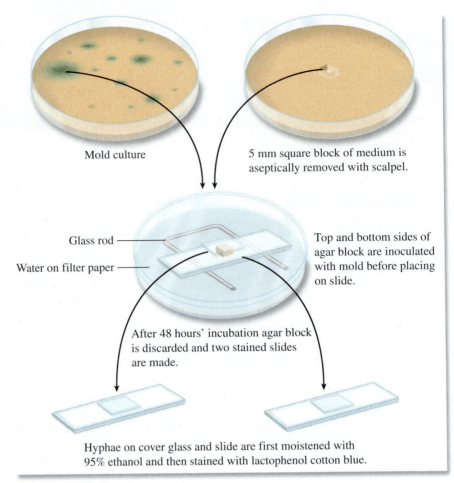

Mold culture

5 mm square block of medium is aseptically removed with scalpel.

Glass rod

Water on filter paper

Top and bottom sides of agar block are inoculated with mold before placing on slide.

After 48 hours' incubation agar block is discarded and two stained slides are made.

Hyphae on cover glass and slide are first moistened with 95% ethanol and then stained with lactophenol cotton blue.

4. With forceps, place a sterile slide on the U-shaped rod.
5. *Gently* flame a scalpel to sterilize, and cut a 5 mm square block of the medium from the plate of Sabouraud's agar or Emmons' medium.

6. Pick up the block of agar by inserting the scalpel into one side, as illustrated in figure 20.2. Inoculate both top and bottom surfaces of the cube with spores from the mold colony. Be sure to flame and cool the loop prior to picking up spores.

Figure 20.2 **Inoculation technique.**

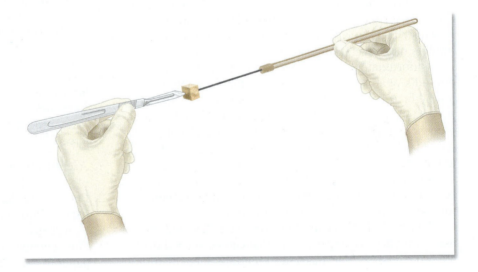

7. Place the inoculated block of agar in the center of a microscope slide. Be sure to place one of the inoculated surfaces down.
8. Aseptically, place a sterile cover glass on the upper inoculated surface of the agar cube.
9. Place the cover on the petri dish and incubate at room temperature for 48 hours.
10. After 48 hours, examine the slide under low power. If growth has occurred, you should see hyphae and spores. If growth is inadequate and spores are not evident, allow the fungus to grow another 24 to 48 hours before making the stained slides.

Second Period

(Application of Stain)

As soon as there is evidence of spores on the slide, prepare two stained slides from the slide culture, using the following procedure:

Materials

- microscope slides and cover glasses
- 95% ethanol
- lactophenol cotton blue stain
- forceps

1. Place a drop of lactophenol cotton blue stain on a clean microscope slide.

2. Remove the cover glass from the slide culture and discard the block of agar in a container provided.
3. Add a drop of 95% ethanol to the hyphae on the cover glass. As soon as most of the alcohol has evaporated, place the cover glass, mold side down, on the drop of lactophenol cotton blue stain on the slide. This slide is ready for examination.
4. Remove the slide from the petri dish, add a drop of 95% ethanol to the hyphae, and follow this with a drop of lactophenol cotton blue stain. Cover the entire preparation with a clean cover glass.
5. Compare both stained slides under the microscope; one slide may be better than the other one.

Laboratory Report

There is no Laboratory Report for this exercise.

Questions

1. Why is a block of agar inoculated rather than streaking the fungal culture on a plate?
2. Why is chloramphenicol added to the fungal culture?
3. At what pH do most fungi grow?

Bacterial Viruses

Viruses differ from bacteria in being much smaller and therefore below the resolution of the light microscope. The smallest virus is one million times smaller than a typical eukaryotic cell. Viruses are obligate intracellular parasites that require a host cell in order to replicate and reproduce, and hence they cannot be grown on laboratory media. Despite these obstacles, we can detect their presence by the effects that they have on their host cells.

Viruses infect all types of cells, eukaryotic and prokaryotic. They are composed of RNA or DNA but never both, and a protein coat, or capsid, that surrounds the nucleic acid. Their dependence on cells is due to their lack of metabolic machinery necessary for the synthesis of viral components. By invading a host cell, they can utilize the metabolic systems of the host cells to achieve their replication.

The study of viruses that parasitize plant and animal cells is time-consuming and requires special tissue culture techniques. Bacterial viruses are relatively simple to study, utilizing ordinary bacteriological techniques. It is for this reason that bacterial viruses will be studied here. However, the principles learned from studying the viruses that infect bacteria apply to viruses of eukaryotic cells.

Viruses that infect bacterial cells are called bacteriophages, or phages. They are diverse in their morphology and size. Some of the simplest ones have single-stranded DNA. Most phages are composed of head, sheath, and tail fibers as seen in figure 21.1. The capsid (head) may be round, oval, or polyhedral and is composed of individual protein subunits called capsomeres. It forms a protective covering around the viral genome. The tail structure or sheath is composed of a contractile protein that surrounds a hollow core, which is a conduit for the delivery of viral nucleic acid into the host cell. At the end of the tail is a base plate with tail fibers and spikes attached to it. The tail fibers bind to chemical groups on the surface of the bacterial cell and are responsible for recognition. Lysozyme associated with the tail portion of the virus erodes and weakens the cell wall of the host cell. This facilitates the injection of the viral nucleic acid by the sheath contracting and forcing the hollow core through the weakened area in the cell wall.

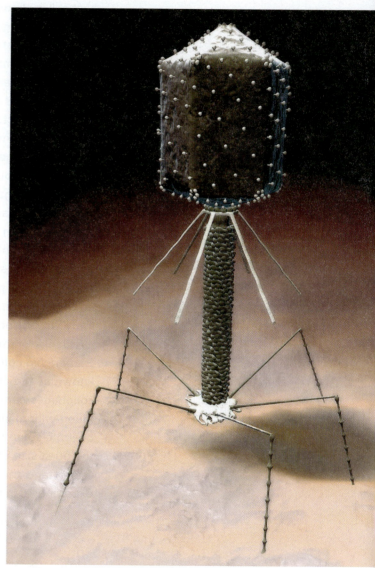
© MedicalRF.com RF

Infections by viruses can have two outcomes. The lytic cycle involves virulent phages that cause lysis and death of the host cell. The lysogenic cycle involves temperate phages, which can either lyse the host or integrate their DNA into host cell DNA and alter the genetics of the host cell.

In the lytic cycle, the virus assumes control of cell metabolism and uses the cell's metabolic machinery to manufacture phage components (i.e., nucleic acid, capsid, sheath, tail fibers, spikes, and base plates). Mature phage particles are assembled and released from the cell by lysis and they can in turn invade new host cells. The result of a lytic infection for the host cell is almost always death (fig. VI.1).

In the lysogenic cycle, the viral DNA of the temperate virus is integrated into host DNA, and no mature phages are made. Cells grow normally and are immune to further infections by the same phage. There is no visible evidence to indicate that a virus is even present in the cell. In some cases, the virus can carry genes that confer new genetic capabilities on the virally infected cell, or lysogen. For example, when *Corynebacterium diphtheriae* is infected with a certain lysogenic phage, because the phage carries a toxin gene in its genome, the host cells begin to produce a potent toxin responsible for the symptoms of diphtheria. This phenomenon is known as lysogenic conversion and is responsible for some of the toxins produced by various pathogens. Periodically, the lysogenic phage DNA can excise from the host DNA and initiate the lytic cycle and the production of mature phages. This results in the lysis of the host cell.

Visual evidence for lysis can be demonstrated by mixing phages with host cells and plating them onto media. The bacteria form a confluent lawn of growth, and where the phages cause lysis of the bacterial cells, there will be seen clear areas called plaques.

Some of the most studied bacteriophages are those that infect *Escherichia coli,* such as the T-even phages and lambda phage. They are known as the **coliphages**. Because *E. coli* is an intestinal bacterium, the coliphages can readily be isolated from raw sewage and coprophagous (dung-eating) insects such as flies. The exercises in this section will demonstrate some of the techniques for isolating, assaying, and determining the burst size of bacteriophages. It is recommended that you thoroughly understand the various stages in the lytic cycle before you begin the experiments in this section.

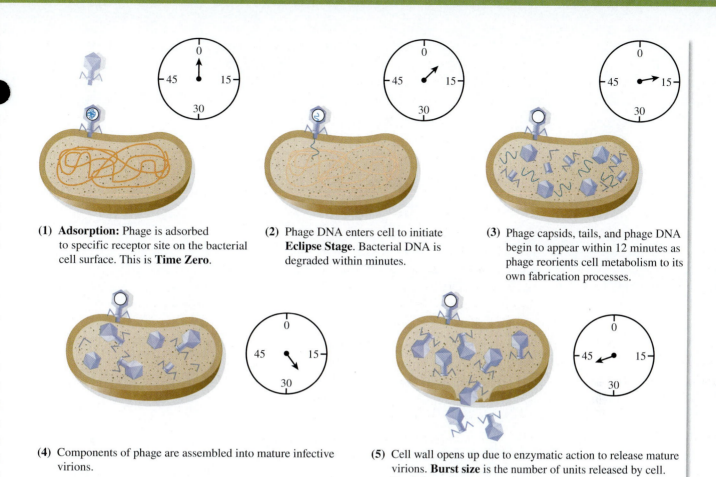

(1) Adsorption: Phage is adsorbed to specific receptor site on the bacterial cell surface. This is **Time Zero**.

(2) Phage DNA enters cell to initiate **Eclipse Stage**. Bacterial DNA is degraded within minutes.

(3) Phage capsids, tails, and phage DNA begin to appear within 12 minutes as phage reorients cell metabolism to its own fabrication processes.

(4) Components of phage are assembled into mature infective virions.

(5) Cell wall opens up due to enzymatic action to release mature virions. **Burst size** is the number of units released by cell. Total time: 40 minutes.

Figure VI.1 **The lytic cycle of a virulent bacteriophage.**

Determination of a Bacteriophage Titer

Bacteriophages are viruses that infect bacterial cells. They were first described by Twort and d'Herelle in 1915 when they both noted that bacterial cultures spontaneously cleared and the bacteria-free liquid that remained could cause new cultures of bacteria to also clear. Because it appeared that the cultures were being "eaten" by some unknown agent, d'Herelle coined the term *bacteriophage,* which means "bacterial eater." Like all viruses, bacteriophages, or phages, for short, are **obligate intracellular parasites,** that is, they must invade a host cell in order to replicate and reproduce. This is due to the fact that viruses are composed primarily of only a single kind of nucleic acid molecule encased in a protein coat, or **capsid,** that protects the nucleic acid. All viruses lack metabolic machinery, such as energy systems, and protein synthesis components necessary for independent replication. In order to replicate and reproduce, they must use the host cell's metabolic machinery to synthesize their various component parts.

Viruses also exhibit specificity for their hosts. For example, a certain bacteriophage may only infect a specific strain of a bacterium. Examples are the T-even bacteriophages that infect *Escherichia coli* B, whereas other phages infect *E. coli* K12, a different strain of the organism. A phage that infects *Staphylococcus aureus* does not infect *E. coli* and vice versa. This specificity can be used in phage typing of pathogens (Exercise 23).

The structure of a T4 bacteriophage is shown in figure 21.1. A phage consists of a **nucleocapsid,** which is the nucleic acid and protein capsid. The nucleocapsid is attached to a protein **sheath** that is contractile and contains a hollow tube in its center.

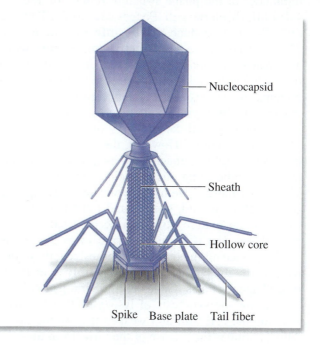

Figure 21.1 **T4 Bacteriophage.**

The sheath sits on a **base plate** to which **tail fibers** and **spikes** are attached. Most of the phage structure is necessary for delivery of the phage nucleic acid into its host. A single virus or phage particle is also called a **virion.**

The steps in a lytic phage infection of a bacterial cell are basically as follows:

Adsorption   A bacteriophage recognizes its host by its tail fibers binding to chemical groups associated with receptors on the surface of the host cell. These receptors perform other functions for the bacterial cell. Examples of receptors are pili and lipopolysaccharide in gram-negative bacteria and teichoic acids in gram-positive bacteria. The specificity in phage infections resides in differences in the receptors. If the groups that are recognized by tail fibers are not present on a bacterial cell, a phage cannot bind to that cell and cause an infection.

Entry of Viral Genome   The phage particle settles onto the surface of the host bacterial cell, and **lysozyme** that is associated with the phage begins to erode a localized area of the cell wall, thus weakening the wall. Then the sheath contracts, forcing a hollow core that is connected to the phage capsid through the weakened area of the cell wall. This results in the viral genome being injected into the bacterial cytoplasm.

Synthesis of Phage Components   Only the phage nucleic acid enters the host cell, while the capsid and remainder of the phage structure remain on the outside of the bacterial cell. Phages such as the T phages (T2, T4, T6) produce a **lytic infection** in *E. coli*. These phages cause a **productive infection** in which new phages are made and released from the cell. To incapacitate the host cell and ensure that phage components are synthesized, a viral nuclease injected with the viral DNA begins to degrade *E. coli* DNA. As a result, the host cell cannot carry out any of its own metabolic functions. However, the virus leaves intact host cell metabolic machinery for producing energy and for synthesizing nucleic acids and proteins. The virus uses these systems to replicate its component parts such as capsid, sheath, and tail fibers. During this time no mature virus can be detected by assay procedures, and it is thus referred to as the **eclipse period**. Once all the parts are synthesized, they come together by the process of **self-assembly** to form mature phage particles. Genes on the viral genome also encode for the synthesis of lysozyme that begins to degrade the cell wall and weaken it from the inside. This facilitates the release of the virus from the cell.

In contrast to lytic phages, when **lysogenic** or **temperate** bacteriophages infect a bacterial cell, the phage genome can integrate into host DNA to become a stable genetic element in the host cell genome. In this instance, the genes for the synthesis of viral components are not usually expressed, and hence no mature phages are produced in these cells. However, each time the bacterial genome replicates during cell division, phage DNA is also replicated, and all bacterial progeny contain phage DNA as part of their genetic makeup. Bacterial cells that have temperate phage DNA integrated into their DNA are called **lysogens,** and the integrated phage genome is known as a **prophage** or **provirus.** An example of a temperate bacteriophage that causes this type of infection is λ (lambda) phage that infects *Escherichia coli*.

Occasionally, the viral DNA in a lysogenic infection can excise from the host DNA, which results in the expression of viral genes responsible for the synthesis of phage components. Mature phages are made by the same steps that are responsible for the production of lytic phages and the host cell is lysed.

Release of Virus   The combination of the weakened cell wall resulting from the action of lysozyme plus the pressure exerted by the phage particles in the cell causes the cell to burst, releasing mature phages into the environment where they can infect susceptible cells. Lytic phages will produce new lytic infections, whereas lysogenic phages will produce new lysogenic infections. One phage particle infecting a single host cell can produce as many as 200 virions. This number is called the **burst size** and will vary for each virus.

If bacterial cells are mixed with bacteriophage in soft agar, the bacteria will first grow to produce a **confluent lawn** of cells. Phages will infect the cells, causing them to undergo lysis and form clear areas in the confluent lawn called **plaques.** Each plaque is formed by the progeny of a single virion that has replicated and lysed the bacterial cells. Like colony-forming units, **plaque-forming units** (PFUs) can be counted to determine the number of viral particles in a suspension of phage.

In the following exercise, you will work in pairs and determine the number of phage particles or PFUs in a suspension of T4 bacteriophage. You will use *E. coli* B as the host for this experiment.

Materials

- 18- to 24-hour broth culture of *Escherichia coli* B
- 2 ml suspension of T4 bacteriophages with a titer of at least 10,000 phages/ml
- 5 trypticase soy agar (TSA) plates. These should be warmed to 37°C before use.
- 5 tubes of soft agar (0.7% agar). Prior to use, melt and hold at 50°C in a water bath.
- 5 tubes of 9.0 ml trypticase soy (TSB) broth
- 1 ml sterile pipettes
- pipette aids

1. Label the 5 TSA plates with your name and dilutions from 1:10 to 1:100,000.
2. Label 5 TS broth tubes with the dilutions 1:10 to 1:100,000 (figure 21.2).
3. Prepare serial tenfold dilutions of the phage stock suspensions by transferring 1 ml of the phage suspension to the first dilution blank. Mix well and transfer 1 ml of the first dilution to the second dilution blank (1:100). Repeat this same procedure until the original phage stock has been diluted 1:100,000 (figure 21.2).
4. Aseptically transfer 2 drops of *E. coli* B broth culture to each of the 5 soft agar overlay tubes.
5. Transfer 1 ml of the first (1:10) phage dilution tube to a soft agar overlay and mix thoroughly but gently. After mixing, pour the contents of the soft agar tube onto the respective TSA plate. Make sure that the soft agar completely covers the surface of the TSA plate. This can be accomplished by gently swirling the plate several times after pouring and while the soft agar is still liquid.

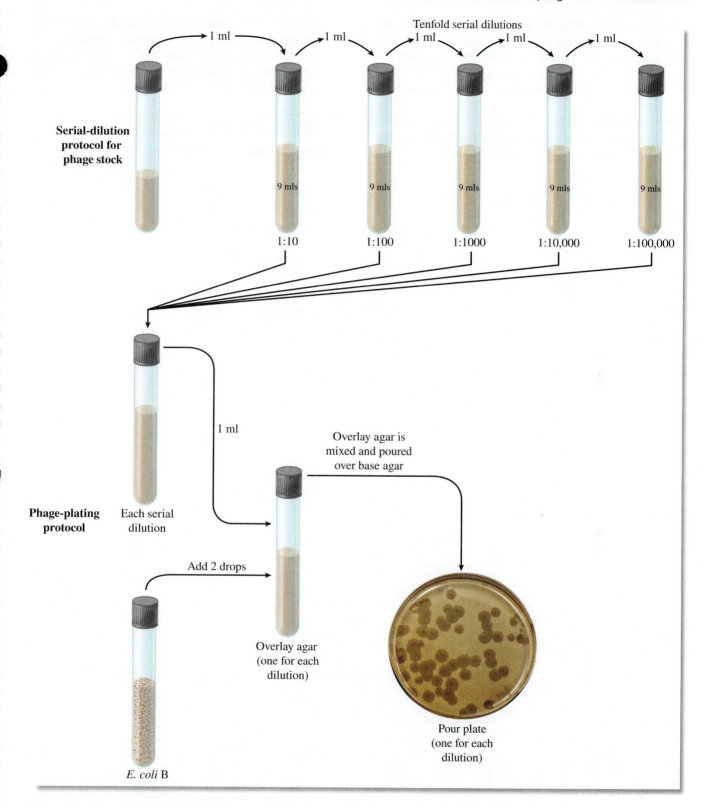

Figure 21.2 **Procedure for determining the titer of bacteriophage.**
© McGraw-Hill Education. Lisa Burgess, photographer

6. Repeat this procedure for each dilution of the phage suspension.
7. Incubate the plates at 37°C for 24 hours. If the exercise cannot be completed at this time, refrigerate the plates until the next laboratory period.
8. Observe the plates. Plaques will appear as clear areas in the bacterial lawn. Count the plaques on the plates. Only include counts between 25 and 250 plaques. This can be facilitated with a bacterial colony counter. Multiply the number of plaques times the dilution factor to determine the number of phage particles in the original suspension of phages.

Laboratory Report

Record the phage titer in Laboratory Report 21.

Laboratory Report

21

21 Determination of a Bacteriophage Titer

A. Results

DILUTION	PLAQUE NUMBER	PFU/ML
1:10		
1:100		
1:1000		
1:10,000		
1:100,000		

For the PFU/ml calculation, only plates giving 25–250 plaques should be used. For the plaque number, indicate when the bacterial cells are cleared due to too many phage particles in the infection.

What is the number of phage particles in the original phage stock? _____

B. Short-Answer Questions

1. To what chemical groups can bacteriophage specific for gram-positive cells attach?

   _____

   _____

2. Why are viruses called obligate intracellular parasites?

   _____

   _____

3. What type of infection is caused by a temperate phage?

   _____

4. What is lysogenic conversion?

   _____

   _____

5. Compare and contrast bacteriophage plaques with bacterial colonies.

   _____

   _____

6. Describe the four steps in a lytic phage infection.

   _____

   _____

7. What part of the host cell is degraded after phage nucleic acid has entered a host cell?

_____

_____

8. Name two stages in a lytic infection where lysozyme is used. Describe the role of lysozyme during each of these stages.

_____

_____

Isolation of Phages from Flies

Learning Outcomes

After completing this exercise, you should be able to

1. Understand where bacteriophages that infect *E. coli* naturally occur in the environment.
2. Isolate bacteriophages that infect *E. coli* from flies.

As stated earlier, coprophagous insects (insects that feed on fecal material and dung, as well as raw sewage) contain various kinds of bacterial viruses. Houseflies are coprophagous because they deposit their eggs in fecal material where the young larvae feed, grow, pupate, and emerge as adult flies. This type of environment is heavily populated by *E. coli* and the various bacteriophages that infect this bacterium.

Fly Collection

To increase the probability of success in isolating phages, it is desirable that one use 20 to 24 houseflies. A smaller number might be sufficient; the larger number, however, increases the probability of initial success. Houseflies should not be confused with the smaller blackfly or the larger blowfly. An ideal spot for collecting these insects is a barnyard or riding stable. Flies should be kept alive until just prior to crushing and placing them in the growth medium. There are many ways that one might use to capture them—use your ingenuity!

Enrichment

Within the flies' digestive tracts are several different strains of *E. coli* and bacteriophages. Our first concern is to enhance the growth of both organisms to ensure an adequate supply of phages. To accomplish this the flies must be ground up with a mortar and pestle and then incubated in a special growth medium for a total of 48 hours. During the last 6 hours of incubation, a lysing agent, sodium cyanide, is included in the growth medium to augment the lysing properties of the phage.

Figures 22.1 and 22.2 illustrate the procedure.

Materials

- bottle of phage growth medium* (50 ml)
- bottle of phage lysing medium* (50 ml)
- Erlenmeyer flask (125 ml capacity) with cap
- mortar and pestle (glass)
  *see Appendix C for composition

1. Into a clean, nonsterile mortar place 24 freshly killed houseflies. Pour half of the growth medium into the mortar and grind the flies to a fine pulp with the pestle.
2. Transfer this fly-broth mixture to an empty flask. Use the remainder of the growth medium to rinse out the mortar and pestle, pouring all the medium into the flask.
3. Wash the mortar and pestle with soap and hot water before returning them to the cabinet.
4. Incubate the fly-broth mixture for 42 hours at 37°C.
5. At the end of the 42-hour incubation period, add 50 ml of lysing medium to the fly-broth mixture. Incubate this mixture for another 6 hours.

Centrifugation

Before attempting filtration, you will find it necessary to separate the fly fragments and miscellaneous bacteria from the culture medium. If centrifugation is incomplete, the membrane filter will clog quickly and filtration will progress slowly. To minimize filter clogging, a triple centrifugation procedure will be used. To save time in the event filter clogging does occur, an extra filter assembly and an adequate supply of membrane filters should be available. These filters have a maximum pore size of 0.45 μm, which holds back all bacteria, allowing only the phage virions to pass through.

Materials

- centrifuge
- 6–12 centrifuge tubes
- 2 sterile membrane filter assemblies (funnel, glass base, clamp, and vacuum flask)
- package of sterile membrane filters (0.45 μm)
- sterile Erlenmeyer flask with cap (125 ml size)
- vacuum pump and rubber hose

Inoculation and Incubation

To demonstrate the presence of bacteriophages in the fly-broth filtrate, a strain of phage-susceptible *E. coli* will be used. To achieve an ideal proportion of phages to bacteria, a proportional dilution method will be used. The phages and bacteria will be added to tubes of soft nutrient agar that will be layered over plates of hard nutrient agar. Soft nutrient agar contains only half as much agar as ordinary nutrient agar. (This medium and *E. coli* provide an ideal "lawn" for phage growth.) Its jellylike consistency allows for better diffusion of phage particles; thus, more even development of plaques occurs.

Figure 22.2 illustrates the overall procedure. It is best to perform this inoculation procedure in the morning so that the plates can be examined in late afternoon. As plaques develop, one can watch them increase in size with the multiplication of phages and simultaneous destruction of *E. coli.*

Materials

- nutrient broth cultures of *Escherichia coli* (ATCC #8677 phage host)
- flask of fly-broth filtrate
- 10 tubes of soft nutrient agar (5 ml per tube) with metal caps
- 10 plates of nutrient agar (15 ml per plate, and prewarmed at 37°C)
- 1 ml serological pipettes, sterile

1. Liquefy 10 tubes of soft nutrient agar and cool to 50°C. Keep tubes in water bath to prevent solidification.

2. With a Sharpie marking pen, number the tubes of soft nutrient agar 1 through 10. Keep the tubes sequentially arranged in the test-tube rack.

3. Label 10 plates of prewarmed nutrient agar 1 through 10. Also, label plate 10 "negative control." Prewarming these plates will allow the soft agar to solidify more evenly.

4. With a 1 ml serological pipette, deliver 0.1 ml of fly-broth filtrate to tube 1, 0.2 ml to tube 2, etc., until 0.9 ml has been delivered to tube 9. Refer to figure 22.2 for sequence. **Note that no fly-broth filtrate is added to tube 10.** This tube will be your negative control.

5. With a fresh 1 ml pipette, deliver 0.9 ml of *E. coli* to tube 1, 0.8 ml to tube 2, etc., as shown in figure 22.2. **Note that tube 10 receives 1.0 ml of *E. coli.*** Make sure to gently but thoroughly mix all the tubes.

6. After flaming the necks of each of the tubes, pour them into similarly numbered plates.

7. When the agar has cooled completely, put the plates, inverted, into a 37°C incubator.

8. **After about 3 hours** of incubation, examine the plates, looking for plaques. If some are visible, measure them and record their diameters in Laboratory Report 22.

9. If no plaques are visible, check the plates again in another **2 hours.**

10. Check the plaque size again at **12 hours,** if possible, recording your results. Incubate a total of 24 hours.

Laboratory Report

Complete Laboratory Report 22.

22 Isolation of Phages from Flies

A. Results

1. Plaque Size Increase
    With a Sharpie, circle and label three plaques on one of the plates and record their sizes in millimeters at 1-hour intervals.

TIME	PLAQUE SIZE (millimeters)		
	Plaque No. 1	Plaque No. 2	Plaque No. 3
3 hours			
5 hours			
12 hours			
24 hours			

    a. Were any plaques seen on the negative control plate? _____

    b. Do the plates show a progressive increase in number of plaques with increased amount of fly-broth filtrate? _____

    c. Did the phage completely "wipe out" all bacterial growth on any of the plates? _____
       If so, which plates? _____

2. Observations
    Count all the plaques on each plate and record the counts in the following table. If the plaques are very numerous, use a colony counter and hand counting device. If this exercise was performed as a class project with individual students doing only one or two plates from a common fly-broth filtrate, collect counts from your classmates to complete the table.

Plate Number	1	2	3	4	5	6	7	8	9	10
E. coli (ml)	0.9	0.8	0.7	0.6	0.5	0.4	0.3	0.2	0.1	1.0
Filtrate (ml)	0.1	0.2	0.3	0.4	0.5	0.6	0.7	0.8	0.9	0
Number of plaques										

B. Short-Answer Questions

1. How does the life cycle of houseflies contribute to the presence of *E. coli* bacteriophages in their guts?

    _____

    _____

2. From what other environments might *E. coli* bacteriophages be readily isolated?

    _____

    _____

3. What is the purpose of including a tube in the phage assay in which only the *E. coli* culture is inoculated? Explain.

    _____

    _____

4. Why would phages pass through the filters used to separate fragments of bacterial cells after the centrifugation step in the exercise?

    _____

    _____

Phage Typing

Learning Outcomes

After completing this exercise, you should be able to

1. Differentiate strains of *S. aureus* based on their susceptibility to different bacteriophage types.

2. Understand how phage typing is used in epidemiological investigations.

The host specificity of bacteriophages is such that it is possible to differentiate strains of individual species of bacteria based on their susceptibility to various kinds of bacteriophages. In epidemiological studies, where it is important to discover the source of a specific infection, determining the phage type of the causative organism can be an important tool in solving the riddle. For example, if it can be shown that the phage type of *S. typhi* in a patient with typhoid fever is the same as the phage type of an isolate from a suspected carrier, chances are excellent that the two cases are epidemiologically related. Since most bacteria are probably infected by bacteriophages, it is theoretically possible to classify each species into strains based on their phage susceptibility. Such phage-typing groups have been determined for *Staphylococcus aureus, Salmonella typhi,* and several other pathogens. The following table illustrates the lytic phage groups for *S. aureus.*

LYTIC GROUP	PHAGES IN GROUP
I	29, 52, 52A, 79, 80
II	3A, 3B, 3C, 55, 71
III	6, 7, 42E, 47, 53, 54, 75, 77, 83A
IV	42D
not allotted	81, 187

In bacteriophage typing, a suspension of the organism to be typed is uniformly swabbed over an agar surface. The bottom of the plate is marked off into squares and each square labeled to indicate which phage type is applied to the square. A small drop of each bacteriophage type is added to its respective square. After incubation, the plate is examined to determine which

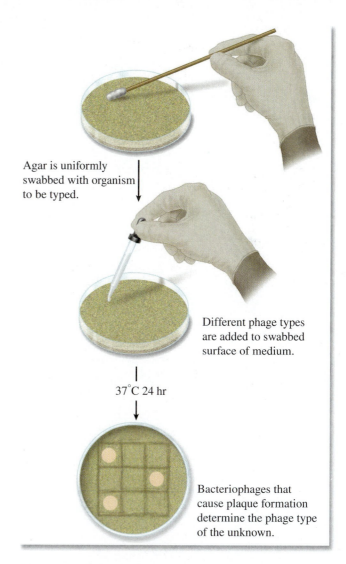

Agar is uniformly swabbed with organism to be typed.

Different phage types are added to swabbed surface of medium.

37°C 24 hr

Bacteriophages that cause plaque formation determine the phage type of the unknown.

Figure 23.1 **Bacteriophage typing.**

phage caused lysis of the test organism. In this exercise, you will determine which phage causes the lysis of *S. aureus* (see figure 23.1).

Materials

- 1 petri plate of tryptone yeast extract agar or trypticase soy agar
- bacteriophage cultures (available types)
- nutrient broth cultures of *S. aureus* with cotton swabs

1. Mark the bottom of a plate of tryptone yeast extract agar with as many squares as there are phage types to be used. Label each square with the phage type numbers.
2. Uniformly swab the entire surface of the agar with the organisms. Please note that *S. aureus* is a BSL 2 organism, and all transfers should be performed in a biosafety hood.
3. Deposit 1 drop of each phage in its respective square.
4. Incubate the plate at 37°C for 24 hours and record the lytic group and phage type of the culture.

Laboratory Report

Record your results in Laboratory Report 23.

23 Phage Typing

A. Results

1. To which phage types was this strain of *S. aureus* susceptible?

   _____

2. To what lytic group does this strain of staphylococcus belong?

   _____

B. Short-Answer Questions

1. *S. aureus* is gram positive. What differences in strains might account for differences in phage susceptibilities?

   _____

   _____

2. Why is phage typing an important clinical tool?

   _____

   _____

Environmental Influences and Control of Microbial Growth

The exercises in the following section demonstrate the effects that factors such as oxygen, temperature, pH, water activity, UV light, antibiotics, disinfectants, antiseptics, and hand washing have on the growth of bacteria. The microbiologist is concerned with providing the optimum conditions for the growth of microorganisms. In contrast, the medical practitioner is concerned with limiting microbial growth to prevent disease. Understanding how environmental factors can affect bacterial growth allows each person to make good occupational decisions.

In Part 5, the primary concern was in formulating a medium that contained all the essential nutrients to support the growth of a microorganism. However, very little emphasis was placed on other limiting factors such as temperature, oxygen, or pH. Even though all its nutritional needs are provided, an organism may fail to grow if these other factors are not considered. The total environment must be considered to achieve the desired growth of microorganisms.

© O. Dimier/PhotoAlto RF

Effects of Oxygen on Growth

Learning Outcomes

After completing this exercise, you should be able to

1. Define the groups of bacteria based on their sensitivity to and metabolic need for oxygen.

2. Explain the different methods of growing anaerobic bacteria.

3. Differentiate the growth patterns for obligate aerobes, facultative aerobes, microaerophiles, aerotolerant anaerobes, and obligate anaerobes in fluid thioglycollate medium, TGYA shake tubes, and anaerobic agar plates.

Bacteria can be classified as either aerobes or anaerobes based upon their metabolic need for oxygen, which comprises approximately 20% of the atmospheric gases. After they have been classified as aerobes or anaerobes, they can be further separated into categories based on their sensitivity to oxygen.

Obligate (Strict) Aerobes: These bacteria must grow in oxygen because their metabolism requires oxygen. They carry out aerobic respiration in which oxygen is utilized as the terminal electron acceptor in the electron transport chain. Examples are *Pseudomonas, Micrococcus,* and many *Bacillus*.

Microaerophiles: These aerobic bacteria prefer to grow in oxygen concentrations of 2–10% rather than the 20% found in the atmosphere. The lower concentration of oxygen is necessary for their respiratory metabolism. Their sensitivity to the higher concentrations of oxygen is not, however, completely understood. *Helicobacter pylori* is a microaerophile that causes stomach ulcers in humans. The oxygen concentration in the stomach is less than the 20% that occurs in the atmosphere.

Facultative Anaerobes: These bacteria grow very well aerobically but also have the capacity to grow anaerobically if oxygen is not present. Their metabolism is flexible because under aerobic conditions they can carry out respiration to produce energy, but if oxygen is absent they can switch to fermentation that does not require oxygen for energy production. *Escherichia coli* is a facultative anaerobe.

Aerotolerant Anaerobes: These anaerobes can tolerate oxygen and even grow in its presence, but they do not require oxygen for energy production. Because they produce their energy strictly by fermentation and not by respiratory means, they are also called **obligate fermenters**. Examples are the streptococci that produce many food products by fermentation such as cheese, yogurt, and sour cream. Other examples are *Enterococcus faecalis* found in the human intestinal tract and *Streptococcus pyogenes,* a pathogen that causes several diseases in humans, such as strep throat and heart and kidney infections.

Obligate (Strict) Anaerobes: Obligate anaerobes cannot tolerate oxygen and must be cultured under conditions in which oxygen is completely eliminated; otherwise, they are harmed or killed by its presence. These organisms carry out fermentation or **anaerobic respiration**, in which inorganic compounds, such as nitrates and sulfate, replace oxygen as the terminal electron acceptor in the electron transport chain. Obligate anaerobes are only found among the prokaryotes and in some protozoa. These anaerobes occur in environments such as the soil, the rumen (stomach) of cattle, and in anaerobic sewage digesters. *Clostridium, Methanococcus,* and *Bacteroides*, a bacterium found in the human intestine, are examples of obligate anaerobes.

The reason for the sensitivity of strict anaerobes to oxygen is not completely understood. Toxic forms of oxygen such as hydrogen peroxide and superoxide are generated by various chemical mechanisms, and these toxic compounds are abundant in most environments. Toxic forms of oxygen are highly reactive compounds that can damage biological molecules such as nucleic acids, proteins, and small molecules such as coenzymes. Most aerobes possess enzyme systems that will convert the toxic forms of oxygen

2. Growth Curves
   Once you have recorded all the O.D. values in the two tables, plot them on the following graph. Use different-colored lines for each species.

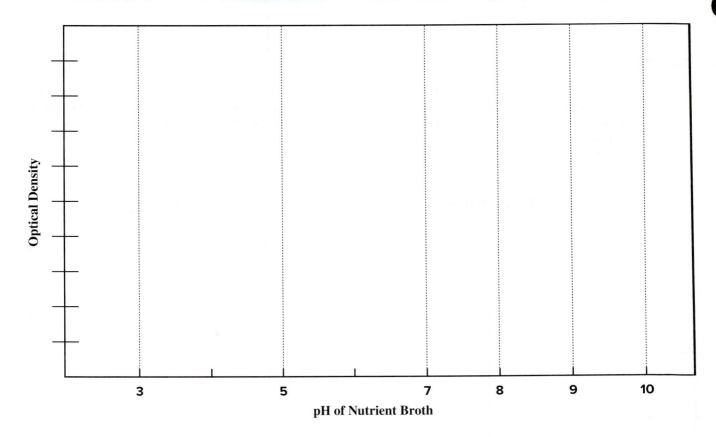

pH of Nutrient Broth

Optical Density

3. Which organism(s) would likely be classified as a neutrophile? _____

4. Which organism(s) would likely be classified as an acidophile? _____

5. Which organism(s) would likely be classified as an alkaliphile? _____

6. Which organism seems to tolerate the broadest pH range? _____

B. Short-Answer Questions

1. How does pH negatively affect the metabolism of microorganisms?

   _____

   _____

2. How would the pH of the culture medium be influenced by sugar fermentation? By urea hydrolysis?

   _____

   _____

3. *Helicobacter pylori* survives in the acidic environment of the human stomach, but this organism is actually a neutrophile. How is this possible?

   _____

   _____

Water Activity and Osmotic Pressure

Learning Outcomes

After completing this exercise, you should be able to

1. Understand the association of water activity to osmotic pressure and solute concentration.

2. Define the terms *plasmolysis, halophile, halotolerant,* and *osmophile.*

3. Demonstrate how different salt concentrations affect the growth of bacteria.

The growth of bacteria can be profoundly affected by the availability of water in an environment. The availability of water is defined by a physical parameter called the water activity, A_w. It is determined by measuring the ratio of the water vapor pressure of a solution to the water vapor pressure of pure water. The values for water activity vary between 0 and 1.0, and the closer the value is to 1.0, the more water is available to a cell for metabolic purposes. Water activity and hence its availability decrease with increases in the concentration of solutes such as salts. This results because water becomes involved in breaking ionic bonds and forming solvation shells around charged species to maintain them in solution.

In the process of **osmosis,** water diffuses from areas of low solute concentration where water is more plentiful to areas of high solute concentration where water is less available. Because there is normally a high concentration of nutrients in the cytoplasm relative to the outside of the cell, water will naturally diffuse into a cell. A medium where solute concentrations on the outside of the cell are lower than the cytoplasm is designated as **hypotonic** (figure 27.1).

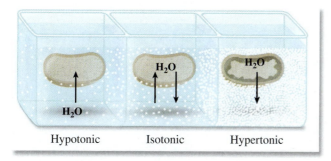

Figure 27.1 Osmotic variabilities.

In general, bacteria are not harmed by hypotonic solutions because the rigid cell wall protects the membrane from being damaged by the osmotic pressure exerted against it. It also prevents the membrane from disrupting when water diffuses across the cell membrane into the cytoplasm.

Environments where the solute concentration is the same inside and outside the cell are termed **isotonic.** Animal cells require isotonic environments or else cells will undergo lysis because only the fragile cell membrane surrounds the cell. Tissue culture media for growing animal cells provides an isotonic environment to prevent cell lysis.

Hypertonic environments exist when the solute concentration is greater on the outside of the cell relative to the cytoplasm, and this causes water to diffuse out of the cytoplasm. When this develops, the cell undergoes **plasmolysis,** resulting in a loss of water, dehydration of the cytoplasm, and shrinkage of the cell membrane away from the cell wall. In these situations, considerable and often irreversible damage can occur to the metabolic machinery of the cell. Low water activity and hypertonic environments have been used by humans for centuries to preserve food. Salted meat and fish, and jams and jellies with high sugar content resist contamination because very little water is available for cells to grow. Most bacteria that might contaminate these foods would undergo immediate plasmolysis.

Microorganisms can be grouped based on their ability to cope with low water activity and high osmotic pressure. Most bacteria grow best when the water activity is around 0.9 to 1.0. In contrast, **obligate halophiles** require high concentrations of sodium chloride to grow. Examples are the halophilic bacteria that require 15–30% sodium chloride to grow and maintain the integrity of their cell walls. These bacteria, which belong to the Archaea, are found in salt lakes and brine solutions, and occasionally growing on salted fish. Some microorganisms are **halotolerant,** which means these organisms do not require salt but are capable of growth in moderate concentrations. For example, *Staphylococcus aureus* can tolerate sodium chloride concentrations that approach 3 *M* or 11%.

Another group of organisms is the **osmophiles,** which are able to grow in environments where sugar concentrations are excessive. An example is *Xeromyces,* a yeast that can contaminate and spoil jams and jellies.

In this exercise, we will test the degree of inhibition of organisms that results with media containing different concentrations of sodium chloride. To accomplish this, you will streak three different organisms on four plates of media. These specific organisms differ in their tolerance of salt concentrations. You will test their tolerance of salt concentrations of 0.5, 5, 10, and 15%. After incubation, comparisons will be made of growth differences to determine each organism's degree of salt tolerance.

Materials

per student:
- 1 nutrient agar plate (0.5% NaCl)
- 1 nutrient agar plate (5% NaCl)
- 1 nutrient agar plate (10% NaCl)
- 1 milk salt agar plate (15% NaCl)

cultures:
- *Escherichia coli* (nutrient broth)
- *Staphylococcus epidermidis* (nutrient broth)
- *Halobacterium salinarium* (slant culture)

1. Mark the bottoms of the four agar plates, as indicated in figure 27.2.
2. Streak each organism in a straight line on the agar, using a wire loop.
3. Incubate all the plates for 48 hours at 37°C. Observe the amount of growth on each plate, and record your results in Laboratory Report 27.

4. Continue the incubation of the milk salt agar (15% NaCl) plate for several more days in the same manner. Observe all four plates again, and record your results in Laboratory Report 27.

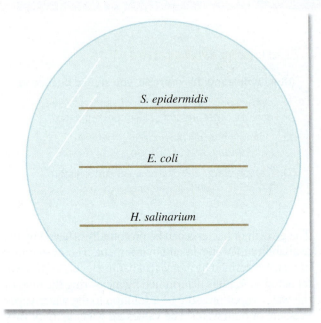

S. epidermidis

E. coli

H. salinarium

Figure 27.2 Streak pattern.

Laboratory Report

27

Student: _____

Date: _____ Section: _____

27 Water Activity and Osmotic Pressure

A. Results

1. Record the amount of growth of each organism at the different salt concentrations, using 0, +1, +2, and +3 to indicate degree of growth.

ORGANISM	SODIUM CHLORIDE CONCENTRATION							
	0.5%		5%		10%		15%	
	48 hr	96 hr	48 hr	96 hr	48 hr	96 hr	48 hr	96 hr
Escherichia coli								
Staphylococcus epidermidis								
Halobacterium salinarium								

2. Evaluate the salt tolerance of the above organisms.

    a. Tolerates very little salt: _____

    b. Tolerates a broad range of salt concentrations: _____

    c. Grows only in the presence of high salt concentration: _____

3. How would you classify *Halobacterium salinarium* as to salt needs? Check one.

    _____ Obligate halophile   _____ Halotolerant   _____ Osmotolerant

B. Short-Answer Questions

1. Why are bacteria generally resistant to hypotonic environments, whereas animal cells are not?

    _____

    _____

2. For each salt concentration used in this exercise, indicate whether it represents an isotonic, hypotonic, or hypertonic environment for the bacterial cells.

    0.5% _____   5% _____

    10% _____   15% _____

3. How do hypertonic environments negatively affect most bacterial cells?

    _____

    _____

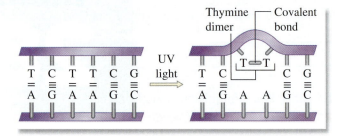

Figure 28.2 **Formation of thymine dimers by UV light.**

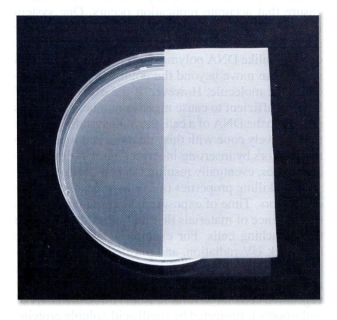

Figure 28.3 **Plates are exposed to UV light with 50% covered by a 3" × 5" card.**

© McGraw-Hill Education/Gabriel Guzman, photographer

Exposure to ultraviolet light may be accomplished with a lamp or with a UV box that has built-in ultraviolet lamps. The UV exposure effectiveness varies with the type of setup used. The exposure times given in table 28.1 work well for a specific type of mercury arc lamp. Note that you may have to use different times based on the UV source you will be using in this exercise. Your instructor will inform you as to whether you should write in new times that will be more suited to the equipment in your lab. Proceed as follows to do this experiment.

Materials

- nutrient agar plates (one or more per student)
- ultraviolet lamp or UV exposure box
- timers
- cards (3" × 5")
- nutrient broth cultures of *Staphylococcus epidermidis* with swabs
- 4-day-old nutrient broth cultures of *Bacillus megaterium* with swabs

1. Refer to table 28.1 to determine which organism you will work with. You may be assigned more than one plate to inoculate. If different times are to be used, your instructor will inform you what times to write in. Since there are only 16 assignment numbers in the table, more student assignment numbers can be written in as designated by your instructor.
2. Label the bottoms of the plates with your assignment number and your initials.
3. Using a cotton-tipped swab, obtain a sample of the bacterial culture and swab the entire surface of the agar in each plate. Before swabbing, express the excess culture from the swab against the inner wall of the tube.
4. Put on the protective goggles provided. Place the plates under the ultraviolet lamp *with the lids removed.* Cover one-half of each plate with a 3" × 5" card, as shown in figure 28.3. Label which side of the plate was covered. **Note:** If your number is 8 or 16, you will not remove the lid from your plate. The purpose of this exposure is to determine to what extent plastic protects cells from the effects of UV light.

Caution

Before exposing the plates to UV light, put on protective goggles. Avoid looking directly into the UV light source. These rays can cause cataracts and eye injury.

5. After exposing the plates for the correct time durations, replace the lids, and incubate them inverted at 37°C for 48 hours.

Laboratory Report

Record your observations in Laboratory Report 28 and answer all the questions.

Table 28.1   Student Inoculation Assignments

	EXPOSURE TIMES (STUDENT ASSIGNMENTS)							
S. epidermidis	1	2	3	4	5	6	7	8
	10 sec	20 sec	40 sec	80 sec	2.5 min	5 min	10 min	10 min*
B. megaterium	9	10	11	12	13	14	15	16
	1 min	2 min	4 min	8 min	15 min	30 min	60 min	60 min*

*These agar plates will be covered with lids during exposure.

28 Ultraviolet Light: Lethal Effects

A. Results

1. Record the results for your assigned organism and exposure time. If substantial growth is present in the exposed area, record your results as $+ + +$. If three or fewer colonies survived, record $+$. Moderate survival should be indicated as $+ +$. No growth should be recorded as $-$. Complete the following table by collecting and recording all class data.

ORGANISMS	EXPOSURE TIMES							
S. epidermidis	10 sec	20 sec	40 sec	80 sec	2.5 min	5 min	10 min	10*min
Survival								
B. megaterium	1 min	2 min	4 min	8 min	15 min	30 min	60 min	60*min
Survival								

* plates covered during exposure

2. How many times more resistant were *B. megaterium* spores than *S. epidermidis* vegetative cells?

   _____

   _____

3. Why was half of each plate covered with an index card?

   _____

   _____

4. What was the purpose of leaving the cover on one set of agar plates? How did the growth on these plates differ from the growth on the uncovered plates that were exposed to UV light for the same amount of time?

   _____

   _____

   _____

B. Short-Answer Questions

1. Describe the damaging effects of UV radiation on living cells.

   _____

   _____

bacteria have **teichoic acids** in their cell walls. Teichoic acids are polymers of ribitol-phosphate or glycerol-phosphate that can bind covalently to peptidoglycan or to the cytoplasmic membrane. They are responsible for the net negative charge on gram-positive cells. They may function in the expansion of peptidoglycan during cell growth. Bacterial cell enzymes called **autolysins** partially open up the peptidoglycan polymer by breaking β (1–4) bonds between the N-acetyl-muramic acid and N-acetyl-glucosamine molecules so that new monomeric subunits of peptidoglycan can be inserted into the existing cell wall and allow for expansion. This process must be limited because if large gaps in peptidoglycan are produced by the autolysins, the cell membrane can rupture through the gaps. Teichoic acids may play a role in limiting the amount of degradation of peptidoglycan by the autolysins. It is interesting to note that teichoic acids are found in *Staphylococcus* but not in *Micrococcus*, even though both are gram-positive bacteria.

In contrast to the gram-positive bacterial cell wall, the cell wall of gram-negative bacteria is composed of a thin layer of peptidoglycan and an outer membrane that encloses the peptidoglycan. Thus, the peptidoglycan of gram-negative bacteria is not accessible from the external environment. The outer membrane of gram-negative bacteria forms an additional permeability barrier in these organisms. Gram-negative bacteria also lack teichoic acids in their cell walls.

Humans are born with intrinsic non-immune factors that protect them from infection by bacteria. One of these factors is **lysozyme,** which is found in most body fluids such as tears and saliva. Lysozyme, like the autolysins, degrades the β (1–4) bond between the amino sugar molecules in peptidoglycan, thus causing breaks in the lattice and weakening the cell wall. As a result, the solute pressure of the cytoplasm can cause the cell membrane to rupture through these breaks, resulting in cell lysis and death.

The presence of lysozyme in tears, for example, can protect our eyes from infection because bacteria that try to establish an infection undergo lysis as a result of the lysozyme in tears that washes the surface of the eye. The egg white of hen's eggs also contains lysozyme where it also may play a protective function for the developing embryo.

Gram-negative bacteria are usually more resistant to lysozyme than are gram-positive bacteria because the gram-negative outer membrane prevents lysozyme from reaching the peptidoglycan layer. However, some gram-positive bacteria, such as *Staphylococcus aureus,* are resistant to lysozyme, presumably because of the presence of teichoic acids in their cell walls, which limits the action of autolysins and

lysozyme. In the following exercise, you will study the effects of lysozyme on gram-negative bacteria and gram-positive bacteria with and without teichoic acids. The source of lysozyme will be human saliva and the egg whites of hen's eggs.

Materials

per class of 30–40
- 2–3 hen's eggs

per pair of students:
- 3 plates of nutrient agar
- small vessels for collecting saliva
- sterile cotton swabs
- small sterile test tubes
- sterile Pasteur pipettes
- pipette bulbs

broth cultures of:
- *Escherichia coli*
- *Micrococcus luteus*
- *Staphylococcus aureus*

1. Using a Sharpie marking pen on the bottom of the plates, divide each plate into three sectors and label them 1, 2, and 3.
2. Using a sterile cotton swab, uniformly inoculate one plate with *E. coli*. Do this by first swabbing the plate in one direction with a cotton swab dipped in the culture. Then turn the plate 90 degrees and swab at a right angle to the first direction.
3. Repeat this procedure for the second and third plates, using *M. luteus* and *S. aureus*.
4. The instructor will break an egg and separate the white portion from the yolk. Using a sterile Pasteur pipette, collect a small amount of egg white and transfer it to a small test tube.
5. Collect saliva by expelling some into the small vessel provided.
6. Using a sterile Pasteur pipette, transfer one drop of egg white to sector 1 on the plate inoculated with *E. coli*. To sector 2, deliver 1 drop of saliva on this same plate. Sector 3 will serve as a growth control for the organism (figure 29.2).
7. Repeat this same procedure for the remaining plates inoculated with *M. luteus* and *S. aureus*.
8. Allow the liquids to absorb into the agar medium and then incubate the plates at 37°C for 24 hours.
9. In the next laboratory period, observe the plates. If the organism is affected by lysozyme, the cells exposed to the enzyme will undergo lysis, forming a clear area in the sector. Compare the effects of the enzyme on the three bacteria and record your results in Laboratory Report 29.

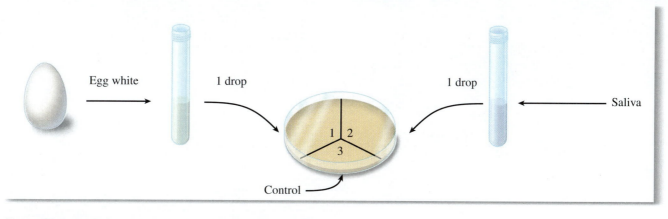

Figure 29.2 **Procedure for setting up the lysozyme plates.**

29 The Effects of Lysozyme on Bacterial Cells

A. Results

1. Record whether cell lysis occurred in each sector of the plates using +++ for a large clear zone, + for a smaller clear zone, and – for no lysis.

	SECTOR		
	Egg White	**Saliva**	**Control**
	1	2	3
Staphylococcus aureus			
Escherichia coli			
Micrococcus luteus			

2. Why did lysozyme affect the test organisms differently?

   _____

   _____

3. What was the purpose of not adding a source of lysozyme to one sector of each plate?

   _____

   _____

B. Short-Answer Questions

1. What is the function of peptidoglycan in bacterial cells?

   _____

   _____

2. How does lysozyme specifically affect peptidoglycan?

   _____

   _____

3. Where in nature can lysozyme be found? Why is it produced in these environments?

   _____

   _____

4. Based on the results, what might one conclude about the types of bacteria that are involved in eye infections?

   _____

   _____

Evaluation of Alcohol:
Its Effectiveness as an Antiseptic

Learning Outcomes

After completing this exercise, you should be able to

1. Understand how alcohols specifically affect bacterial cells.

2. Demonstrate the effectiveness of alcohol in killing bacteria on the human skin.

As an antiseptic, alcohol is widely used to swab the skin before inoculations or the drawing of blood from veins. This is to prevent the introduction of skin bacteria, especially pathogens, into tissue where they could cause disease. The number of bacteria on the skin will vary depending on location and available moisture. For example, approximately 1000 organisms per square centimeter may be present on the back, whereas greater than 10 million per square centimeter can be found in the moist armpits and groin.

Ethyl alcohol and isopropyl alcohol are effective antiseptics for preventing inoculation of skin bacteria but only when used in aqueous solutions of 60% to 80%. These concentrations specifically kill bacterial cells by denaturing proteins and damaging cell membranes. Proteins are more soluble in aqueous solutions of alcohols, and hence they denature more readily when water is present. Alcohols are not, however, effective against endospores and some naked viruses (i.e., viruses that lack a lipid envelope). An advantage of alcohols is that they are relatively nontoxic to the human skin, but a disadvantage is that they are volatile and can lose their effectiveness due to evaporation.

Some important questions concerning alcohols include: How effective is alcohol in routine use? When the skin is swabbed prior to penetration, are all or mostly all of the surface skin bacteria killed? To determine alcohol effectiveness as it might be used in routine skin disinfection, we are going to perform a very simple experiment that utilizes four thumbprints and a plate of enriched agar. Class results will be pooled to arrive at a statistical analysis.

Figure 30.1 illustrates the various steps in this test. Note that the agar plate is divided into four quadrants. On the left side of the plate, an unwashed left thumb is first pressed down on the agar in the lower-left quadrant of the plate. Next, the left thumb is pressed down on the upper-left quadrant. With the left thumb, we are trying to establish the percentage of bacteria that are removed by simple contact with the agar.

On the right side of the plate, an unwashed right thumb is pressed down on the lower-right quadrant of the plate. The next step is to either dip the right thumb into alcohol or to scrub it with an alcohol swab and air-dry it. Half of the class will use the dipping method and the other half will use alcohol swabs. Your instructor will indicate what your assignment

❷
Without touching any other surface the left thumb is pressed against the agar in quadrant B.

❻
The pad of the treated right thumb is pressed against the agar in the D quadrant.

❺
The alcohol-treated right thumb is allowed to completely air-dry.

❹
The pad of the right thumb is immersed in 70% alcohol or scrubbed with an alcohol swab for 10 seconds.

❶
The pad of the unwashed left thumb is momentarily pressed against the agar in quadrant A.

❸
The pad of the unwashed right thumb is momentarily pressed against the agar in quadrant C.

Figure 30.1 **Procedure for testing the effectiveness of alcohol on the skin.**

will be. The last step is to press the air-dried right thumb on the upper-right quadrant of the plate.

After the plate is inoculated, it is incubated at 37°C for 24 to 48 hours. Colony counts will establish the effectiveness of the alcohol.

Materials

- 1 veal infusion agar plate
- small beaker
- 70% ethanol
- alcohol swab

1. Perform this experiment with unwashed hands.
2. With a Sharpie marking pen, mark the bottom of the agar plate with two perpendicular lines that divide it into four quadrants. Label the left quadrants **A** and **B** and the right quadrants **C** and **D**, as shown in figure 30.1. *(Keep in mind that when you turn the plates over to label them, the A and B quadrants will be on the right and C and D will be on the left.)*
3. Press the pad of your left thumb against the agar surface in the A quadrant.
4. Without touching any other surface, press the left thumb into the B quadrant.
5. Press the pad of your right thumb against the agar surface of the C quadrant.
6. According to your instructor's assignment, disinfect your right thumb by one of the two following methods:
   - dip the thumb into a beaker of 70% ethanol for 5 seconds, or
   - scrub the entire pad surface of the right thumb with an alcohol swab.
7. Allow the alcohol to completely evaporate from the skin.
8. Press the right thumb against the agar in the D quadrant.
9. Incubate the plate at 37°C for 24 to 48 hours.
10. Follow the instructions in Laboratory Report 30 for evaluating the plate and answer all of the questions.

30   Evaluation of Alcohol: Its Effectiveness as an Antiseptic

A. Results

1. Count the number of colonies that appear on each of the thumbprints and record them in the following table. If the number of colonies has increased in the second press, record a 0 in percent reduction. Calculate the percentages of reduction and record these data in the appropriate column. Use this formula:

$$\text{Percent reduction} = \frac{(\text{Colony count 1st press}) - (\text{Colony count 2nd press})}{(\text{Colony count 1st press})} \times 100$$

Collect data from your classmates, and then determine an average reduction percentage for each type of treatment at the bottom of the table.

LEFT THUMB (Control)			RIGHT THUMB (Dipped)			RIGHT THUMB (Swabbed)		
Colony Count 1st Press (A)	Colony Count 2nd Press (B)	Percent Reduction	Colony Count 1st Press (C)	Colony Count 2nd Press (D)	Percent Reduction	Colony Count 1st Press (C)	Colony Count 2nd Press (D)	Percent Reduction
Av. % Reduction, Control			Av. % Reduction, Dipped			Av. % Reduction, Swabbed		

2. In general, what effect does alcohol have on the level of skin contaminants? _____

3. Is there any difference between the effects of dipping versus swabbing? _____

    Which method appears to be more effective? _____

4. There is definitely survival of some microorganisms even after alcohol treatment. Without staining or microscopic scrutiny, predict what types of microbes are growing on the medium where you made the right thumb impression after treatment. _____

    _____

    _____

    _____

    _____

B. Short-Answer Questions (see Exercise 32 for a definition of *antiseptic*)

1. For what purposes is alcohol a useful antiseptic?

    _____

    _____

2. What advantages does alcohol have over hand soap for antisepsis of the skin?

    _____

    _____

3. Why does treatment of human skin with alcohol not create a completely sterile environment?

    _____

    _____

4. Explain how the left thumb acted as a control for the experiment.

    _____

    _____

    _____

Antimicrobic Sensitivity Testing:
The Kirby-Bauer Method

Antimicrobials and antibiotics have been our first line of defense in the battle to conquer infectious diseases. Diseases caused by *Staphylococcus, Streptococcus, Pseudomonas,* and *Mycobacterium* that once were fatal to humans became manageable and curable with the discovery of antibiotics. In 1929, Alexander Fleming noticed that *Penicillium,* a fungus, inhibited growth of the bacterium *Staphylococcus aureus* when both organisms were present on the same agar plate. This discovery eventually led to the purification of an antibacterial substance produced by the fungus. Penicillin, purified by the efforts of Howard Florey, Norman Heatley, and Ernst Chain, became the first antibiotic to be used in the clinical treatment of bacterial infections. Other antibiotics were subsequently found and added to the list that could be used to treat bacterial diseases. Prior to the discovery of antibiotics, very few options were available to treat infectious agents except for some antimicrobial chemical compounds such as salvarsan, an arsenic compound, and the sulfa drugs, which were used with some success. Today we have a vast number of compounds that are effective against bacterial, fungal, and some viral agents. By definition, **antimicrobials,** also known as antimicrobics, are any compounds that kill or inhibit microorganisms. **Antibiotics** are antimicrobials, usually of low molecular weight, naturally produced by microorganisms to inhibit or kill other microorganisms. Two common examples are penicillin and streptomycin. Many times antibiotics are chemically altered to make them more effective in their mode of action. These are referred to as **semi-synthetics.** Some antimicrobials are chemically synthesized in the laboratory and are not produced by microbial biosynthesis at all. Examples of these **synthetics** include the sulfa drugs

that were used to treat some bacterial diseases before the discovery of penicillin.

Our perceived defeat of the infectious microbe, however, has been short-lived. The development of antibiotic-resistant strains of bacteria such as *Staphylococcus aureus, Pseudomonas aeruginosa, Mycobacterium tuberculosis, Neisseria gonorrhoeae,* and *Enterococcus* that do not respond to the antibiotics normally used to treat these bacteria is presenting an increasingly difficult challenge in medicine. *Staphylococcus aureus* has become especially problematic because both hospital and community strains of MRSA (MRSA: methicillin-resistant *S. aureus*) have been isolated. These strains do not respond to penicillin, and hence other and more expensive drugs such as vancomycin must be used to manage infections caused by these organisms. As might be expected, strains of *Staphylococcus* and *Enterococcus* have also developed resistance to vancomycin. In hospitals MRSA is responsible for many **healthcare acquired infections (HAIs),** formerly known as nosocomial infections because they were acquired in hospitals. In 2011, the Centers for Disease Control (CDC) estimated that there were 722,000 HAIs at acute care hospitals in the United States. Penicillin has long been the drug of choice to treat gonorrhea. However, penicillin-resistant *N. gonorrhoeae* strains have been isolated that require more expensive drugs to treat this sexually transmitted disease. Strains of *Mycobacterium tuberculosis* have been isolated from street people and drug addicts that do not respond to any antimicrobials available. Needless to say, antibiotic resistance is a major public health problem affecting the treatment and spread of disease.

When a patient seeks medical assistance for a bacterial infection, the physician will normally take a conservative approach and prescribe an antimicrobial without isolating the pathogen. The physician bases the choice of antimicrobial on symptoms presented by the patient, probable causative agents associated with the symptoms, and drugs known to be effective against these pathogens. Usually this approach is sufficient to treat the infection, but occasionally a pathogen will not respond to the prescribed antimicrobial. It is then crucial that the causative agent be isolated in pure culture, specifically identified, and tested for its sensitivity and resistance to various antimicrobial agents. Antimicrobial agents can vary in their effectiveness against various pathogenic bacteria. Some antimicrobial agents are narrow in their spectrum and may be more effective against gram-positive bacteria, while others are more effective against gram-negative bacteria. Broad-spectrum antimicrobials are effective against both kinds of organisms. Based on test results, an antimicrobial that the pathogen is sensitive to would then be prescribed to treat the infection.

Whether an antimicrobial has a broad or narrow spectrum depends appreciably on its mode of action and its ability to be transported into the cell. Antimicrobials have different modes of action, and they affect different aspects of bacterial cell metabolism. They can target cell wall synthesis (penicillin), DNA and RNA synthesis (cipro, rifampin), protein synthesis (tetracyclines, streptomycin), and vitamin synthesis (sulfa drugs). For example, antimicrobials that target 70S ribosomes tend to be broad spectrum because all bacteria have 70S ribosomes, which are necessary for protein synthesis. Sulfa drugs are more narrow in spectrum because not all bacterial cells can synthesize the vitamin folic acid, whose synthesis is inhibited by this drug. Another factor that affects the ability of an antimicrobial to function is permeability. The outer membrane of gram-negative bacteria acts as a permeability barrier and can restrict the entry of antimicrobials into the cell. Table 31.1 summarizes the modes of action of selected antimicrobials.

The **Kirby-Bauer method** (figure 31.1) is used to determine the sensitivity or resistance of a bacterium to an antimicrobial. This is a standardized test procedure that is reliable, relatively simple, and yields results in as short a time as possible. It is performed by uniformly streaking a standardized inoculum of the test organism on Mueller-Hinton medium, and then paper disks containing specific concentrations of an antimicrobial are deposited on the agar surface. The antimicrobial diffuses out from the disk into the agar, forming a concentration gradient. If the agent inhibits or kills the test organism, there will be a zone around the disk where no growth occurs called the **zone of inhibition** (figure 31.2). This zone can vary, however, with the diffusibility of the agent, the size of the inoculum, the type of medium, and other factors. All of these factors were taken into consideration in developing this test. The Kirby-Bauer method is sanctioned by the U.S. FDA and the Subcommittee on Antimicrobial Susceptibility Testing of the National Committee for Clinical Laboratory Standards. Although time is insufficient here to consider all facets of this test, its basic procedure will be followed to test antimicrobials.

The recommended medium in this test is Mueller-Hinton II agar. Its pH should be between 7.2 and 7.4, and it should be poured to a uniform thickness of 4 mm in the plate. This requires 60 ml in a 150 mm plate and 25 ml in a 100 mm plate. For certain fastidious microorganisms, 5% defibrinated sheep's blood is added to the medium.

Inoculation of the surface of the medium is made with a cotton swab from a broth culture. In clinical applications, the broth turbidity has to match a defined standard. Care must also be taken to express excess broth from the swab prior to inoculation.

High-potency disks are used that may be placed on the agar with a mechanical dispenser or sterile forceps. To secure the disks to the medium, it is necessary to press them down gently onto the agar.

Table 31.1 Modes of Action of Antimicrobial Drugs

ANTIMICROBIAL DRUG	CELLS AFFECTED	CELL TARGET/SPECIFIC SITE
Penicillin	G+; G−	Cell wall/β-lactamase, peptidoglycan synthesis—amino acid side chain
Vancomycin	G+	Cell wall/Peptidoglycan synthesis
Bacitracin	G+	Cell wall/Transport of peptidoglycan monomer
Isoniazid	*Mycobacterium tuberculosis*	Cell wall/Mycolic acid synthesis in *Mycobacterium*
Fluroquinolones	G+; G−	DNA/Topoisomerase unwinding of DNA in DNA synthesis
Rifamycins	G+; some G−	RNA/RNA polymerase in RNA synthesis
Tetracyclines	G+; G−	Protein synthesis/30S subunit of 70S ribosome
Streptomycin	G+; G−	Protein synthesis/30S subunit of 70S ribosome
Chloramphenicol	G+; G−	Protein synthesis/50S subunit of 70S ribosome
Sulfa drugs	G+; G−	Structural analogue of para-amino benzoic acid (PABA)—inhibit enzyme linking pteridine to PABA in folic acid synthesis

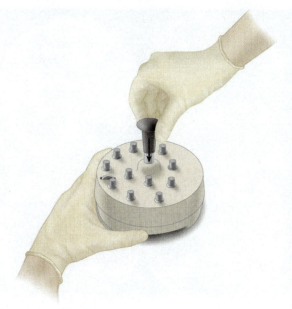

(1) The entire surface of a plate of nutrient medium is swabbed with organism to be tested.

(2) Handle of dispenser is pushed down to place multiple disks on the medium. In addition to dispensing disks, this dispenser also tamps disks onto medium.

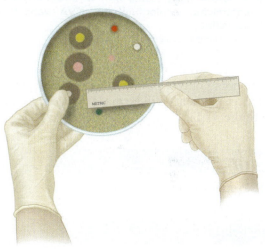

(3) Individual cartridges (Difco) or a sterile forceps can be used to dispense single disks. Only 4 or 5 disks should be placed on small (100 mm) plates.

(4) After 18 hours of incubation, the zones of inhibition (diameters) are measured in millimeters. Significance of zones is determined from the Kirby-Bauer chart (table 31.2).

Figure 31.1 **Antimicrobic sensitivity testing.**

After 16 to 18 hours of incubation, the plates are examined and the diameters of the zones of inhibition are measured to the nearest millimeter. Diameters for tested bacteria are compared to those in a table that are based on values obtained for ATCC (American Type Culture Collection) reference cultures that are sensitive to the specific antibiotic. Cultures are designated as resistant, sensitive, or intermediate in their response to the antibiotic. These designations are derived by comparison to the response of the reference culture. (To determine the significance of the zone diameters, consult table 31.2.)

In this exercise, we will work with four microorganisms: *Staphylococcus aureus, Escherichia coli, Proteus vulgaris,* and *Pseudomonas aeruginosa.* Each student will inoculate one plate with one of the four organisms and place the disks on the medium by whichever method is available. Since each student

Figure 31.2 Kirby-Bauer plates for *Pseudomonas aeruginosa*, top left; *Staphylococcus aureus*, top right; *Escherichia coli*, bottom left; and *Proteus vulgaris*, bottom right, tested with: S - streptomycin; GM - gentamicin; C - chloramphenicol; VA - vancomycin; TE - tetracycline.
© McGraw-Hill Education/Lisa Burgess, photographer

will be doing only a portion of the total experiment, student assignments will be made. Proceed as follows (see figure 31.1):

⏰ First Period

(Plate Preparation)

Materials

- 1 Mueller-Hinton II agar plate
- nutrient broth cultures (with swabs) of *S. aureus, E. coli, P. vulgaris,* and *P. aeruginosa*
- disk dispenser (BBL or Difco)
- cartridges of disks (BBL or Difco)
- forceps and Bunsen burner
- zone interpretation charts (Difco or BBL)
- metric ruler

1. Select the organism you are going to work with from the following table.

ORGANISM	STUDENT NUMBER
S. aureus	1, 5, 9, 13, 17, 21, 25
E. coli	2, 6, 10, 14, 18, 22, 26
P. vulgaris	3, 7, 11, 15, 19, 23, 27
P. aeruginosa	4, 8, 12, 16, 20, 24, 28

2. Label your plate with the name of your organism.
3. Inoculate the surface of the medium with the swab after expressing excess fluid from the swab by pressing and rotating the swab against the inside walls of the tube above the fluid level. Cover the surface of the agar evenly by swabbing in three directions. A final sweep should be made of the agar rim with the swab.
4. Allow **3 to 5 minutes** for the agar surface to dry before applying disks.
5. Dispense disks as follows:
   a. If an automatic dispenser is used, remove the lid from the plate, place the dispenser over the plate, and push down firmly on the plunger. With the sterile tip of forceps, tap each disk lightly to secure it to medium.
   b. If forceps are used, sterilize them first by flaming before picking up the disks. Keep each disk at least 15 mm from the edge of the plate. Place no more than 13 on a 150 mm plate, and no more than 5 on a 100 mm plate. Apply light pressure to each disk on the agar with the tip of a sterile forceps or inoculating loop to secure it to the medium.
6. Invert and incubate the plate for 16 to 18 hours at 37°C.

⏱ Second Period

(Interpretation)

After incubation, measure the zone diameters with a metric ruler to the nearest whole millimeter (figure 31.3). The disk is included in the measurement. If growth occurs against the border of the disk, record the zone measurement as zero. The zone of inhibition is determined without magnification. Ignore faint growth or tiny colonies that can be detected by very close scrutiny. Large colonies growing within the clear zone might represent resistant variants or a mixed inoculum and may require retesting in clinical situations. Ignore the "swarming" characteristics of *Proteus,* measuring only to the margin of heavy growth.

Record the zone measurements in the table in Laboratory Report 31. Use table 31.2 for identifying the various disks.

To determine which antimicrobials your organism is sensitive to (S), or resistant to (R), or intermediate (I), consult table 31.2. It is important to note that the significance of a zone of inhibition varies with the type of organism. A large zone does not necessarily mean the organism is sensitive; the standardized Kirby-Bauer chart must be consulted to make the susceptibility determinations. If you cannot find your antimicrobial on the chart, consult a chart supplied by your instructor. Table 31.2 is incomplete.

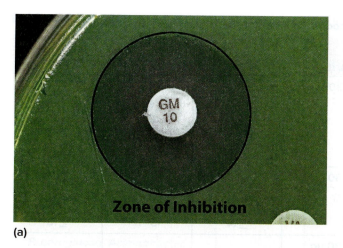

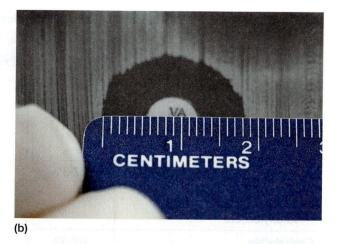

(a)

(b)

Figure 31.3 (a) Zone of inhibition for gentamicin on *Pseudomonas aeruginosa*. (b) Measuring the zone of inhibition.
© McGraw-Hill Education/Lisa Burgess, photographer

Table 31.2   Zones of Inhibition in the Kirby-Bauer Method of Antimicrobic Sensitivity Testing

ANTIBIOTIC	CODE	POTENCY	Zone of Inhibition (mm)		
			RESISTANT	INTERMEDIATE	SENSITIVE
Amikacin	AN-30	30 µg			
Enterobacteriaceae					
P. aeruginosa, Acinetobactor					
staphylococci			≤14	15–16	≥17
Amoxicillin/Clavulinic acid	AmC-30	20/10 µg			
Enterobacteriaceae			≤13	14–17	≥18
Staphylococcus spp.			≤19	—	≥20
Haemophilus spp.			≤19	—	≥20
Ampicillin	AM-10	10 µg			
Enterobacteriaceae			≤13	14–16	≥17
Staphylococcus spp.			≤28	—	≥29
Enterococcus spp.			≤16	—	≥17
Listeria monocytogenes			≤19	—	≥20
Haemophilus spp.			≤18	19–21	≥22
β-hemolytic streptococci			—	—	≥24
Aziocillin	AZ-75	75 µg			
P. aeruginosa			≤17	—	≥18
Bacitracin	B-10	10 units	≤8	9–12	≥13
Carbenicillin	CB-100	100 µg			
Enterobacteriaceae and					
Acinetobacter			≤19	20–22	≥22
P. aeruginosa			≤13	14–16	≥17

(continued)

32   Evaluation of Antiseptics and Disinfectants: The Filter Paper Disk Method

A. Results

1. With a metric ruler, measure the zones of inhibition on all four plates, and record this information in the table below.

DISINFECTANT/ANTISEPTIC	Zone of Inhibition (mm)	
	Staphylococcus aureus	*Pseudomonas aeruginosa*
3% Hydrogen peroxide		
5% Lysol		
5% Iodine		
1:10 Bleach		
Mouthwash		
Betadine		
Isopropyl alcohol		

2. Which chemical was the most effective for inhibiting the growth of *S. aureus?* Of *P. aeruginosa?*

_____

_____

3. Which chemical was the least effective for inhibiting the growth of *S. aureus?* Of *P. aeruginosa?*

_____

_____

4. What do your results indicate about the relative chemical resistances of these two species?

_____

_____

B. Short-Answer Questions

1. Differentiate between antiseptic and disinfectant. Include examples of each in your answer. Indicate whether any chemicals can be used as both.

_____

_____

231

in healthcare-acquired infections because of its antibiotic resistance. Recently it has been shown that two different genetic strains of MRSA exist, a hospital-acquired strain and a community-acquired strain. The latter accounts for about 79% of MRSA infections and has been especially prevalent in infections that occur in young athletes using school gyms and training facilities.

Fungi: The fungi are represented by *Malassezia furfur,* a small, nonpathogenic yeast that utilizes fatty substances for growth. It often grows on the face, especially around the nose, where it can cause a harmless flaking of the skin. The spores of transient saprophytic fungi can collect on the skin and be grown if the skin is sampled for microorganisms. Some fungi and yeasts can cause opportunistic infections. The dermatophytes are fungi that cause infections of the hair, skin, and nails such as athlete's foot.

In addition to the normal microbiota, there are transient bacteria that can occur on the skin. Organisms such as endospore formers may be cultured by swabbing the skin because their endospores are present. However, they are not part of the persistent population that inhabits the skin. Other bacteria may be present because the skin has become temporarily contaminated with them. Most are easily removed by washing because they are contaminants on the skin, and antiseptic soaps are effective in killing them. The normal microbiota are much more difficult to remove by washing because these organisms reside in hair follicles and are entrenched in the skin, making them very difficult to remove or kill.

In this exercise, the entire class will work together to evaluate the effectiveness of length of time in the removal of organisms from the hands using a surgical scrub technique. One member of the class will be selected to perform the scrub. Another student will assist by supplying the soap, brushes, and basins, as needed. During the scrub, at 2-minute intervals, the hands will be scrubbed into a basin of sterile water. Bacterial counts will be made of these basins to determine the effectiveness of the previous 2-minute scrub in reducing the bacterial microbiota of the hands. Members of the class not involved in the scrub procedure will make the inoculations from the basins for the plate counts.

Scrub Procedure

The two members of the class who are chosen to perform the surgical scrub will set up their materials near a sink for convenience. As one student performs the scrub, the other will assist in reading the instructions and providing materials as needed. The basic steps, which are illustrated in figure 33.1, are also described in detail below. Before beginning the scrub, both students should read all the steps carefully.

Materials

- 5 sterile surgical scrub brushes, individually wrapped
- 5 basins (or 2000 ml beakers), containing 1000 ml each of sterile water. These basins should be covered to prevent contamination.
- 1 dispenser of soap
- 1 tube of hand lotion

Step 1 To get some idea of the number of transient organisms on the hands, the scrubber will scrub all surfaces of each hand with a sterile surgical scrub brush for 30 seconds in basin A. No soap will be used for this step. The successful performance of this step will depend on:

- spending the same amount of time on each hand (30 seconds),
- maintaining the same amount of activity on each hand, and
- scrubbing under the fingernails, as well as working over their surfaces.

After completion of this 60-second scrub, notify Group A that their basin is ready for the inoculations.

Step 2 Using the *same* brush as above, begin scrubbing with soap for 2 minutes, using cool tap water to moisten and rinse the hands. One minute is devoted to each hand.

The assistant will make one application of soap to each hand as it is being scrubbed.

Rinse both hands for 5 seconds under tap water at the completion of the scrub.

Discard the brush.

Note: This same procedure will be followed exactly in steps 4, 6, and 8 of figure 33.1.

Step 3 With a *fresh* sterile brush, scrub the hands in basin B in a manner that is identical to step 1. Don't use soap. Notify Group B when this basin is ready.

Note: Exactly the same procedure is used in steps 5, 7, and 9 of figure 33.1, using basins C, D, and E.

Remember: It is important to use a fresh sterile brush for the preparation of each of these basins.

After Scrubbing After all scrubbing has been completed, the scrubber should dry his or her hands and apply hand lotion.

Making the Pour Plates

While the scrub is being performed, the rest of the class will be divided into five groups (A, B, C, D, and E)

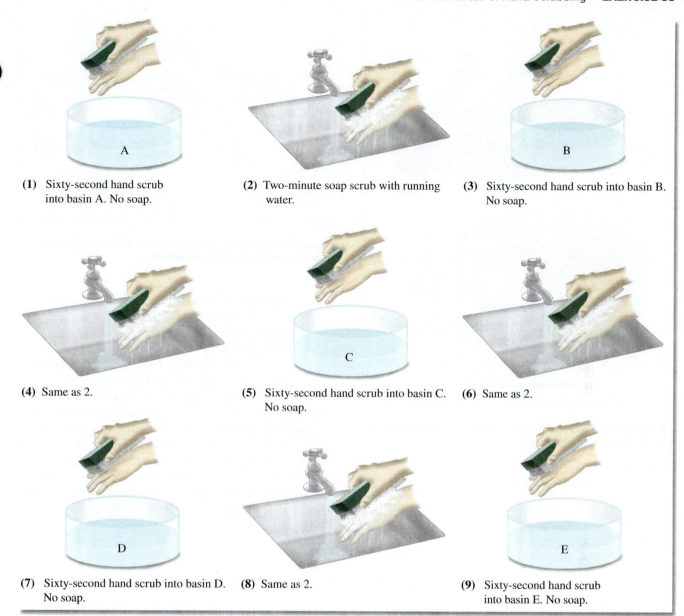

(1) Sixty-second hand scrub into basin A. No soap.

(2) Two-minute soap scrub with running water.

(3) Sixty-second hand scrub into basin B. No soap.

(4) Same as 2.

(5) Sixty-second hand scrub into basin C. No soap.

(6) Same as 2.

(7) Sixty-second hand scrub into basin D. No soap.

(8) Same as 2.

(9) Sixty-second hand scrub into basin E. No soap.

Figure 33.1 **Hand scrubbing routine.**

by the instructor. Each group will make six plate inoculations from one of the five basins (A, B, C, D, or E). It is the function of these groups to determine the bacterial count per milliliter in each basin. In this way, we hope to determine, in a relative way, the effectiveness of scrubbing in bringing down the total bacterial count of the skin.

Materials

- 30 veal infusion agar pours—6 per group
- 1 ml pipettes
- 30 sterile empty plates—6 per group
- 70% alcohol
- L-shaped glass stirring rod (optional)

1. Liquefy six pours of veal infusion agar and cool to 50°C. While the medium is being liquefied, label two sets of plates each: 0.1 ml, 0.2 ml, and 0.4 ml. Also, indicate your group designation on the plate.
2. As soon as the scrubber has prepared your basin, take it to your table and make your inoculations as follows:
   a. Stir the water in the basin with a pipette or an L-shaped stirring rod for 15 seconds. If the stirring rod is used (figure 33.2), sterilize it before using by immersing it in 70% alcohol and flaming. *For consistency of results, all groups should use the same method of stirring.*
   b. Deliver the proper amounts of water from the basin to the six empty plates with a sterile

Figure 33.2 An alternative method of stirring utilizes an L-shaped glass stirring rod.

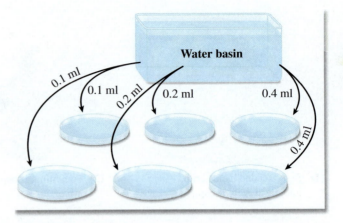

Figure 33.3 **Scrub water for count is distributed to six petri plates in amounts as shown.**

serological pipette. Refer to figure 33.3. If a pipette was used for stirring, it may be used for the deliveries.

 c. Pour a tube of veal infusion agar, cooled to 50°C, into each plate, rotate to get good distribution of organisms, and allow to cool.

 d. Incubate the plates at 37°C for 24 hours.

3. After the plates have been incubated, select the pair that has the best colony distribution with no fewer than 30 or more than 300 colonies. Count the colonies on the two plates and record your counts on the classroom data chart.

4. After all class data are recorded, complete the table and graph in Laboratory Report 33.

33

33 Effectiveness of Hand Scrubbing

A. Results

1. The instructor will provide a class data table similar to the one below. Examine the six plates that your group inoculated from the basin of water. Select the two plates of a specific dilution that have approximately 30 to 300 colonies and count all of the colonies of each plate with a colony counter. Record the counts for each plate and their averages in the class data table. Once all the groups have recorded their counts, record the dilution factors for each group in the proper column. To calculate the organisms per milliliter, multiply the average count by the dilution factor. The dilution factor depends on the volume used in the inoculation. These dilution factors are given below the table for each volume used in this exercise. Choose the one that matches the volume for the plates you counted, and use that to determine the number of organisms per milliliter in the last column.

GROUP	0.1 ml COUNT		0.2 ml COUNT		0.4 ml COUNT		DILUTION FACTOR*	ORGANISMS PER MILLILITER
	Per Plate	Average	Per Plate	Average	Per Plate	Average		
A								
B								
C								
D								
E								

*Dilution factors: 0.1 ml = 10; 0.2 ml = 5; 0.4 ml = 2.5

Identification of Unknown Bacteria

One of the most interesting experiences in introductory microbiology is to attempt to identify an unknown microorganism that has been assigned to you. The next six exercises pertain to this phase of microbiological work. You will be given one or more cultures of bacteria to identify. The only information that might be given to you about your unknowns will pertain to their sources and habitats. All of the other information needed for identification will have to be acquired by you through laboratory study.

Although you will be focused on trying to identify an unknown organism, there is a more fundamental underlying objective of this series of exercises that goes far beyond simply identifying an unknown. That objective is to gain an understanding of the cultural and physiological characteristics of bacteria. Physiological characteristics will be determined with a series of biochemical tests that you will perform on the organisms. Although correctly identifying the unknowns that are given to you is very important, it is just as important that you understand the chemistry of the tests that you perform on the organisms.

The first step in the identification procedure is to accumulate information that pertains to the organ-

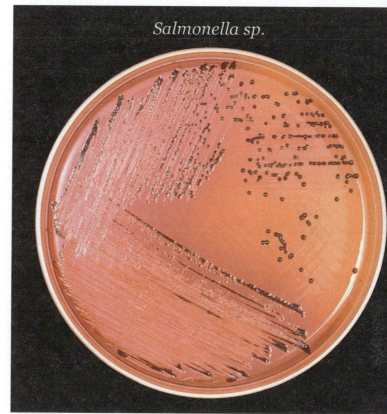

Salmonella sp.

Source: Centers for Disease Control

isms' morphological, cultural, and physiological (biochemical) characteristics. This involves making different kinds of slides for cellular studies and the inoculation of various types of media to note the growth characteristics and types of enzymes produced. As this information is accumulated, it is recorded in an orderly manner on descriptive charts, which are located at the end of Exercise 34.

After sufficient information has been recorded, the next step is to consult a taxonomic key, which enables one to identify the organism. For this final step, *Bergey's Manual of Systematic Bacteriology* or *Bergey's Manual of Determinative Bacteriology* will be used.

Success in this endeavor will require meticulous techniques, intelligent interpretation, and careful record keeping. Your mastery of aseptic methods in the handling of cultures and the performance of inoculations will show up clearly in your results. Contamination of your cultures with unwanted organisms will yield false results, making identification hazardous speculation. If you have reason to doubt the validity of the results of a specific test, repeat it; ***don't rely on chance!*** As soon as you have made an observation or completed a test, record the information on the descriptive chart. Do not trust your memory—record data immediately.

Morphological Study of an Unknown Bacterium

Learning Outcomes

After completing this exercise, you should be able to

1. Establish working and reserve stock cultures of your unknown organism(s).
2. Determine the optimum growth temperature for your unknown.
3. Determine the Gram reaction, cell shape and arrangement, motility, and other important morphological characteristics of your unknown.

The first step in the study of your unknown bacterium is to set up stock cultures that will be used in the subsequent exercises. Your reserve stock culture will not be used for making slides or inoculating tests. It will be stored in the refrigerator in case your working stock becomes contaminated and you need to make a fresh working stock. The working stock will be used to inoculate the various tests that you will perform to identify your unknown bacterium. It is crucial that you practice good aseptic technique when inoculating from your working stock in order to avoid contaminating the culture. If it becomes contaminated or loses viability, you can prepare a fresh culture from the reserve stock culture that you have maintained in the refrigerator.

Identifying your unknown will be a kind of "microbiological adventure" that will test the skills and knowledge that you have acquired thus far. You will gather a great deal of information regarding your unknown by performing staining reactions and numerous metabolic tests. The Gram stain will play a very critical role in the process because it will eliminate thousands of possible organisms. The results of these tests will be compared to flowcharts provided in this manual and to information in *Bergey's Manual*. From your "detective" work, you will be able to ascertain the identity of the unknown that you were given. To set up the stock cultures, proceed in the following way (see figure 34.1).

Stock Cultures

You will receive a broth culture or an agar slant of your unknown bacterium. From this culture, you will prepare your working stock and your reserve stock cultures.

From the working stock, you will be able to determine such things as cell morphology, the Gram reaction of the unknown, and, in some cases, whether the culture forms any pigment. You can also determine other morphological characteristics such as the presence of a glycocalyx, endospores, or cytoplasmic granules.

Materials

- nutrient agar or tryptone agar slants

🕐 First Period

1. Label the agar slants with the code number of the unknown, your name, lab section, and date.
2. Inoculate both slants with your unknown organism. Begin your streak at the bottom of the slant and move the inoculating loop toward the top of the slant in a straight motion. Remember to practice good aseptic technique.
3. Place the respective tubes in the appropriate test tube racks labeled with the two incubation temperatures, 20°C and 37°C (figure 34.1). Also label the test tubes with the incubation temperature. Incubate the slants for 18 to 24 hours.

🕐 Second Period

1. Examine the slants. Look for growth. Some organisms produce sparse growth, and you must examine the cultures closely to determine if growth is present. Is either culture producing a pigment and, if so, is the pigment associated with the cells or has it diffused into the agar? Remember, however, that pigment production could require longer incubation times.
2. Determine which incubation temperature produced the best growth. If no growth occurred on either slant, your original culture could be nonviable or more time is needed for growth of the culture to occur. A third possibility is that neither temperature supported growth. Think through the possibilities and decide what course of action you need to take.
3. If growth occurred on the slant, pick the tube with the best growth and designate it as your **reserve stock culture.** Store the reserve stock in the refrigerator. Cultures stored in this manner are

Characteristics of Unknown Bacterium			Observations	Interpretation
Physiological Characteristics		O/F Glucose Test		
	Fermentative Tests	Glucose Fermentation		
		Lactose Fermentation		
		Sucrose Fermentation		
		Mannitol Fermentation		
	Oxidative Tests	Oxidase Test		
		Catalase Test		
		Nitrate Reduction Test		
	IMViC Tests	Indole Test (tryptone broth)		
		Methyl Red Test		
		Voges-Proskauer Test		
		Citrate Test		
	Hydrolysis Tests	Starch Agar		
		Skim Milk Agar		
		Spirit Blue Agar		
		Urea Agar		
		Phenylalanine agar		
	Multiple Test Media	SIM Medium (H_2S)		
		SIM Medium (Indole)		
		SIM Medium (Motility)		
		Kliger's Iron Agar (Carbohydrate Fermentations)		
		Kliger's Iron Agar (H_2S)		
		EXAMPLE: Catalase Test	bubbles	catalase +

Cultural Characteristics

Learning Outcomes

After completing this exercise, you should be able to

1. Determine the cultural characteristics of your unknown grown on nutrient agar, gelatin, and in thioglycollate broth.

2. Define the colony characteristics of your unknown such as elevation, edge, and any pigment production.

The cultural characteristics of an organism pertain to its macroscopic appearance on different kinds of media. In *Bergey's Manual,* you will find descriptive terms used by bacteriologists for recording cultural characteristics. For the general description of colonies, nutrient agar or any complex, rich medium is useful for this purpose. The nature of the growth in a nutrient broth can vary, and this too can be a source of certain information about an organism. Thioglycollate medium (Exercise 24) can be used to determine the oxygen requirements of an organism: Where do the cells grow in a tube of this medium? Some media, such as blood agar, are "differential," demonstrating the hemolytic capability of an organism. In the following exercise, you will inoculate your unknown into different media to determine its cultural characteristics in the various media.

First Period

(Inoculations)

During this period, one nutrient agar plate, one nutrient gelatin deep, two nutrient broths, and one tube of fluid thioglycollate medium will be inoculated. Inoculations will be made with the original broth culture of your unknown. The reason for inoculating two tubes of nutrient broth here is to recheck the optimum growth temperature of your unknown. In Exercise 34, you incubated your nutrient agar slants at 20°C and 37°C. It may well be that the optimum growth temperature is closer to 30°C. It is to check out this intermediate temperature that an extra nutrient broth is being inoculated. Proceed as follows:

Materials

for each unknown:
- 1 nutrient agar plate
- 1 nutrient gelatin deep

- 2 nutrient broths
- 1 fluid thioglycollate medium (FTM)

1. Obtain or pour a nutrient agar plate for each unknown and streak it with a method that will give good isolation of colonies. Use the original broth culture for streaking.
2. Inoculate the tubes of nutrient broth with a loop.
3. Make a stab inoculation into the gelatin deep by stabbing the inoculating needle (straight wire) directly down into the medium to the bottom of the tube and pulling it straight out. The medium must not be disturbed laterally.
4. Inoculate the tube of FTM with a loopful of your unknown. Mix the organisms throughout the tube by rolling the tube between your palms.
5. Place all tubes except one nutrient broth into a test tube rack and incubate for 24 hours at the temperature that seemed optimal in Exercise 34. Incubate the remaining tube of nutrient broth separately at 30°C. Incubate the agar plate, inverted, at the presumed best temperature.

Second Period

(Evaluation)

After the cultures have been properly incubated, *carry them to your desk in a careful manner* to avoid disturbing the growth pattern in the nutrient broths and FTM. Before studying any of the tubes or plates, place the tube of nutrient gelatin in an ice water bath. It will be studied later. Proceed as follows to study each type of medium and record your observations in the descriptive chart at the end of Exercise 34.

Materials

- reserve stock agar slant of unknown
- spectrophotometer and cuvettes
- magnification lens
- ice water bath

Nutrient Agar Slant (Reserve Stock)

Examine your reserve stock agar slant of your unknown that has been stored in the refrigerator since

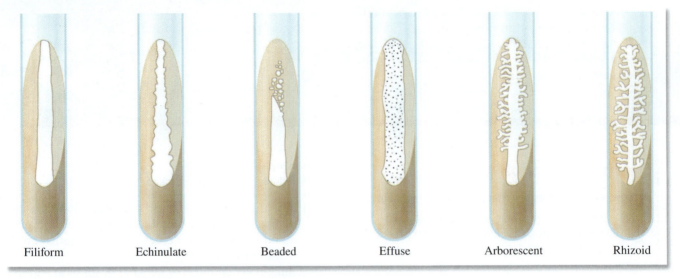

| Filiform | Echinulate | Beaded | Effuse | Arborescent | Rhizoid |

Figure 35.1 **Types of bacterial growth on nutrient agar slants.**

the last laboratory exercise. Evaluate it in terms of the following criteria:

Amount of Growth   The abundance of growth may be described as *none, slight, moderate,* and *abundant.*

Color   Pigments can be associated with a colony, for example, prodigiosin, the red pigment made by *Serratia marcescens* when grown at 27°C. However, pigments can be produced by an organism that diffuses into the medium, causing the medium to be colored, such as the case for the green fluorescent pigment produced by *Pseudomonas fluorescens.* To check for diffusable pigments, hold your plate up to the light and observe the color of the medium in the plate. Most bacteria, however, do not produce pigments, and their colonies are white or buff colored.

Opacity   Organisms that grow prolifically on the surface of a medium will appear more opaque than those that exhibit a small amount of growth. Degrees of opacity may be expressed in terms of *opaque, transparent,* and *translucent* (partially transparent).

Form   The gross appearance of different types of growth are illustrated in figure 35.1. The following descriptions of each type will help in differentiation:

> *Filiform:* characterized by uniform growth along the line of inoculation
> *Echinulate:* margins of growth exhibit toothed appearance
> *Beaded:* separate or semiconfluent colonies along the line of inoculation
> *Effuse:* growth is thin, veil-like, unusually spreading
> *Arborescent:* branched, treelike growth
> *Rhizoid:* rootlike appearance

Nutrient Broth

The nature of growth on the surface, subsurface, and bottom of the tube is significant in nutrient broth cultures. Describe your cultures as thoroughly as possible on the descriptive chart with respect to these characteristics:

Surface   Figure 35.2 illustrates different types of surface growth. A *pellicle* type of surface differs from the *membranous* type in that the latter is much thinner. A *flocculent* surface is made up of floating adherent masses of bacteria.

Subsurface   Below the surface, the broth may be described as *turbid* if it is cloudy, *granular* if specific small particles can be seen, *flocculent* if small masses are floating around, and *flaky* if large particles are in suspension.

Sediment   The amount of sediment in the bottom of the tube may vary from none to a great deal. To

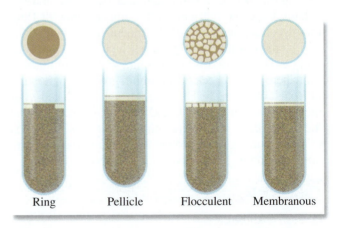

| Ring | Pellicle | Flocculent | Membranous |

Figure 35.2 **Types of surface growth in nutrient broth.**

describe the type of sediment, agitate the tube, putting the material in suspension. The type of sediment can be described as *granular, flocculent, flaky,* and *viscid.* Test for viscosity by probing the bottom of the tube with a sterile inoculating loop.

Amount of Growth To determine the amount of growth, it is necessary to shake the tube to disperse the organisms. Terms such as *slight, moderate,* and *abundant* adequately describe the amount.

Temperature Requirements To determine which temperature produces better growth, transfer the contents of the nutrient broth tubes to separate cuvettes and measure the optical density (absorbance) with a spectrophotometer. Because the cultures may be too turbid to measure, you may have to dilute the cultures with water before taking the readings. Record in the descriptive chart which temperature produces better growth for your organism. This temperature will be closer to the one needed for optimum growth of your organism.

Fluid Thioglycollate Medium

The growth pattern of your bacterium in fluid thioglycollate medium will give some indication of the oxygen requirement of your organism. Examine your FTM tube and compare the growth pattern of your organism with that of figure 24.5. More than likely, your bacterium will be either aerobic, microaerophilic, or a facultative anaerobe. Strict anaerobes such as *Clostridium* require special culture conditions for growth.

Gelatin Stab

Some bacteria produce **proteases,** enzymes that degrade proteins. Determine if your unknown produces proteases by examining the nutrient gelatin tube that you inoculated with your unknown. After incubation, place the culture in an ice bath and allow it to stand for several minutes. Remove the tube and tilt it several times from side to side to ascertain if liquefaction has occurred. If gelatin is degraded by proteases, the culture will remain liquid after being placed in the ice bath. If proteases are not produced by the bacterium, the contents of the tube will be a solid. Check the liquefaction configuration with figure 35.3 to see if any of the illustrations match your tube. A description of each type follows:

Crateriform: saucer-shaped liquefaction
Napiform: turnip-like
Infundibular: funnel-like or inverted cone
Saccate: elongate sac, tubular, cylindrical
Stratiform: liquefied to the walls of the tube in the upper region

Note: The configuration of liquefaction is not as significant as the mere fact that liquefaction takes place.

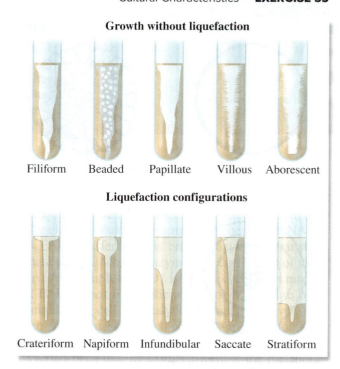

Growth without liquefaction

Filiform   Beaded   Papillate   Villous   Aborescent

Liquefaction configurations

Crateriform   Napiform   Infundibular   Saccate   Stratiform

Figure 35.3 **Growth in gelatin stabs.**

If your organism liquefies gelatin, but you are unable to determine the exact configuration, don't worry about it. However, be sure to record on the descriptive chart the *presence* or *absence* of protease production.

Another important point: Some organisms produce protease at a very slow rate. Tubes that are negative should be incubated for another 4 or 5 days to see if protease is produced slowly.

Type of Growth (No Liquefaction) If no liquefaction has occurred, check the tube to see if the organism grows in nutrient gelatin (some do, some don't). If growth has occurred, compare the growth with the top of the illustration in figure 35.3. It should be pointed out, however, that the pattern of growth in gelatin is not very important for bacterial identification.

Nutrient Agar Plate

Colonies grown on plates of nutrient agar should be studied with respect to size, color, opacity, form, elevation, and margin. With a magnifying lens, study individual colonies carefully. Refer to figure 35.4 for descriptive terminology. Record your observations in the descriptive chart in Exercise 34.

Laboratory Report

There is no Laboratory Report for this exercise. Record all information in the descriptive chart at the end of Exercise 34.

to snapping division of cells. Metachromatic granules formed. Facultative anaerobic. Catalase-positive. Most species produce acid from glucose and some other sugars. Often produce pellicle in broth.

Propionibacterium: Pleomorphic rods, often diphtheroid or club-shaped with one end rounded and the other tapered or pointed. Cells may be coccoid, bifid (forked, divided), or even branched. Nonmotile. Some produce clumps of cells with "Chinese character" arrangements. Anaerobic to aerotolerant. Generally catalase-positive. Produce large amounts of propionic and acetic acids. All produce acid from glucose.

Arthrobacter: Gram-positive rod and coccoid forms. Pleomorphic. Growth often starts out as rods, followed by shortening as growth continues, and finally becoming coccoidal. Some V-shaped and angular forms; branching by some. Rods usually nonmotile; some motile. Oxidative, never fermentative. Catalase-positive. Little or no gas produced from glucose or other sugars. Type species is *Arthrobacter globiformis*.

Endospore-Forming Cocci This group consists of unusual cocci.

Sporosarcina: Cells are spherical or oval when single. Cells may adhere to each other after division, producing tetrads or packets of eight or more. Endospores are formed. Strictly aerobic. Generally motile.

Catalase-Positive, Gram-Positive Cocci Our concern here is with only three genera in this group. Oxygen requirements and cellular arrangement are the principal factors in differentiating the genera. Most of these genera are not closely related.

Micrococcus: Spheres, occurring as singles, pairs, irregular clusters, tetrads, or cubical packets. Usually nonmotile. Strict aerobes (one species is facultative anaerobic). Catalase- and oxidase-positive. Most species produce carotenoid pigments. All species will grow in media containing 5% NaCl.

Planococcus: Spheres, occurring singly, in pairs, in groups of three cells, occasionally in tetrads. Although cells are generally gram-positive, they may be gram-variable. Motility is present. Catalase- and gelatinase-positive. Carbohydrates not utilized. Do not hydrolyze starch or reduce nitrate.

Staphylococcus: Spheres, occurring as singles, pairs, and irregular clusters. Nonmotile. Facultative anaerobes. Usually

catalase-positive. Most strains grow in media with 10% NaCl. Susceptible to lysis by lysostaphin. Glucose fermentation: acid, no gas. Coagulase production by some. Refer to Exercise 52 for species differentiation.

Catalase-Negative, Gram-Positive Cocci Note that the single genus of this group is included in the same section of *Bergey's Manual* as other gram-positive cocci.

Streptococcus: Spherical to ovoid cells that occur in pairs or chains when grown in liquid media. Some species, notably *S. mutans*, will develop short rods when grown under certain circumstances. Facultative anaerobes. Catalase-negative. Carbohydrates are fermented to produce lactic acid without gas production. Many species are commensals or parasites of humans or animals. Refer to Exercise 53 for species differentiation of pathogens.

Aerobic Gram-Negative Rods Although there are many genera of gram-negative aerobic rod-shaped bacteria, only four genera are likely to be encountered in this exercise.

Pseudomonas: Generally motile. Strict aerobes. Catalase-positive. Some species produce soluble fluorescent pigments that diffuse into the agar of a slant.

Alcaligenes: Rods, coccal rods, or cocci. Motile. Obligate aerobes with some strains capable of anaerobic respiration in presence of nitrate or nitrite.

Halobacterium: Cells may be rod- or disk-shaped. Cells divide by constriction. Most are strict aerobes; a few are facultative anaerobes. Catalase- and oxidase-positive. Colonies are pink, red, or red to orange. Gelatinase not produced. Most species require high NaCl concentrations in media. Cell lysis occurs in hypotonic solutions.

Flavobacterium: Gram-negative rods with parallel sides and rounded ends. Nonmotile. Oxidative. Catalase-, oxidase-, and phosphatase-positive. Growth on solid media is typically pigmented yellow or orange. Nonpigmented strains do exist.

Facultatively Anaerobic Gram-Negative Rods If your unknown appears to fall into this group, use the separation outline in figure 39.3 to determine the genus. Another useful separation outline is provided in figure 54.1. *Keep in mind, when using these separation outlines, that there are some minor exceptions in the applications of these tests. The diversity of species within a particular genus often presents some problematical exceptions to the rule.*

Laboratory Report   **39**

39   Use of *Bergey's Manual*

A. Identification of an Unknown Bacterium (from Exercises 34–38)

1.   Unknown bacterial species: _____

2.   What tests/characteristics were the most helpful in your identification
_____
_____

3.   What is the significance of your unknown organism? Use your text
learn more about its relevance.
_____
_____
_____

4.   Recent advances in DNA technology have provided other techniques
the methods you used in these exercises still valuable, and if so, when
_____
_____

B. Challenge Exercise

1.   Record initial information you obtained for your unknown bacterium.

a. Preferred temperature: _____

b. Cell shape/arrangement: _____

c. Gram reaction: _____

2.   Draw your separation outline in the space below or on a separate shee

---

Gram-Negative Cocci   These genera are morphologically quite similar, yet physiologically quite different.

> *Neisseria:* Cocci, occurring singly, but more
> often in pairs (diplococci); adjacent sides
> are flattened. One species (*N. elongata*)
> consists of short rods. Nonmotile. Except
> for *N. elongata*, all species are oxidase- and
> catalase-positive. Aerobic.
>
> *Veillonella:* Cocci, appearing as diplococci,
> masses, and short chains. Diplococci have
> flattening at adjacent surfaces. Nonmotile.
> All are oxidase- and catalase-negative.
> Nitrate is reduced to nitrite. Anaerobic.

Problem Analysis

If you have identified your unknown by following the above procedures, congratulations! Not everyone succeeds at their first attempt. If you are having difficulty, consider the following possibilities:

- You may have been given the wrong unknown! Although this is a remote possibility, it does happen at times. Occasionally, clerical errors are made when unknowns are put together.
- Your organism may be giving you a false negative result on a test. This may be due to an incorrectly prepared medium, faulty test reagents, or improper testing technique.
- Your unknown organism may not match the description exactly as stated in *Bergey's Manual*. The information given in *Bergey's Manual* is based on the fact that 90% of organisms within a species conform, whereas 10% of a species can deviate from the given descriptions. *In other words, test results, as stated in the manual, may not always apply to your unknown.*
- Your culture may be contaminated. If you are not working with a pure culture, all tests are unreliable.
- You may not have performed enough tests. Check the various tables in *Bergey's Manual* to see if there is some other test that will be helpful. In addition, double-check the tables to make sure that you have read them correctly.

Confirmation of Results

There are several ways to confirm your presumptive identification. One method is to apply serological techniques, if your organism is one for which typing serum is available. Another alternative is to use one of the miniature multitest systems that are described in the next section of this manual. Your instructor will indicate which of these alternatives, if any, will be available.

Challenge Exercise

To further practice the techniques and tests you have already learned to use in identifying an unknown bacterium, here is another challenging activity. In this part of the exercise, your instructor will give you a broth culture of one of the organisms listed in table 39.1. Your task is to correctly identify this unknown bacterium in the least number of steps, using the least amount of materials. Use the following general procedure to complete this assignment.

1.   Streak your unknown for isolation onto two agar plates. Incubate one at 25°C and one at 37°C.
2.   After incubation, record the preferred temperature for your organism in part B of the Laboratory Report, and use this temperature when incubating all further tests.
3.   From one colony on your plate, make two pure culture slants to use throughout this exercise. One can serve as your working culture, and the other can serve as a reserve culture in case of contamination during the process.
4.   From that same colony, prepare a Gram stain of the bacterium and observe microscopically. Record the Gram reaction and the cellular morphology of your organism.
5.   Using the relevant information provided in table 39.1, construct a dichotomous key to use as you attempt to identify your unknown. First, find a test that divides your type of organism (gram-positive or gram-negative) into two groups. Then, continue by dividing each group into two groups until all remaining candidates for the unknown organism are separated from each other on your outline. Complete this task on a separate sheet of paper, and ask your instructor to review your separation outline before you conduct any further tests.
6.   Run the tests indicated on your outline one-by-one until you have eliminated all but one organism. This should be your unknown. Consider using multiple test media as appropriate to limit the amount of materials and time for this process. As you work, record all of the tests that you conduct and the results in the Laboratory Report table.
7.   Conduct a confirmatory test on your organism. This should be a test that you have not already used, and it is always best to choose one where you expect a positive result for your unknown.
8.   Record the identity of your unknown species and check your answer with your instructor.

Laboratory Report

Record the identity of your unknown in part B of Laboratory Report 39 and answer all of the questions.

Table 39.1 Characteristics of Common Gram-Positive and Gram-Negative Bacterial Species

SPECIES	SHAPE	GRAM REACTION	OXIDASE	CATALASE	METHYL RED	VOGES-PROSKAUER	LACTOSE FERMENTATION	GLUCOSE FERMENTATION	CITRATE	H₂S	NITRATE REDUCTION	INDOLE	UREA	MOTILITY
Alcaligenes faecalis	Rod	Neg	Pos	Pos	Neg	Neg	Neg	Neg	Pos	Neg	Neg	Neg	Neg	Pos
Enterobacter aerogenes	Rod	Neg	Neg	Pos	Neg	Pos	AG	AG	Pos	Neg	Pos	Neg	Neg	Pos
Escherichia coli	Rod	Neg	Neg	Pos	Pos	Neg	AG	AG	Neg	Neg	Pos	Pos	Neg	Pos
Klebsiella pneumoniae	Rod	Neg	Neg	Pos	Neg	Pos	AG	AG	Pos	Neg	Pos	Neg	Pos	Neg
Proteus vulgaris	Rod	Neg	Neg	Pos	Pos	Neg	Neg	A or AG	Neg	Pos	Pos	Pos	Pos	Pos
Pseudomonas aeruginosa	Rod	Neg	Pos	Pos	Neg	Neg	Neg	Neg	Pos	Neg	Pos	Neg	Neg	Pos
Salmonella typhimurium	Rod	Neg	Neg	Pos	Pos	Neg	Neg	A or AG	Pos	Pos	Pos	Neg	Neg	Pos
Serratia marcescens	Rod	Neg	Neg	Pos	Neg	Pos	Neg	A or AG	Pos	Neg	Pos	Neg	Neg	Pos

SPECIES	SHAPE	GRAM REACTION	OXIDASE	CATALASE	METHYL RED	VOGES-PROSKAUER	LACTOSE FERMENTATION	GLUCOSE FERMENTATION	CITRATE	H₂S	NITRATE REDUCTION	INDOLE	UREA	MOTILITY
Enterococcus faecalis	Coccus	Pos	Neg	Neg	Pos	Neg	A or AG	A or AG	Neg	Neg	Neg	Neg	Neg	Neg
Micrococcus luteus	Coccus	Pos	Pos	Pos	Neg	Neg	Neg	Neg	Neg	Neg	Pos	Neg	Neg	Neg
Staphylococcus aureus	Coccus	Pos	Neg	Pos	Pos	Pos	A or AG	A or AG	Neg	Neg	Pos	Neg	Pos	Neg
Staphylococcus epidermidis	Coccus	Pos	Neg	Pos	Neg	Neg	Neg	A or AG	Neg	Neg	Neg	Neg	Pos	Neg

---

B. Short-Answer Questions

1. What are the advantages and disadvantages of multitest systems for bacterial identification?

2. Before using the EnteroPluri-*Test* System, what test must be performed to confirm the identity of your unknown as a member of the family *Enterobacteriaceae*? What is the expected result?

3. For each of the following aspects, compare and contrast the EnteroPluri-*Test* System to the API® 20E System (Exercise 40) for *Enterobacteriaceae* identification.

    a. time requirement

    b. specimen preparation

    c. tests utilized

    d. anaerobic conditions

    e. interpretation of results

4. The five-digit code for all members of the family *Enterobacteriaceae* starts with the number 2 or 3. What does this indicate about their common biochemistry?

5. If the first number of the five-digit code is 0, what does this indicate? What bacterial genera are likely to give this result? What should you do next?

6. The VP test is a confirmatory test. In what situations would this test be utilized?

7. If the five-digit code that is tabulated cannot be found in the EnteroPluri-*Test* Codebook, what might that indicate about the bacterial culture?

   _____

   _____

8. What biochemical characteristics are commonly found in *Enterobacteriaceae*?

   _____

   _____

9. List and describe at least three medically important members of family *Enterobacteriaceae*. What biochemical characteristics can be used to distinguish between these different members?

   _____

   _____

   _____

   _____

Staphylococcus Identification:
The API® Staph System

Learning Outcomes

After completing this exercise, you should be able to

1. Inoculate an API® Staph strip with an unknown staphylococcus.

2. Evaluate the test results to generate a seven-digit profile number for the unknown bacterium.

3. Use the seven-digit profile number and *Staph Profile Register* to correctly identify the unknown bacterium.

The API® Staph System, produced by bioMérieux, is a reliable method for identifying 23 species of gram-positive cocci, including 20 clinically important species of staphylococci. This system consists of 19 microampules that contain dehydrated substrates and/or nutrient media. Except for the coagulase test, all the tests are important in the differentiation of *Staphylococcus, Kocuria,* and *Micrococcus.*

Figure 42.1 illustrates two inoculated strips: the upper one just after inoculation and the lower one with positive reactions. Note that the appearance of each microcupule undergoes a pronounced color change when a positive reaction occurs.

Figure 42.2 illustrates the overall procedure. The first step is to make a saline suspension of the organism from an isolated colony. An API® Staph strip is then placed in a tray that has a small amount of water added to it to provide humidity during incubation. Next, a sterile Pasteur pipette is used to dispense 2–3 drops of the bacterial suspension to each microcupule. The

inoculated tray is covered and incubated aerobically for 18 to 24 hours at 35–37°C. Finally a seven-digit profile number is obtained and used to determine the identity of the organism in Appendix D.

As simple as this system might seem, there are a few limitations that one must keep in mind. Final species determination by a competent microbiologist must take into consideration other factors such as the source of the specimen, the catalase reaction, colony characteristics, and antimicrobial susceptibility pattern. Very often there are confirmatory tests that must also be made.

If you have been working with an unknown that appears to be one of the staphylococci, use this system to confirm your conclusions. If you have already done the coagulase test and have learned that your organism is coagulase-negative, this system will enable you to identify one of the numerous coagulase-negative species that are not identifiable by the procedures in Exercise 52.

First Period

(Inoculations and Coagulase Test)

Materials

- API® Staph test strip, incubation tray, and cover
- blood agar plate culture of unknown (must not have been incubated over 30 hours)
- sterile blood agar plate (if needed for purity check)
- serological tube of 2 ml sterile saline
- squeeze bottle of tap water
- tubes containing McFarland No. 3 (BaSO₄) standard
- sterile Pasteur pipette (5 ml size)

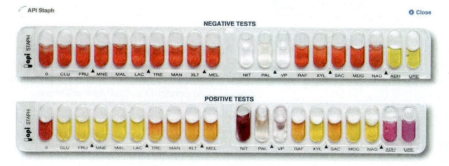

Figure 42.1 Negative and positive results on API® Staph test strips.
Courtesy of bioMérioux, Inc.

0.85%
saline

(1) Use several loopfuls of organisms to make saline suspension of unknown. Turbidity of suspension should match McFarland No. 3 barium sulfate standard.

(2) After labeling the end tab of a tray with your name and unknown number, dispense approximately 5 ml of tap water into the bottom of the tray.

(3) Place an API® Staph test strip into the bottom of the moistened tray. Take care not to contaminate the microcupules with fingers when handling test strip.

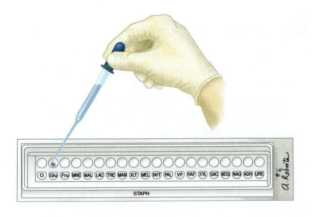

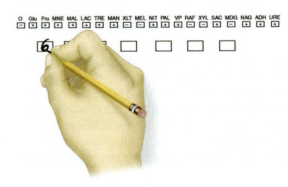

(4) With a Pasteur pipette, dispense 2–3 drops of the bacterial suspension into each of the 20 microcupules. Cover the tray with the lid and incubate at 35–37°C for 18 to 24 hours.

(5) Once all results are recorded in the Laboratory Report, total up the positive values in each group to determine the seven-digit profile. Consult the *Staph Profile Register* or chart V in Appendix D to find your unknown.

Figure 42.2 The API® Staph procedure.
Courtesy of bioMerioux, Inc.

Refer to figure 42.2 as you complete the following steps.

1. If the coagulase test has not been performed, refer to Exercise 52 for the procedure and perform it on your unknown.
2. Prepare a saline suspension of your unknown by transferring organisms to a tube of sterile saline from one or more colonies with a loop or sterile swab. Turbidity of the suspension should match a tube of No. 3 McFarland barium sulfate standard (see Appendix B).
   Important: Do not allow the bacterial suspension to go unused for any great length of time. Suspensions older than 15 minutes become less effective.
3. Label the end strip of the tray with your name and unknown number. See illustration 2, figure 42.2.
4. Dispense about 5 ml of tap water into the bottom of the tray with a squeeze bottle. Note that the bottom of the tray has numerous depressions to accept the water.
5. Remove the API® test strip from its sealed envelope and place the strip in the bottom of the tray (see illustration 3).
6. After shaking the saline suspension to disperse the organisms, fill a sterile Pasteur pipette with the bacterial suspension.
7. Inoculate each of the microcupules with 2 or 3 drops of the suspension (see illustration 4). If a purity check is necessary, use the excess suspension to inoculate another blood agar plate.
8. Place the plastic lid on the tray and incubate the strip aerobically for 18 to 24 hours at 35–37°C.

⏲ Second Period

(Evaluation of Test)

During this period, the results will be recorded on the Laboratory Report, the profile number will be determined, and the unknown will be identified by looking up the number on the *Staph Profile Register* (or chart V, Appendix D).

Materials

- API® Staph test strip (incubated 18 to 24 hours)
- *Staph Profile Register*
- Barritt's reagents A and B
- nitrate reagents A and B
- ZYM reagent A
- ZYM reagent B

1. After 18 to 24 hours of incubation, read and interpret the test results (chart IV in Appendix D or figure 42.3).
   a. For the Voges-Proskauer test, add 1 drop each of Barritt's A and Barritt's B to the VP ampule. Incubate 10 minutes. A positive test is indicated by a violet-pink color; a negative test shows no color change.
   b. For the nitrate test, add 1 drop each of nitrate A and nitrate B test reagents to the NIT ampule. Incubate 10 minutes. A positive test is a violet-pink color; a negative test shows no color change.
   c. For the alkaline phosphatase (PAL) test, add 1 drop each of ZYM A and ZYM B to the PAL ampule. Incubate 10 minutes. A positive test is violet in color; a negative test is yellow.
2. Record the results in the Laboratory Report.
3. Construct the profile number according to the instructions in the Laboratory Report and determine the name of your unknown. Use chart V in Appendix D or look up the seven-digit profile in a *Staph Profile Register* provided by your instructor.

Disposal

Once all the information has been recorded, be sure to place the entire incubation unit in a receptacle that is to be autoclaved.

Laboratory Report

Complete Laboratory Report 42 by recording your results, identifying your unknown, and answering the questions.

Figure 42.3 Test results of a positive strip.
Courtesy of bioMérieux, Inc.

42 Staphylococcus Identification: The API® Staph System

A. Results

1. Tabulation

   By referring to figure 42.3 and chart IV in Appendix D, determine the results of each test, and record these results as positive (+) or negative (−) in the Profile Determination Table below. Note there are two more tables on the next page for tabulation of additional organisms.

	O 1	GLU 2	FRU 4	MNE 1	MAL 2	LAC 4	TRE 1	MAN 2	XLT 4	MEL 1	NIT 2	PAL 4	VP 1	RAF 2	XYL 4	SAC 1	MDG 2	NAG 4	ADH	URE	Lysost*
RESULTS																					

PROFILE NUMBER ☐ ☐ ☐ ☐ ☐ ☐ ☐

GRAM STAIN ☐   COAGULASE ☐   Additional Information ☐   Identification ☐

MORPHOLOGY ☐   CATALASE ☐

*Lysostaphin-lysostatin resistance to be done as a separate test as described by the manufacturer at the bottom of chart IV in Appendix D. To be included to determine the seven-digit ID number.

2. Construction of Seven-Digit Profile

   Note in the above table that each test has a value of 1, 2, or 4. To compute the seven-digit profile for your unknown, total up the positive values for each group.

3. Final Determination

   Refer to the *Staph Profile Register* (or chart V, Appendix D) to find the organism that matches your profile number. Write the name of your unknown in the space below and list any additional tests that are needed for final confirmation. If the materials are available for these tests, perform them.

   Name of Unknown: _____

   Additional Tests: _____

	O 1	GLU 2	FRU 4	MNE 1	MAL 2	LAC 4	TRE 1	MAN 2	XLT 4	MEL 1	NIT 2	PAL 4	VP 1	RAF 2	XYL 4	SAC 1	MDG 2	NAG 4	ADH	URE	Lysost*
RESULTS																					

PROFILE NUMBER

GRAM STAIN ☐   COAGULASE ☐   Additional Information   Identification

MORPHOLOGY ☐   CATALASE ☐

*Lysostaphin-lysostatin resistance to be done as a separate test as described by the manufacturer at the bottom of chart IV in Appendix D. To be included to determine the seven-digit ID number.

	O 1	GLU 2	FRU 4	MNE 1	MAL 2	LAC 4	TRE 1	MAN 2	XLT 4	MEL 1	NIT 2	PAL 4	VP 1	RAF 2	XYL 4	SAC 1	MDG 2	NAG 4	ADH	URE	Lysost*
RESULTS																					

PROFILE NUMBER

GRAM STAIN ☐   COAGULASE ☐   Additional Information   Identification

MORPHOLOGY ☐   CATALASE ☐

*Lysostaphin-lysostatin resistance to be done as a separate test as described by the manufacturer at the bottom of chart IV in Appendix D. To be included to determine the seven-digit ID number.

B. Short-Answer Questions

1. What are the advantages and disadvantages of multitest systems for bacterial identification?

   _____

   _____

2. Before using the API® Staph System, what tests should be performed on bacteria to confirm that they are staphylococci?

   _____

   _____

3. What single test differentiates *Staphylococcus aureus* from other species of staphylococci? What is the expected result?

   _____

   _____

4. If the seven-digit code that is tabulated cannot be found in the Profile Determination Table, what might that indicate about the bacterial culture?

   _____

   _____

Applied Microbiology

Applied microbiology encompasses many aspects of modern microbiology. We use microorganisms to produce many of the foods we eat such as cheese, yogurt, bread, sauerkraut, and a whole list of fermented beverages. Microorganisms are important in industrial applications where they are involved in producing antibiotics, pharmaceuticals, and even solvents and starting materials for the manufacture of plastics. Their presence and numbers in our foods and drinking water determine if it is safe to consume these substances as they could cause us harm and disease. In the following exercises, you will explore some of the applications of microbiology by determining bacterial numbers and/or kinds in food and water. You will also study the process of alcohol fermentation as an example of food production.

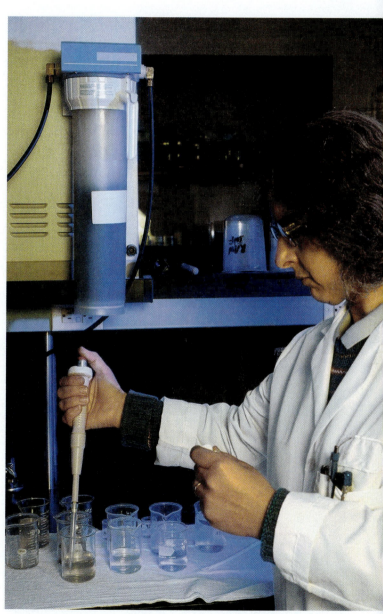

© Brand X Pictures/PunchStock RF

Bacterial Counts of Foods

Learning Outcomes

After completing this exercise, you should be able to

1. Understand the role that microorganisms have in food production and in food spoilage.

2. Perform a standard plate count on a food sample to determine the number of bacteria in the sample.

The presence of microorganisms in food does not necessarily indicate that the food is spoiled or that it has the potential to cause disease. Some foods can have high counts because microorganisms are used in their production. Yogurt, sauerkraut, and summer sausage are examples of foods prepared by microbial fermentation and, therefore, they have high bacterial counts associated with them during production. However, postproduction treatments such as pasteurization or smoking will significantly reduce the numbers of bacteria present. During processing and preparation, food can become contaminated with bacteria, which naturally occur in the environment. These bacteria may not be necessarily harmful or pathogenic. Bacteria are naturally associated with some foods when they are harvested. For example, green beans, potatoes, and beets have soil bacteria associated with them when harvested. Even after washing, some bacteria can remain and will be preserved with the food when it is frozen. The chalky appearance of grapes is due to yeasts that are naturally associated with grapes. This also true of other fruits. Milk in the udders of healthy cows is sterile, but bacteria such as *Streptococcus* and *Lactobacillus* are introduced during milking and processing because they are bacteria that are associated with the animal, especially on the outside of the udder. Pasteurization kills many of the bacteria that are introduced during processing, and any pathogens that may be present. However, it does not kill all the bacteria in milk as some bacteria can survive pasteurization temperatures, and these bacteria can eventually cause spoilage and souring of milk. Hamburger can also have high counts of bacteria that are introduced during processing and grinding of the meat. Many bacteria in hamburger are harmless saprophytes (organisms that live on decaying plant and animal material) that come from the environment where processing occurs. For example, endospore-forming

bacteria and others can be introduced into ground beef during its preparation.

We must also bear in mind that food can be an important means for the transmission of disease. The Centers for Disease Control estimates that 48 million people per year in the United States become sick; 128,000 are hospitalized; and 3,000 people die from foodborne illnesses. Foodborne illnesses usually result because pathogenic bacteria or their toxins are introduced into food products during processing, handling, or preparation. Food handlers can transmit pathogens associated with the human body, like *Staphylococcus aureus* or intestinal bacteria, such as *Salmonella typhi*. Transmission occurs because of unsanitary practices such as failure to wash their hands before preparing or handling food. Botulism food poisoning results from ingesting a toxin produced by *Clostridium botulinum* when its endospores grow in improperly home-canned foods. The endospores occur in the soil and the environment and contaminate the prepared vegetables. *Salmonella* and *Campylobacter* are associated with poultry and eggs and can cause illness if these foods are not properly prepared. *Escherichia coli* O157:H7 is found in the intestines of cattle and can become associated with meat if fecal material from the animal's intestines contaminates meat during the butchering process. This pathogen is then incorporated into hamburger during grinding and processing. Serious illness results from eating improperly cooked hamburger because cooking temperatures are insufficient to kill the organism. Transmission of this pathogen has also occurred when fecal material of cattle contaminated fruits and vegetables such as apples, lettuce, and spinach.

Although high bacterial counts in food do not necessarily mean that the food is spoiled or that it harbors disease-causing organisms, it can suggest the potential for rapid spoilage of the food. Thus, high counts can be important indicators of potential problems. One method to ascertain if food is contaminated with fecal bacteria and, therefore, has the potential to spread disease is to perform coliform counts. Coliforms are bacteria such *Escherichia coli* that normally occur in the intestines of humans and warm-blooded animals. Their presence in food or water indicates that fecal contamination has occurred and therefore, there is the high potential for the spread of serious disease such as typhoid

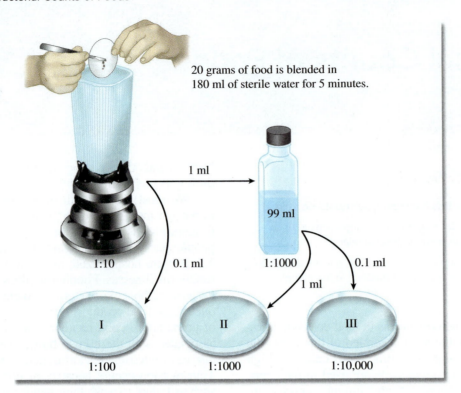

20 grams of food is blended in 180 ml of sterile water for 5 minutes.

1 ml

99 ml

1:10   0.1 ml   1:1000   0.1 ml

1 ml

I   II   III

1:100   1:1000   1:10,000

Figure 43.1 Dilution procedure for bacterial counts of food.

fever, bacillary dysentery, cholera, and intestinal viral diseases.

The standard plate count and the coliform count can be used to evaluate foods in much the same manner that they are used for milk and water to determine total bacterial counts and the number of coliforms. However, because most foods are solid in nature, organisms associated with a food sample must be put into suspension by using a food blender before counts can be performed.

In this exercise, samples of ground beef, dried fruit, and frozen food will be tested for total numbers of bacteria. This procedure, however, will not determine the numbers of coliforms. Your instructor will indicate the specific kinds of foods to be tested and make individual assignments. Figure 43.1 illustrates the general procedure.

Materials

per student:
- 3 petri plates
- 1 bottle (45 ml) of Plate Count agar or Standard Methods agar
- one 99 ml sterile water blank
- two 1.0 ml dilution pipettes

per class:
- food blender
- sterile blender jars (one for each type of food)
- sterile weighing paper
- 180 ml sterile water blanks (one for each type of food)
- samples of ground meat, dried fruit, and frozen vegetables, thawed 2 hours

1. Using aseptic technique, weigh out on sterile weighing paper 20 grams of food to be tested.
2. Add the food and 180 ml of sterile water to a sterile blender jar. Blend the mixture for 5 minutes. This suspension will provide a 1:10 dilution.
3. With a sterile 1.0 ml pipette, dispense from the blender 0.1 ml to plate 1 and 1.0 ml to the water blank. See figure 43.1.
4. Shake the water blank 25 times in an arc for 7 seconds with your elbow on the table, as done in Exercise 19 (Enumeration of Bacteria).
5. Using a fresh pipette, dispense 0.1 ml to plate III and 1.0 ml to plate II.
6. Pour agar (50°C) into the three plates and incubate them at 35°C for 24 hours.
7. Count the colonies on the best plate and record the results in Laboratory Report 43.

43

43 Bacterial Counts of Foods

A. Results

Record your count and the bacterial counts of various other foods made by other students.

TYPE OF FOOD	PLATE COUNT	DILUTION	ORGANISMS PER ML

B. Short-Answer Questions

1. Which type of food had the highest bacterial count? Explain.

_____

_____

2. Which type of food had the lowest bacterial count? Explain.

_____

_____

3. Coliform bacteria are common contaminants of meats.

a. Why might one expect to find coliforms in samples of meat?

_____

_____

b. High coliform counts in food indicate the potential for finding which intestinal pathogens?

_____

_____

c. In terms of food safety, why is it suggested to cook hamburgers medium-well to well-done, whereas steaks can be cooked rare?

_____

_____

4. What considerations should be made to safely thaw frozen foods for later consumption?

   _____

   _____

5. Why is refrigeration not always an effective means for preventing food spoilage?

   _____

   _____

6. Why can milk become sour due to bacteria even though it has been pasteurized?

   _____

   _____

7. Give two examples of foods that have high bacterial counts but are not spoiled.

   _____

   _____

Bacteriological Examination of Water:

Most Probable Number Determination

Water is an essential compound for all living organisms. Water is the solvent that keeps in solution the nutrients that are necessary for metabolism in cells. As a result, the metabolic reactions essential for growth and survival can occur. Prior to the modern age of public health, however, water was also a means for the spread of many infectious diseases. With the emergence of large cities, diseases such as cholera, dysentery, and typhoid fever were common because sewage would often contaminate sources of drinking water. In the 1840s, an English physician, John Snow, showed that a cholera epidemic in London was the result of cesspool overflow into the Thames River, also a source of drinking water for the population. The overflow caused a pump at Broad Street in a section of London to become contaminated with sewage and therefore the means by which cholera was spread to the population in the area. Snow's solution was to remove the handle on the pump and the epidemic stopped. Water safety is still a major concern of municipalities today, and it is important that they make sure that water does not become contaminated with pathogenic bacteria and viruses so that the water drawn from our faucets is safe and disease free.

From a microbiological standpoint, it is not the numbers of bacteria in water that are a problem, but rather the specific kinds of bacteria. Water in rivers, lakes, and streams can contain a variety of bacteria that come from the surrounding ecosystem. These are usually harmless non-pathogenic bacteria. It is important for cities and municipalities to ensure that pathogens which cause cholera, dysentery, and typhoid fever do not enter the water supply. In modern cities, sewage is treated and then discharged into receiving waters such as rivers, lakes, and streams, and this constitutes a major problem because these same bodies of water are also a source of drinking water. As a result, methods have been developed to treat water to eliminate pathogens, both bacteria and viruses, and microbiological tests are done to determine if water is potable and safe for consumption.

Since the pathogens that contaminate water and cause disease are known, it might seem reasonable to directly test for the presence of each specific pathogen *Vibrio cholerae, Salmonella typhi,* and *Shigella dysenteriae.* However, this not practical because the bacterial pathogens are fastidious and could be easily overgrown by other bacteria in the water, causing them to be missed in an analysis. It is even more complicated for viruses because they cannot be grown in culture like bacteria. A much easier approach is to test for the presence of an indicator bacterium that is routinely found in the human intestine but is not usually found in soil or water. Its presence would then signal the likelihood that fecal contamination is present and could cause serious disease.

E. coli is a good indicator bacterium for such testing, for several reasons: (1) it is present in the intestines of humans and several warm-blooded animals but is not routinely found in water or soil; (2) the organism can be identified using specific microbiological tests; and (3) it is not as fastidious as the other pathogens and can survive longer in water samples. By definition, intestinal bacteria such as *Escherichia coli* and *Enterobacter aerogenes* are designated as **coliforms**. These are gram-negative, facultative anaerobic, non-endospore-forming rods that ferment lactose to produce acid and gas in 48 hours when grown at 35°C. Lactose fermentation with acid and gas formation is the basis for determining the total coliform count in water samples in the United States and therefore the purity of water. *Enterococcus faecalis* is a gram-positive coccus that also inhabits the human intestine. The presence of this bacterium is not tested for in the United States, but it has been used as an indicator in Europe to test for fecal contamination.

Three different tests are done to determine the coliform count (figure 44.1): presumptive, confirmed, and completed. Each test is based on one or more of the characteristics of a coliform. A description of each test follows.

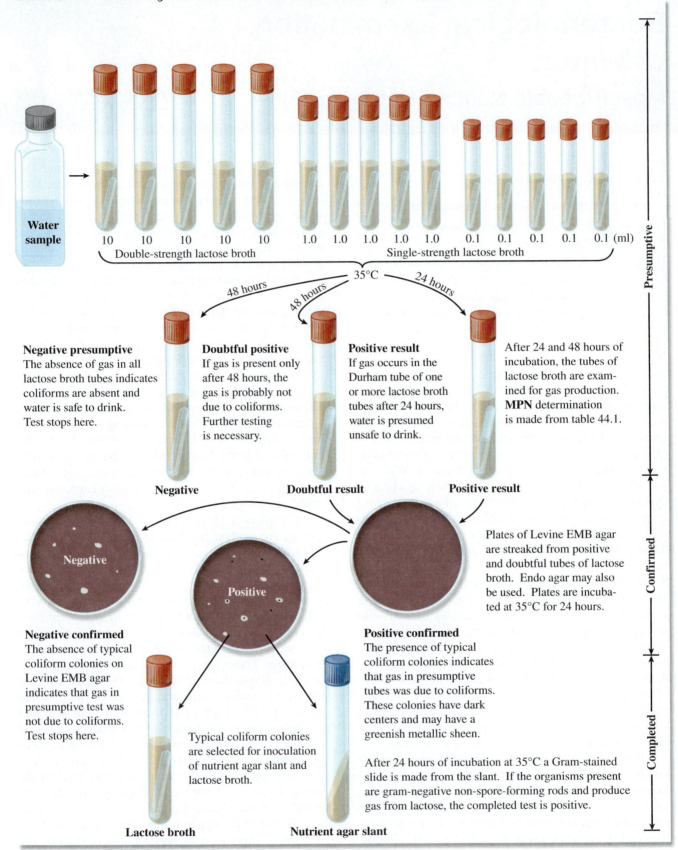

Water sample

10   10   10   10   10   1.0   1.0   1.0   1.0   1.0   0.1   0.1   0.1   0.1   0.1   (ml)

Double-strength lactose broth   Single-strength lactose broth

Presumptive

35°C

48 hours   48 hours   24 hours

Negative presumptive
The absence of gas in all lactose broth tubes indicates coliforms are absent and water is safe to drink. Test stops here.

Doubtful positive
If gas is present only after 48 hours, the gas is probably not due to coliforms. Further testing is necessary.

Positive result
If gas occurs in the Durham tube of one or more lactose broth tubes after 24 hours, water is presumed unsafe to drink.

After 24 and 48 hours of incubation, the tubes of lactose broth are examined for gas production. **MPN** determination is made from table 44.1.

Negative   **Doubtful result**   **Positive result**

Plates of Levine EMB agar are streaked from positive and doubtful tubes of lactose broth. Endo agar may also be used. Plates are incubated at 35°C for 24 hours.

Confirmed

Negative

Positive

Negative confirmed
The absence of typical coliform colonies on Levine EMB agar indicates that gas in presumptive test was not due to coliforms. Test stops here.

Positive confirmed
The presence of typical coliform colonies indicates that gas in presumptive tubes was due to coliforms. These colonies have dark centers and may have a greenish metallic sheen.

After 24 hours of incubation at 35°C a Gram-stained slide is made from the slant. If the organisms present are gram-negative non-spore-forming rods and produce gas from lactose, the completed test is positive.

Typical coliform colonies are selected for inoculation of nutrient agar slant and lactose broth.

Completed

Lactose broth   **Nutrient agar slant**

Figure 44.1 **Bacteriological analysis of water.**

Presumptive Test   In the presumptive test, 15 tubes of lactose broth are inoculated with measured amounts of water to see if the water contains any lactose-fermenting bacteria that produce gas. After incubation, if gas is seen in any of the lactose broths, it is *presumed* that coliforms are present in the water sample. This test is also used to determine the **most probable number (MPN)** of coliforms present per 100 ml of water.

Confirmed Test   In this test, plates of Levine EMB agar or Endo agar are inoculated from positive (gas-producing) tubes to see if the organisms that are producing the gas are gram-negative (another coliform characteristic). Both of these media inhibit the growth of gram-positive bacteria. Either medium allows coliform colonies to be distinguished from noncoliforms. On EMB agar, coliforms produce large blue-black colonies with a green metallic sheen for *E. coli*. On Endo agar, coliforms produce pink to rose-red colonies. The presence of coliform colonies indicates the presence of a gram-negative lactose-fermenting bacterium.

Completed Test   In the completed test, the concern is to determine if the isolate from the agar plates truly matches our definition of a coliform. The media for this test include a nutrient agar slant and a Durham tube of lactose broth. If gas is produced in the lactose tube and a slide from the agar slant reveals that we have a gram-negative, non-spore-forming rod, the presence of a coliform is certain.

The completion of these three tests with positive results establishes that coliforms are present; however, there is no certainty that *E. coli* is the coliform present. The organism might be *E. aerogenes*. Of the two, *E. coli* is the better sewage indicator since *E. aerogenes* can be of nonsewage origin. To differentiate these two species, one must perform the **IMViC tests,** which are described in Exercise 38.

In this exercise, water will be tested from local ponds, streams, swimming pools, and other sources supplied by students and instructor. Enough known positive samples will be evenly distributed throughout the laboratory so that all students will be able to see positive test results. All three tests in figure 44.1 will be performed. If time permits, the IMViC tests may also be performed.

The Presumptive Test

As stated earlier, the presumptive test is used to determine if gas-producing lactose fermenters are present in a water sample. The American Public Health Association's *Standard Methods for Examination of Water and Waste Water,* 22nd edition, employs a 15-tube method for the analysis of water samples which is more statistically valid than previous procedures using fewer tubes. This 15-tube procedure is most applicable to clear surface waters but can be used for brackish water and waters with sediments. Figure 44.1 shows the procedure for determining the MPN of a water sample using sample volumes of 10 ml, 1.0 ml, and 0.1 ml each delivered to five tubes of test medium. The figure also describes the subsequent procedures for performing the confirmed and completed tests for tubes in the presumptive test that show gas production. In addition to detecting the presence or absence of coliforms in the presumptive test, the pattern of positive tubes is used to ascertain the most probable number (MPN) of coliforms in 100 ml of water. See table 44.1 to determine that value from the number of positive tubes.

Set up the test for your water sample using the procedure outlined in figure 44.1. As stated, the method will be used for clear surface waters as well as turbid waters with sediments.

Materials

- 5 Durham tubes of DSLB
- 10 Durham tubes of SSLB
- one 10 ml pipette
- one 1.0 ml pipette

  Note: DSLB designates double-strength lactose broth. It contains twice as much lactose as SSLB (single-strength lactose broth).

1. Set up 5 DSLB and 10 SSLB tubes as illustrated in figure 44.1. Label each tube according to the amount of water that is to be dispensed to it: *10 ml, 1.0 ml,* and *0.1 ml,* respectively.
2. Mix the bottle of water to be tested by shaking it 25 times.
3. With a 10 ml pipette, transfer 10 ml of water to each of the DSLB tubes.
4. With a 1.0 ml pipette, transfer 1.0 ml of water to each of the middle set of tubes, and 0.1 ml to each of the last five SSLB tubes.
5. Incubate the tubes at 35°C for 24 hours.
6. Examine the tubes and record the number of tubes in each set that have 10% gas or more.
7. Determine the MPN by referring to table 44.1. Consider the following:

   Example: If gas was present in the first five tubes (10 ml samples), and detected in only one tube of the second series (1.0 ml samples), but was not present in any of the last five tubes (0.1 ml samples), you would read across the row that has 5-1-0 for the three sets of five tubes. Table 44.1 indicates that the MPN for this pattern of tubes is 33. This means that this water sample would have approximately 33 organisms per 100 ml. *Keep in mind that the MPN of 33 is only a statistical probability figure.*

8. Record the data in Laboratory Report 44.

Table 44.1 MPN Index for Different Combinations of Positive and Negative Results Using the 15-Tube Method for Water Analysis

NO. OF TUBES GIVING POSITIVE REACTION OUT OF				NO. OF TUBES GIVING POSITIVE REACTION OUT OF			
5 OF 10 ML EACH	5 OF 1 ML EACH	5 OF 0.1 ML EACH	MPN INDEX PER 100 ML	5 OF 10 ML EACH	5 OF 1 ML EACH	5 OF 0.1 ML EACH	MPN INDEX PER 100 ML
0	0	0	<1.8	4	0	3	25
0	0	1	1.8	4	1	0	17
0	1	0	1.8	4	1	1	21
0	1	1	3.6	4	1	2	26
0	2	0	3.7	4	1	3	31
0	2	1	5.5	4	2	0	22
0	3	0	5.6	4	2	1	26
1	0	0	2.0	4	2	2	32
1	0	1	4.0	4	2	3	38
1	0	2	6.0	4	3	0	27
1	1	0	4.0	4	3	1	33
1	1	1	6.1	4	3	2	39
1	1	2	8.1	4	4	0	34
1	2	0	6.1	4	4	1	40
1	2	1	8.2	4	4	2	47
1	3	0	8.3	4	5	0	41
1	3	1	10	4	5	1	48
1	4	0	10	5	0	0	23
2	0	0	4.5	5	0	1	31
2	0	1	6.8	5	0	2	43
2	0	2	9.1	5	0	3	58
2	1	0	6.8	5	1	0	33
2	1	1	9.2	5	1	1	46
2	1	2	12	5	1	2	63
2	2	0	9.3	5	1	3	84
2	2	1	12	5	2	0	49
2	2	2	14	5	2	1	70
2	3	0	12	5	2	2	94
2	3	1	14	5	2	3	120
2	4	0	15	5	2	4	150
3	0	0	7.8	5	3	0	79
3	0	1	11	5	3	1	110
3	0	2	13	5	3	2	140
3	1	0	11	5	3	3	170
3	1	1	14	5	3	4	210
3	1	2	17	5	4	0	130
3	2	0	14	5	4	1	170
3	2	1	17	5	4	2	220
3	2	2	20	5	4	3	280
3	3	0	17	5	4	4	350
3	3	1	21	5	4	5	430
3	3	2	24	5	5	0	240
3	4	0	21	5	5	1	350
3	4	1	24	5	5	2	540
3	5	0	25	5	5	3	920
4	0	0	13	5	5	4	1600
4	0	1	17	5	5	5	>1600
4	0	2	21				

*Results to two significant figures.

Source: Data from Rice, E.W., R.B. Baird, A.D. Eaton, L.S. Clesceri, eds. *Standard Methods for the Examination of Water and Wastewater,* 22nd Edition. 2012. Washington, DC: APHA, AWWA, WEF; p. 9–71 (Table 9221:IV).

The Confirmed Test

Once it has been established that gas-producing lactose fermenters are present in the water, it is *presumed* to be unsafe. However, gas formation may be due to noncoliform bacteria. Some of these organisms, such as *Clostridium perfringens*, are gram-positive. To confirm the presence of gram-negative lactose fermenters, the next step is to inoculate media such as Levine eosin-methylene blue (EMB) agar or Endo agar from positive presumptive tubes.

Levine EMB agar contains methylene blue, which inhibits gram-positive bacteria. Gram-negative lactose fermenters (coliforms) that grow on this medium will produce blue-black colonies due to the uptake of the dyes eosin and methylene blue. Colonies of *E. coli* and *E. aerogenes* can be differentiated on the basis of the presence of a greenish metallic sheen. *E. coli* colonies on this medium have this metallic sheen, whereas *E. aerogenes* colonies usually lack the sheen. Differentiation in this manner is not completely reliable, however. It should be remembered that *E. coli* is the more reliable sewage indicator since it is not normally present in soil, while *E. aerogenes* has been isolated from soil and grains.

Endo agar contains a fuchsin sulfite indicator that makes identification of lactose fermenters relatively easy. Coliform colonies and the surrounding medium appear red on Endo agar. Nonfermenters of lactose, on the other hand, are colorless and do not affect the color of the medium.

In addition to these two media, there are several other media that can be used for the confirmed test. Brilliant green bile lactose broth, Eijkman's medium, and EC medium are just a few examples that can be used.

To demonstrate the confirmation of a positive presumptive in this exercise, the class will use Levine EMB agar and Endo agar. One-half of the class will use one medium; the other half will use the other medium. Plates will be exchanged for comparisons.

Materials

- 1 plate of Levine EMB agar (odd-numbered students)
- 1 plate of Endo agar (even-numbered students)

1. Select one positive lactose broth tube from the presumptive test and streak a plate of medium according to your assignment. Use a streak method that will produce good isolation of colonies. If all your tubes were negative, borrow a positive tube from another student.
2. Incubate the plate for 24 hours at 35°C.
3. Look for typical coliform colonies on both kinds of media. Record your results in Laboratory Report 44. If coliform colonies are not present, the water is considered bacteriologically safe to drink.

Note: In actual practice, confirmation of all presumptive tubes would be necessary to ensure accuracy of results.

The Completed Test

A final check of the colonies that appear on the confirmatory media is made by inoculating a nutrient agar slant and lactose broth with a Durham tube. After incubation for 24 hours at 35°C, the lactose broth is examined for gas production. A Gram-stained slide is made from the slant, and the slide is examined under oil immersion.

If the organism proves to be a gram-negative, non-spore-forming rod that ferments lactose, we know that coliforms were present in the water sample. If time permits, complete these last tests and record the results in Laboratory Report 44.

The IMViC Tests

Review the discussion of the IMViC tests in Exercise 38. The significance of these tests should be much more apparent at this time. Your instructor will indicate whether these tests should be performed if you have a positive completed test.

44

44 Bacteriological Examination of Water: Most Probable Number Determination

A. Results

1. **Presumptive Test (MPN Determination)**
   Record the number of positive tubes for various water samples tested by you and your classmates in the table below. Determine the MPN according to the instructions and table 44.1 in this exercise.

WATER SAMPLE (SOURCE)	NUMBER OF POSITIVE TUBES			MPN
	5 Tubes DSLB 10 ml	5 Tubes SSLB 1.0 ml	5 Tubes SSLB 0.1 ml	

2. **Confirmed Test**
   Record the results of the confirmed tests for each water sample that was positive on the presumptive test.

WATER SAMPLE (SOURCE)	POSITIVE	NEGATIVE

3. **Completed Test**
   Record the results of completed tests for each water sample that was positive on the confirmed test.

WATER SAMPLE (SOURCE)	LACTOSE FERMENTATION RESULTS	MORPHOLOGY	EVALUATION

B. Short-Answer Questions

1. Does a positive presumptive test mean that the water is absolutely unsafe to drink? _____

   Explain: _____

2. What might cause a false positive presumptive test?_____
   _____

3. List three characteristics required of a good sewage indicator:

   a. _____ b. _____ c. _____

4. The fermentation of what disaccharide is the basis for determining the presence of coliforms?
   _____

5. Describe the appearance of coliforms on Levine EMB agar. _____
   _____

6. Why don't health departments routinely test for pathogens instead of using a sewage indicator?
   _____
   _____

7. List five characteristics of coliform bacteria.
   _____
   _____
   _____

8. How is each of the following media used for the detection of coliforms?

   a. lactose broth with Durham tube

   _____

   _____

   b. Levine EMB agar

   _____

   _____

   c. nutrient agar slant

   _____

   _____

9. Once the completed test establishes the presence of coliforms in the water sample, why might you perform the IMViC tests on these isolates?

   _____

   _____

10. What gram-positive bacterium can give a positive presumptive test?

    _____

    _____

Bacteriological Examination of Water:

The Membrane Filter Method

The most probable number method for determining coliform bacteria in water samples is complicated and requires several days to complete. Furthermore, more than one kind of culture medium is needed for each phase of the test to finally establish the presence of coliforms in a water sample. A more rapid method is the **membrane filter method**, also recognized by the United States Public Health Service as a reliable procedure for determining coliforms. In this test, known volumes of a water sample are filtered through membrane filters that have a pore size of 0.45 μm in diameter. Most bacteria, including coliforms, are larger than the pore diameters, and hence they are retained on the membrane filter. Once the water sample has been filtered, the filter disk containing bacterial cells is placed in a petri dish with an absorbent pad saturated with Endo broth. The plate is then incubated at 35°C for 22 to 24 hours, during which time individual cells on the filter multiply, forming colonies.

Any coliforms that are present on the filter will ferment the lactose in the Endo broth, producing acids. The acids produced from fermentation interact with basic fuchsin, a dye in the medium, causing coliform colonies to have a characteristic metallic sheen. Non-coliform bacteria will not produce the metallic sheen. Gram-positive bacteria are inhibited from growing because of the presence of bile salts and sodium lauryl sulfate, which inhibit these bacteria. Colonies are easily counted on the filter disk, and the total coliform count is determined based on the volume of water filtered.

Figure 45.1 illustrates the procedure we will use in this experiment.

Materials

- vacuum pump or water faucet aspirators
- membrane filter assemblies (sterile)
- side-arm flask, 1000 ml size, and rubber hose
- sterile graduates (100 ml or 250 ml size)

- sterile, plastic petri dishes, 50 mm dia (Millipore #PD10 047 00)
- sterile membrane filter disks (Millipore #HAWG 047 AO)
- sterile absorbent disks (packed with filters)
- sterile water
- 5 ml pipettes
- bottles of *m* Endo MF broth (50 ml)*
- water samples

1. Prepare a small plastic petri dish as follows:
   a. With a flamed forceps, transfer a sterile absorbent pad to a sterile plastic petri dish.
   b. Using a 5 ml pipette, transfer 2.0 ml of *m* Endo MF broth to the absorbent pad.
2. Assemble a membrane filtering unit as follows:
   a. *Aseptically* insert the filter holder base into the neck of a 1-liter side-arm flask.
   b. With a flamed forceps, place a sterile membrane filter disk, grid side up, on the filter holder base.
   c. Place the filter funnel on top of the membrane filter disk and secure it to the base with the clamp.
3. Attach the rubber hose to a vacuum source (pump or water aspirator) and pour the appropriate amount of water into the funnel.

   The amount of water used will depend on water quality. No less than 50 ml should be used. For water samples with few bacteria and low turbidity, 200 ml or more can be filtered. Your instructor will advise you as to the amount of water that you should use. Use a sterile graduate for measuring the water.
4. Rinse the inner sides of the funnel with 20 ml of sterile water.
5. Disconnect the vacuum source, remove the funnel, and aseptically transfer the filter disk with sterile forceps to the petri dish of *m* Endo MF broth. *Keep grid side up*.
6. Incubate at 35°C for 22 to 24 hours. *Do not invert the plate*.
7. After incubation, remove the filter from the dish and dry for 1 hour on absorbent paper.
8. Count the colonies on the disk with low-power magnification, using reflected light. Ignore all colonies that lack the golden metallic sheen. If desired, the disk may be held flat by mounting between two 2″ × 3″ microscope slides after drying. Record your count in the first portion of Laboratory Report 45.

---

*See Appendix C for special preparation method.

(1) Sterile absorbent pad is aseptically placed in the bottom of a sterile plastic petri dish.

(2) Absorbent pad is saturated with 2.0 ml of *m* Endo MF broth.

(3) Sterile membrane filter disk is placed on filter holder base with grid side up.

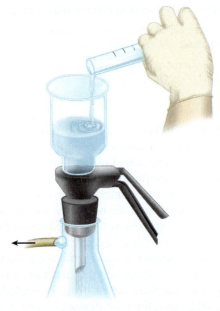

(4) Water sample is poured into assembled funnel, utilizing vacuum. A rinse of 20 ml of sterile water follows.

(5) Filter disk is carefully removed with sterile forceps after disassembling the funnel.

(6) Membrane filter disk is placed on medium-soaked absorbent pad with grid side up. Incubate at 35°C for 24 hours.

Figure 45.1 **Membrane filter routine.**

45   Bacteriological Examination of Water: The Membrane Filter Method

A. Results

Record your coliform count in the table below. Collect data from your classmates to complete the entire table.

SAMPLE	SOURCE	COLIFORM COUNT	AMOUNT OF WATER FILTERED	MPN*
A				
B				
C				
D				
E				
F				
G				
H				

$$*MPN = \frac{\text{Coliform Count} \times 100}{\text{Amount of Water Filtered}}$$

B. Short-Answer Questions

1. In what ways is the membrane filter method for coliform detection superior to the most probable number method (Exercise 44)?

    _____

    _____

2. How is the proper amount of water to be filtered determined? Why is this determination critical to the outcome of the testing?

    _____

    _____

3. Besides pathogens that cause typhoid fever, cholera, and dysentery, what other pathogens would be indicated by the presence of coliforms?

    _____

    _____

4. What is the pore size of the membrane filter used in the membrane filter method for water analysis?

_____

_____

5. On what medium is the membrane filter incubated? Describe the appearance of coliforms on this medium.

_____

_____

Reductase Test

Learning Outcomes

After completing this exercise, you should be able to

1. Understand the basis for the reductase test for determining the quality of a milk sample.

2. Perform the reductase test and determine the quality of a milk sample.

Milk that contains large numbers of actively growing bacteria will have a lowered oxidation-reduction (redox) potential because of the use and exhaustion of dissolved oxygen by the microorganisms. The lowered redox potential can be measured by the addition of the dye methylene blue to a milk sample, which is the basis for the **reductase test** of milk. Methylene will change from blue to clear when the dye is reduced. The test is based on the assumption that bacteria in a milk sample when allowed to grow at 35°C will lower the redox potential, causing the dye to go from blue to colorless. The greater the number of bacteria in the milk, the less time is required for the reduction of the dye.

In the test, 1.0 ml of methylene blue (1:25,000) is added to 10 ml of milk. The tube is sealed with a rubber stopper and slowly inverted three times to mix the sample. It is incubated in a water bath at 35°C and examined at intervals for up to 6 hours. The time required for the methylene blue to become colorless is defined as the **methylene blue reduction time (MBRT)**. The shorter the time, the more bacteria are present and the poorer the quality of the milk. An MBRT of 6 hours indicates a good quality of milk, whereas an MBRT of 30 minutes indicates milk of poor quality (figure 46.1). Bacteria that do not grow at 35°C would not produce a positive test. These would include psychrophiles and thermophiles and bacteria that can survive pasteurization temperatures. Microorganisms can also vary in the extent to which they reduce the dye. This type of test is not applicable to foods other than milk because some foods such as raw meats have natural substances that can reduce the dye.

Unpasteurized milk will contain *Streptococcus lactis* and other lactobacilli, and it can also contain *Escherichia coli*. These organisms are strong reducers, and thus this test is suitable for screening milk that arrives at receiving stations for processing. It is a relatively simple test that does not require extensive training of personnel who would use it.

In this exercise, samples of low- and high-quality raw milk will be tested.

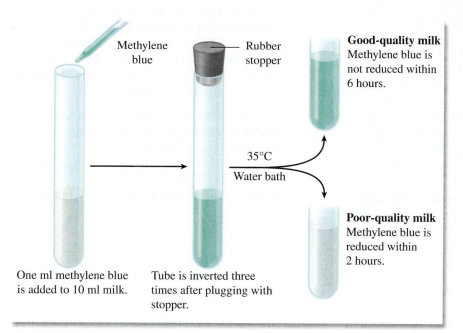

Methylene blue — Rubber stopper

Good-quality milk
Methylene blue is not reduced within 6 hours.

35°C
Water bath

Poor-quality milk
Methylene blue is reduced within 2 hours.

One ml methylene blue is added to 10 ml milk.

Tube is inverted three times after plugging with stopper.

Figure 46.1 Procedure for performing the reductase test on raw milk and interpretation of results.

Figure 46.2 **Procedure for preparing dilutions and plating of milk samples.**

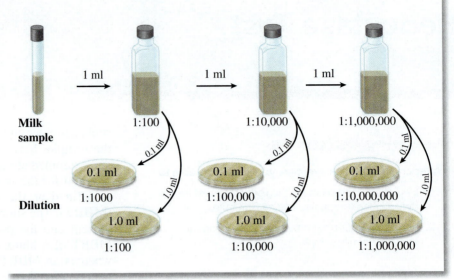

Materials

- 2 sterile test tubes with rubber stoppers per student
- raw milk samples of low and high quality (samples A and B)
- water bath set at 35°C
- methylene blue (0.004%; 4 μl/100 ml)
- 10 ml pipettes
- 1 ml pipettes
- 2 bottles of nutrient agar (100 ml each)
- 10 petri plates
- 6 sterile 99 ml water blanks
- incubator set at 32°C

1. Label two test tubes with your name and the type of milk. Each student will test a good-quality as well as a poor-quality milk.
2. Using separate 10 ml pipettes for each type of milk, transfer 10 ml to each test tube. To the milk in the tubes add 1.0 ml of the methylene blue with a 1 ml pipette. Insert the rubber stoppers and gently invert three times to mix. Record your name and the time on the labels and place the tubes in the water bath set at 35°C. Refrigerate the remaining portions of both types of milk for analysis using the standard plate count.
3. After 5 minutes of incubation, remove the tubes from the water bath and invert once to mix. Return

the tubes to the water bath. This is the last time they should be mixed.
4. To be performed by groups of four students. While waiting to observe the color changes in the tubes at every 30 minutes, perform a standard plate count (Exercise 19) on the two types of milk that you saved in the refrigerator. For each milk sample, make the following dilutions and plate aliquots as shown in figure 46.2: **1:100; 1:10,000; 1:1,000,000.**
5. Incubate the nutrient agar plates at 32°C for 48 hours. Count the bacterial colonies and record the results in Laboratory Report 46.
6. Carefully remove the tubes from the water bath after 30 minutes of incubation and for every half hour until the end of the laboratory period. *When at least four-fifths of the tube has turned white,* the end point of reduction has occurred. Record this time in Laboratory Report 46. The classification of milk quality is as follows:

Class 1: Excellent, not decolorized in 8 hours.
Class 2: Good, decolorized in less than 8 hours, but not less than 6 hours.
Class 3: Fair, decolorized in less than 6 hours, but not less than 2 hours.
Class 4: Poor, decolorized in less than 2 hours.

46 Reductase Test

A. Results

How would you grade the two samples of milk that you tested? Give the MBRT for each one.

Sample A: _____

Sample B: _____

B. Results

Record the standard plate count results (CFU/ml) for each milk sample.

Sample A: _____

Sample B: _____

How do these numbers correlate with MBRT values?

C. Short-Answer Questions

1. Why are the tubes sealed with a rubber stopper rather than left open to the atmosphere?

   _____

   _____

2. Is milk with a short reduction time necessarily unsafe to drink? _____

   Explain: _____

   _____

   _____

3. When oxygen is exhausted in a milk sample, what color is the methylene blue indicator? _____

   _____

4. What advantage do you see in this method over the direct microscopic count method? _____

   _____

   _____

   _____

5. What kinds of organisms may be plentiful in a milk sample, yet give a negative reductase test?

   _____

   _____

   _____

   _____

Temperature:
Lethal Effects

Learning Outcomes

After completing this exercise, you should be able to

1. Understand how temperature kills bacterial cells.

2. Define *thermal death point* and *thermal death time* and their importance in canning of foods.

3. Demonstrate how endospore-forming bacteria are more resistant to the lethal effects of temperature than are vegetative cells.

Most microorganisms are killed by elevated temperatures primarily because of the susceptibility of their macromolecules to heat. Elevated temperatures cause proteins to denature and unfold, resulting in a loss of their tertiary structure and biological activity. Because enzymes are proteins, the metabolic capabilities of an organism are irreversibly damaged by heat. Damage to enzymatic systems in the cell, such as those involved in energy production, protein synthesis, and cell division, invariably results in the death of a cell. Nucleic acids can also be damaged by heat, resulting in the loss of DNA and RNA structure. Loss of nucleic acids prevents cell division and protein synthesis and thus causes cell death. Also, small molecules in the cell, such as NAD^+ and other coenzymes, can be damaged or destroyed by elevated temperatures, and loss of these essential factors contributes to cell death. Some organisms are more resistant to heat than others because they form endospores. *Bacillus* and *Clostridium* form endospores, and these structures are more resistant to heat than vegetative cells for several reasons. Endospores have a much lower water content than vegetative cells. As a result, their macromolecules are less susceptible to denaturation. Because the endospore is dehydrated, water is unavailable for chemical reactions that can damage macromolecules, coenzymes, and other essential small molecules. Endospores, unlike vegetative cells, contain calcium dipicolinate, which plays a vital role in heat resistance by further preventing thermally induced denaturation of proteins. The calcium associated with dipicolinic acid may likewise provide resistance to oxidizing agents that can irreversibly destroy proteins, nucleic acids, and small molecules in the cell. Additionally,

specific proteins found in the endospore can bind to nucleic acids and prevent their denaturation. All of these factors ensure that the endospore remains dormant and undamaged so that it will give rise to viable cells when conditions are reestablished for germination and growth of the organism.

In attempting to compare the susceptibility of different organisms to elevated temperatures, it is necessary to use some metric of comparison. Two methods of comparison are used: the **thermal death point** and the **thermal death time**. The thermal death point (TDP) is the *lowest* temperature at which *a population of a target organism* is killed in 10 minutes. The thermal death time (TDT) is the *shortest* time required to kill a suspension of cells or spores *under defined conditions* at a given temperature. However, various factors such as pH, moisture, medium composition, and age of cells can greatly influence results, and therefore these variables must be clearly stated.

The thermal death time and thermal death point are important in the food industry because canned foods must be heated to temperatures that kill the endospores of *Clostridium botulinum* and *Clostridium perfringens,* two bacteria involved in food-borne illnesses.

In this exercise, you will subject cultures of three different bacteria to temperatures of 60°C, 70°C, 80°C, 90°C, and 100°C. At intervals of 10 minutes, samples of the test bacteria will be removed and plated out to determine the number of survivors. The endospore-former *Bacillus megaterium* will be compared with the non-endospore-formers, *Staphylococcus aureus* and *Escherichia coli*. The overall procedure is illustrated in figure 47.1.

Note in figure 47.1 that *before* the culture is heated, a **control plate** is inoculated with 0.1 ml of the organism. When the culture is placed in the water bath, a tube of nutrient broth with a thermometer inserted into it is placed in the water bath at the same time. Timing of the experiment starts when the thermometer reaches the test temperature.

Due to the large number of plates that have to be inoculated to perform the entire experiment, it will be necessary for each member of the class to be assigned a specific temperature and bacterium. The inoculation assignment chart in this exercise provides

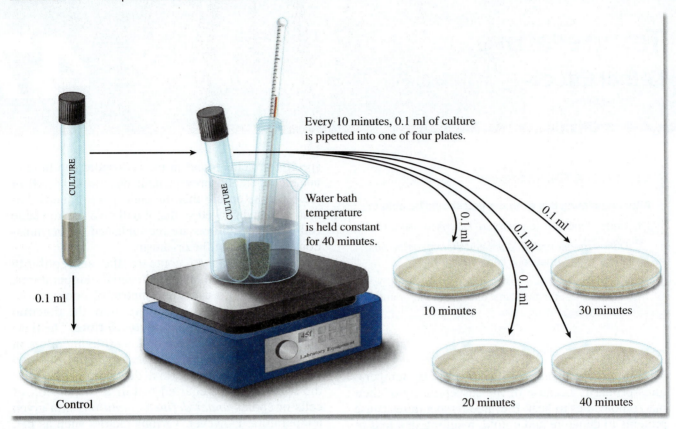

Figure 47.1 Procedure for determining thermal endurance.

suggested assignments by student numbers. After the plates have been incubated, each student's results will be tabulated on a Laboratory Report chart and shared with the entire class.

Although this experiment is not difficult to perform, inappropriate results can occur as a result of errors. Common errors are (1) omission of the control (no-heat) plate inoculation, (2) putting the thermometer in the culture tube instead of in a tube of sterile nutrient broth, and (3) not using fresh sterile pipettes when instructed to do so. Other sources of error include not using similar size test tubes and a temperature gradient between the portion of the tube immersed in the water bath and that portion above the water level.

Materials

per student:
- 5 petri plates
- 5 pipettes (1 ml size)
- 1 tube of nutrient broth
- 1 bottle of nutrient agar (60 ml)
- 1 culture of organisms

class equipment:
- water baths set up at 60°, 70°, 80°, 90°, and 100°C

broth cultures:
- *Staphylococcus aureus, Escherichia coli,* and *Bacillus megaterium* (minimum of 5 cultures of each species per lab section)

1. Consult the chart on the next page to determine what organism and temperature have been assigned to you. Several thermostatically controlled water baths will be provided.

    If your temperature is 100°C, use a hot plate to prepare a beaker of boiling water. When setting up a water bath, use hot tap water to start with to save heating time.
2. Liquefy a bottle containing 60 ml of nutrient agar in a boiling water bath, steamer, etc. When it is completely liquefied, place it in a 50°C water bath to cool.
3. Label 5 petri plates: **control (no heat), 10 min, 20 min, 30 min,** and **40 min.**
4. Shake the culture of organisms and transfer 0.1 ml of culture with a 1 ml pipette to the control (no heat) plate.
5. Place the culture and a tube of sterile nutrient broth into the water bath. Remove the cap from

Chart for Inoculation Assignments

ORGANISM	STUDENT NUMBER				
	60°C	70°C	80°C	90°C	100°C
Staphylococcus aureus	1, 16	4, 19	7, 22	10, 25	13, 28
Escherichia coli	2, 17	5, 20	8, 23	11, 26	14, 29
Bacillus megaterium	3, 18	6, 21	9, 24	12, 27	15, 30

the tube of nutrient broth and insert a thermometer into the tube. *Do not insert the thermometer into the tube containing the test organisms!*

6. As soon as the temperature of the nutrient broth reaches the desired temperature, record the time here: _____.
Watch the temperature carefully to make sure it does not vary appreciably.

7. After 10 minutes have elapsed, remove the culture from the water bath and mix throughly. Using a fresh pipette, quickly transfer 0.1 ml from the culture to the 10-minute plate and immediately return the culture tube to the water bath. Repeat this procedure at 10-minute intervals until all the plates have been inoculated. *Use fresh pipettes each time and be sure to resuspend the culture before each delivery.*

8. Pour liquefied nutrient agar (50°C) into each plate, rotate, and cool.

9. Incubate at 37°C for 24 to 48 hours. After evaluating your plates, record your results in the chart in Laboratory Report 47 and in the chart in the demonstration table.

Laboratory Report

Complete Laboratory Report 47 once you have all class results.

47 Temperature: Lethal Effects

A. Results

Examine your five petri plates, looking for evidence of growth. Record your results in the table below by indicating the presence or absence of growth as (+) or (−). When all members of the class have recorded their results, complete the entire chart.

ORGANISM	60°C					70°C					80°C					90°C					100°C				
	C*	10	20	30	40	C*	10	20	30	40	C*	10	20	30	40	C*	10	20	30	40	C*	10	20	30	40
S. aureus																									
E. coli																									
B. megaterium																									

*control (no-heat) tubes

1. If they can be determined from the above information, record the **thermal death point** for each of the organisms.

   S. aureus: _____ E. coli: _____ B. megaterium: _____

2. From the table shown above, determine the thermal death time for each organism at the tabulated temperatures and *record this information in the table below.*

ORGANISM	THERMAL DEATH TIME				
	60°C	70°C	80°C	90°C	100°C
S. aureus					
E. coli					
B. megaterium					

B. Short-Answer Questions

1. What is the importance of inoculating a control plate in this experiment?

   _____

   _____

2. To measure the culture temperature, why is the thermometer placed in a tube separate from the culture?

   _____

   _____

3. *Bacillus megaterium* has a high thermal death point and a long thermal death time, but it is not classified as a thermophile. Explain.

    _____

    _____

4. Give three reasons why endospores are much more resistant to heat than are vegetative cells.

    a. _____

    b. _____

    c. _____

5. List four diseases caused by spore-forming bacteria.

    a. _____

    b. _____

    c. _____

    d. _____

6. Would heating the culture in a sealed, small-diameter tube that is totally immersed in a water bath produce more accurate results than the use of tubes that are partially submerged in a water bath? Give reasons for your answer.

    _____

    _____

7. Give two reasons why this experiment can fail to give the appropriate results.

    _____

    _____

Microbial Spoilage of Canned Food

Learning Outcomes

After completing this exercise, you should be able to

1. Define how endospore-forming bacteria are involved in the spoilage of canned foods.

2. Explain the types of spoilage that occur in canned foods.

3. Characterize the bacteria in spoiled canned food using the Gram and endospore stains.

Spoilage of inadequately heat-processed, commercially canned foods is confined almost entirely to the action of bacteria that produce heat-resistant endospores. Two genera of endospore-forming bacteria, *Clostridium* and *Bacillus*, are of major concern in canned foods. *Clostridium* species are anaerobic, whereas *Bacillus* is aerobic to facultative anaerobic. Within both genera are strains that are mesophilic or thermophilic, and the heat resistance of their endospores can vary. Mesophilic anaerobic spore-formers, including *Clostridium* species, may be proteolytic or nonproteolytic. The proteolytic strains degrade proteins to produce hydrogen sulfide and gas. Nonproteolytic strains of *Clostridium botulinum* are an insidious problem because they can grow and produce a toxin in canned food without the typical signs of spoilage such as gas production, off-flavors, or hydrogen sulfide production. Canning of foods normally involves heat exposure for long periods of time at temperatures that are adequate to kill bacterial endospores. Particular concern is given to the processing of low-acid foods in which *Clostridium botulinum* can thrive to produce botulism toxin, and thereby cause botulism food poisoning.

Spoilage occurs when the heat processing fails to meet accepted standards. This can occur for several reasons: (1) lack of knowledge on the part of the processor (this is often the case in home canning); (2) materials to be canned may have high levels of associated bacteria that are not killed by the heat processing; (3) equipment malfunction that results in undetected underprocessing; and (4) defective containers that permit contamination after the heat process.

Our concern here will be with the most common types of food spoilage caused by heat-resistant, spore-forming bacteria. There are three types: flat sour, thermophilic anaerobe (T.A.) spoilage, and stinker spoilage.

Flat sour pertains to spoilage in which acids are formed with no gas production; the result is sour food in cans that have flat ends. **T.A. spoilage** is caused by thermophilic anaerobes that produce acid and gases (CO_2 and H_2, but not H_2S) in low-acid foods. Cans swell to various degrees, sometimes bursting. **Stinker spoilage** is due to spore-formers that produce hydrogen sulfide and blackening of the can and contents. Blackening is due to the reaction of H_2S with the iron in the can to form iron sulfide.

In this experiment, you will observe some of the morphological and physiological characteristics of organisms that cause canned food spoilage, including both aerobic and anaerobic endospore formers of *Bacillus* and *Clostridium*, as well as a non-spore-forming bacterium.

Working as a single group, the entire class will inoculate 10 cans of vegetables (corn and peas) with five different organisms. Figure 48.1 illustrates the procedure. Note that the cans will be sealed with solder after inoculation and incubated at different temperatures. After incubation the cans will be opened so that stained microscope slides can be made to determine Gram reaction and presence of endospores. Your instructor will assign individual students or groups of students to inoculate one or more of the 10 cans. One can of corn and one can of peas will be inoculated with each of the organisms. Proceed as follows:

🕐 First Period

(Inoculations)

Materials

- 5 small cans of corn
- 5 small cans of peas
- cultures of *Geobacillus stearothermophilus, B. coagulans, C. sporogenes, Thermoanaerobacterium thermosaccharolyticum,* and *E. coli*

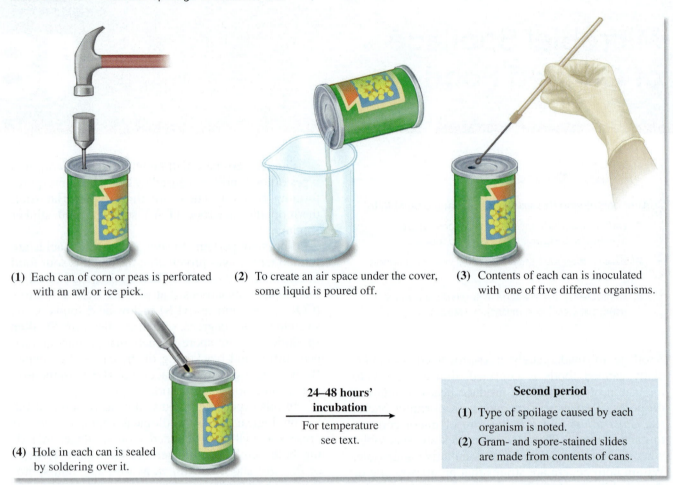

(1) Each can of corn or peas is perforated with an awl or ice pick.

(2) To create an air space under the cover, some liquid is poured off.

(3) Contents of each can is inoculated with one of five different organisms.

(4) Hole in each can is sealed by soldering over it.

24–48 hours' incubation

For temperature see text.

Second period

(1) Type of spoilage caused by each organism is noted.
(2) Gram- and spore-stained slides are made from contents of cans.

Figure 48.1 Canned food inoculation procedure.

- ice picks or awls
- hammer
- solder and soldering iron
- plastic bags
- Sharpies
- bleach solution (0.01%; 10 µl in 100 ml)
- Bunsen burner

1. Label the can or cans with the name of the organism that has been assigned to you. In addition, label one of the plastic bags to be used after sealing of the cans.
2. Pour the bleach solution onto the tops of all the cans. After 10 minutes, drain off the bleach solution and heat the top of each can over a Bunsen burner to remove excess moisture. Use an autoclaved ice pick or awl to punch a hole through a flat area in the top of the can. This can be done with the heel of your hand or a hammer if available.
3. Pour off about 5 ml of the liquid from the can to leave an air space under the lid.
4. Use an inoculating needle to inoculate each can of corn or peas with the organism indicated on the label.

5. Take the cans up to the demonstration table where the instructor will seal the hole with solder.
6. After sealing, place each can in two plastic bags. Each bag must be closed separately with rubber bands, and the outer bag must be labeled.
7. Incubation will be as follows until the next period:
   - **45°C; 72 hrs.**—*T. thermosaccharolyticum*
   - **37°C; 24–48 hrs.**—*C. sporogenes* and *B. coagulans*
   - **37°C; 24–48 hrs.**—*E. coli*
   - **55°C; 24–48 hrs.**—*G. stearothermophilus*

Note: If cans begin to swell during incubation, they should be placed in refrigerator.

⏱ Second Period

(Interpretation)

After incubation, place the cans under a hood to open them. The odors of some of the cans will be very strong due to H_2S production.

Materials

- can opener
- small plastic beakers
- Parafilm
- Gram staining kit
- spore-staining kit

1. Open each can carefully with a can opener. If the can is swollen, hold an inverted plastic funnel over the can during perforation to minimize the effects of any explosive release of contents.
2. Remove about 10 ml of the liquid through the opening, pouring it into a small plastic beaker. Cover with Parafilm. This fluid will be used for making stained slides.

3. Return the cans of food to the plastic bags, reclose them, and dispose in the biohazard bin.
4. Prepare Gram-stained and endospore-stained slides from your canned food extract as well as from the extracts of all the other cans. Examine under brightfield oil immersion.
5. Record your observations in the Results section, and collect data from the entire class.

Laboratory Report

Complete the first portion of Laboratory Report 48.

48 Microbial Spoilage of Canned Food

A. Results

1. Observations

   Record your observations of the effects of each organism on the cans of vegetables. Share results with other students.

ORGANISM	PEAS		CORN	
	Gas Production + or −	Odor	Gas Production + or −	Odor
E. coli				
B. coagulans				
G. stearothermophilus				
C. sporogenes				
T. thermosaccharolyticum				

2. Microscopy

   After making Gram-stained and spore-stained slides of organisms from the canned food extracts, sketch in representatives of each species:

		G. s.	G. s.	
		C. s.	C. s.	

E. coli   B. coagulans   G. stearothermophilus   C. sporogenes   T. thermosaccharolyticum

B. **Short-Answer Questions**

1. Which organisms, if any, caused flat sour spoilage? _____

   _____

2. Which organisms, if any, caused T.A. spoilage? _____

   _____

3. Which organisms, if any, caused stinker spoilage? _____

   _____

4. Does flat sour cause a health problem? _____

   _____

5. What types of foods are usually associated with botulism food poisoning? _____

   _____

   _____

6. Why is spoilage more likely to occur for individuals who do home canning than in a canning factory?

   _____

   _____

Microbiology of Alcohol Fermentation

Fermented beverages are as old as civilization. One of the earliest records of wine making was 7000 B.C. to 6600 B.C. in China. The practice then spread to other regions of the world and is documented in writings from Mesopotamia, Egypt, Phoenicia, and the Greek and Roman civilizations. The importance of wine in Greek and Roman cultures is evidenced by the fact that Dionysus was the Greek god of the grape harvest and wine making and his counterpart was the Roman god Bacchus. An Assyrian tablet even states that Moses took beer aboard the ark.

Fermentation is actually an ancient method to preserve foods. Buttermilk, cheese, sauerkraut, pickles, summer sausage, and yogurt are all produced by microbial fermentations. Bacteria, yeast, and various fungi are used in the production of fermented foods. Fermentation end products such as acids (e.g., lactic, propionic, acetic) and alcohols are produced which inhibit the growth of spoilage microorganisms, thus preserving the food. Fermentation is still a means to produce and preserve some foods.

The starting point for the production of wine is the preparation of **must,** which is the juice produced by crushing the grapes. A yeast, *Saccharomyces cerevisiae,* variety ellopsoideus, is usually added, which then ferments the grape sugars to produce wine which reaches a maximum alcohol content of 17%. Greater concentrations of alcohol will inhibit the growth of the yeast. Although we usually associate wine with the fermentation of grape juice, it may also be made from other starting materials such as berries, dandelions, and rhubarb. The basic conditions necessary for the production of wine are sugar, yeast, and anaerobic conditions. The production of alcohol is shown in the following reaction:

$$C_6H_{12}O_6 \xrightarrow{\text{yeast}} 2C_2H_5OH + 2CO_2$$

Figure 49.1 Alcohol fermentation setup.

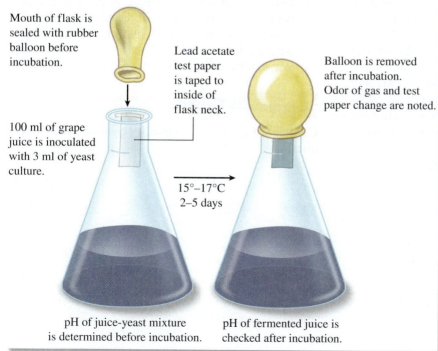

Mouth of flask is sealed with rubber balloon before incubation.

Lead acetate test paper is taped to inside of flask neck.

100 ml of grape juice is inoculated with 3 ml of yeast culture.

15°–17°C
2–5 days

Balloon is removed after incubation. Odor of gas and test paper change are noted.

pH of juice-yeast mixture is determined before incubation.

pH of fermented juice is checked after incubation.

Commercially, wine may be produced as a red, rosé, or white wine. Red wine is produced by allowing the skins and seeds to remain during the initial fermentation process. Components are extracted from the skins and seeds that confer the characteristic color of a red wine. Red wines are fermented at a temperature of 24°C (75°F). Rosé wines are produced by only allowing the skins and seeds of a red grape to remain in the primary fermentation process for 1 to 3 days, after which they are removed. White wines can be made from either red or white grapes, but the skins and seeds are removed before fermentation begins. Fermentation of white wines is carried out at 13°C (55°F).

A serious problem in wine making is the formation of hydrogen sulfide (H_2S). This compound can confer the smell of rotten eggs on the wine. Hydrogen sulfide can result from different sources. A primary source is the yeast used in the fermentation. *Saccharomyces cerevisiae* can synthesize the sulfur-containing amino acids cysteine and methionine by reducing sulfate in the environment to sulfide, which is then incorporated into these amino acids in several enzymatic steps. Excess sulfide that is not utilized by the yeast for the synthesis of the amino acids is converted to hydrogen sulfide, which can contaminate the wine, causing unwanted off-flavors and odors. Wine producers have developed various methods and strategies to prevent the formation of hydrogen sulfide and thus avoid these problems.

In this exercise, you will set up a grape juice fermentation experiment to learn some of the characteristics about the fermentation of sugar to alcohol. To provide the necessary anaerobic conditions, a balloon will be placed over the mouth of the fermentation flask (figure 49.1). The balloon will also trap any gases produced during the fermentation. To detect if any hydrogen sulfide is produced by *Saccharomyces cerevisiae* during fermentation, a lead acetate test paper will be taped to the inside of the flask. Sulfide will cause the test paper to turn black. The pH of the sugar substrate will also be monitored before and after the reaction to determine if any changes have occurred.

First Period

Materials

- 100 ml grape juice (no preservative)
- bottle of juice culture of wine yeast
- 125 ml Erlenmeyer flask
- one 10 ml pipette
- balloon
- hydrogen sulfide (lead acetate) test paper
- tape
- pH meter

1. Label an Erlenmeyer flask with your initials and date.
2. Add about 100 ml of grape juice to the flask.
3. Determine the pH of the juice with a pH meter and record the pH in Laboratory Report 49.
4. Agitate the container of yeast and juice to suspend the culture, remove 5 ml with a pipette, and add it to the flask.
5. Attach a short strip of tape to a piece of lead acetate test paper (3 cm long), and attach it to the inside surface of the neck of the flask. Make certain that neither the tape nor the test paper protrudes from the flask.
6. Cover the flask opening with a balloon.
7. Incubate at 15–17°C for 2 to 5 days.

Second Period

Material

- pH meter

1. Remove the balloon and note the aroma of the flask contents. Describe the odor in Laboratory Report 49.
2. Determine the pH and record it in the Laboratory Report.
3. Record any change in color of the lead acetate test paper in the Laboratory Report. If any H_2S is produced, the paper will darken due to the formation of lead sulfide as hydrogen sulfide reacts with the lead acetate.
4. Wash out the flask and return it to the drain rack.

Laboratory Report

Complete Laboratory Report 49 by answering all the questions.

Laboratory Report

49 Microbiology of Alcohol Fermentation

A. Results

Record your observations of the fermented product:

Aroma: _____

pH: _____

H₂S production: _____

B. Short-Answer Questions

1. What compound in the grape juice is fermented? What other major product is produced during fermentation?

    _____

    _____

2. What is the purpose of sealing the flask with a balloon? What product is captured in the balloon?

    _____

    _____

3. What group of microorganisms is responsible for producing fermented beverages? What microorganism was used in this exercise?

    _____

    _____

4. Why is hydrogen sulfide production monitored during fermentation? Why is this end product undesirable in wine production?

    _____

    _____

5. What process in the microorganism used for wine production contributes to the formation of hydrogen sulfide?

    _____

    _____

6. Fermentation is a means of food preservation. What end products produced in fermentation are antagonistic or inhibitory to spoilage organisms?

    _____

    _____

7. What other foods are produced by microbial fermentations? What organisms are involved?

    _____

    _____

8. What is the must in wine production? Why is the final concentration of alcohol limited to 17% for wine?

    _____

    _____

Bacterial Genetics and Biotechnology

The genetic information encoded in the DNA molecules that make up the genome of the bacterial cell determines its metabolic capabilities and its cellular structure, and thus defines the cell. The information is encoded primarily in the chromosomal DNA, but information is also found in small DNA molecules called plasmids, which can replicate and be transferred between cells independent of the chromosomal DNA. The information in chromosomal DNA and plasmids is arranged into genetic units called genes, which primarily encode for the various enzymes that catalyze the metabolic reactions in a cell and confer upon the cell its unique characteristics that make it different from other cells. The information in nuclear DNA is required for any cell to survive because it defines the cell. However, plasmid DNA supplies accessory information that a bacterial cell might need under special circumstances, and its presence in the cell is not essential for viability. Plasmid DNA can encode for characteristics such as antibiotic resistance, specialized metabolic functions, enhanced virulence, or the production of toxins. Such functions can aid a cell in times when the specialized characteristics are advantageous to the cell but are not essential at other times.

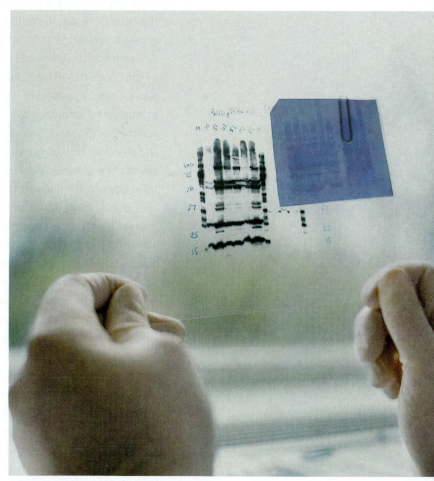

© Getty Images RF

It is crucial that the information in the DNA be replicated accurately when cells divide to ensure that the progeny have the same genotype as the cells from which they are derived. If changes occur in nucleotide bases during replication because of insertion or deletion of nucelotide bases, this results in changes in the sequence of nucleotide bases, or mutations, which are usually damaging to the cell. Mutations also result when nucleotide bases are altered by radiation and mutagenic chemicals, such as ultraviolet light and nitrosamines, respectively. Because bacteria only carry a single copy of a particular gene, any mutation results in immediate changes to the cell. Not every mutation is detrimental, however, and mutation is one way that the genetic information of a bacterial cell can be permanently altered to change the phenotypic characteristics of cells.

The genotype of a cell can also be changed by the reassortment of genes that occurs during the exchange of genetic material by sexual mechanisms. Bacteria are asexual, dividing by binary fission, but they have evolved several horizontal mechanisms for transferring and exchanging DNA. These include transformation, or the uptake of "naked" DNA; conjugation, involving DNA exchange between two living cells; and transduction, in which genes are transferred by a virus. When genetic information is exchanged, it can replace existing DNA sequences in a cell. If the changes confer a selective advantage to the cell and its offspring, they will become a permanent part of the cell's genotype. For example, the new DNA could replace DNA with mutations. The ultimate purpose of genetic exchange is to introduce variability into the gene pool.

In bacteria, variability can also be quickly introduced by the transfer of plasmid DNA. The number of genes carried on a plasmid varies from a few genes to several hundred. Plasmids are exchanged between bacteria primarily by conjugation. As indicated, plasmids encode for specialized functions nonessential to the cell. For example, in a hospital environment, *Staphylococcus aureus* can exchange penicillinase plasmids that confer resistance to penicillin, resulting in resistant strains that can cause serious infections in pediatric and surgical wards.

The knowledge that we have gained from the study of bacterial genetics and extrachromosomal elements such as plasmids has provided new and exciting approaches in molecular biology and generated the new field of biotechnology. Using recombinant DNA technology, we can now introduce "foreign" DNA into bacterial plasmids and cause the expression of desirable products in bacteria. For example, we no longer have to isolate insulin from the pancreases of slaughtered animals to treat diabetics because the genes for human insulin have been genetically engineered into *Escherichia coli,* and strains of this bacterium now produce human insulin or humulin. This is only one example of how pharmaceuticals, such as human growth hormone and interferon, have been produced by genetically engineering bacterial cells for our purposes.

In the following section, you will perform exercises that involve some genetics and molecular biology techniques. These include confirming the presence of a particular gene in a plasmid by PCR amplification, agarose gel electrophoresis, and using transformation to change the genetic makeup of bacterial cells. These techniques are indispensable and form the basis for many of the procedures used in genetic engineering and modern biotechnology.

Polymerase Chain Reaction for Amplifying DNA

Learning Outcomes

After completing this exercise, you should be able to

1. Use the polymerase chain reaction (PCR) to amplify a gene found on the pGLO plasmid for further analysis.

2. Use agarose gel electrophoresis to visualize and confirm the gene amplification and therefore presence on the pGLO plasmid.

Verification of Gene in a Recombinant Plasmid

The recombinant plasmid pGLO used in this exercise contains a gene of interest known as the GFP gene (figure 50.1). This gene was isolated from a marine jellyfish and introduced into a bacterial plasmid using recombinant DNA techniques. The GFP gene codes for a *green fluorescent protein* that is visible when expressed by bacterial colonies that possess the plasmid.

In this exercise, the presence of the GFP gene on the pGLO plasmid will be verified by two molecular methods: amplification of this target gene via the polymerase chain reaction (PCR) and visualization of the amplified DNA using agarose gel electrophoresis. These are both widely used techniques, critical to the field of biotechnology as well as many others. On the first day, students should set up the polymerase chain reaction (PCR). On the second day, students run the PCR product on an agarose gel to visualize the amplified DNA.

Part 1: Polymerase Chain Reaction (PCR)

One of the most important developments of the past 30 years in molecular biology is the polymerase chain reaction invented by Kary Mullins in the 1980s. The ability to rapidly and easily make billions of copies of almost any piece of DNA revolutionized molecular biology and made the new field of genomics possible. This relatively simple technique is used for many important purposes, including microbiology and cellular biology research, infectious disease diagnosis, and even forensics. Figure 50.2 illustrates how PCR

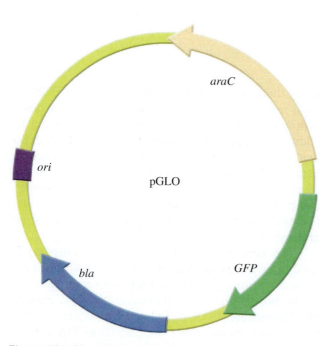

Figure 50.1 **The pGLO plasmid.**

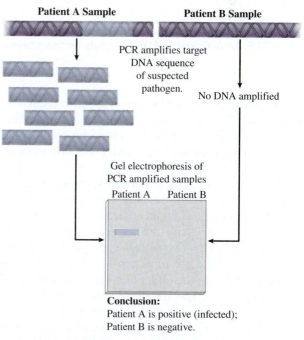

Figure 50.2 **The polymerase chain reaction (PCR) can be reliably used for diagnosis of infectious disease.**

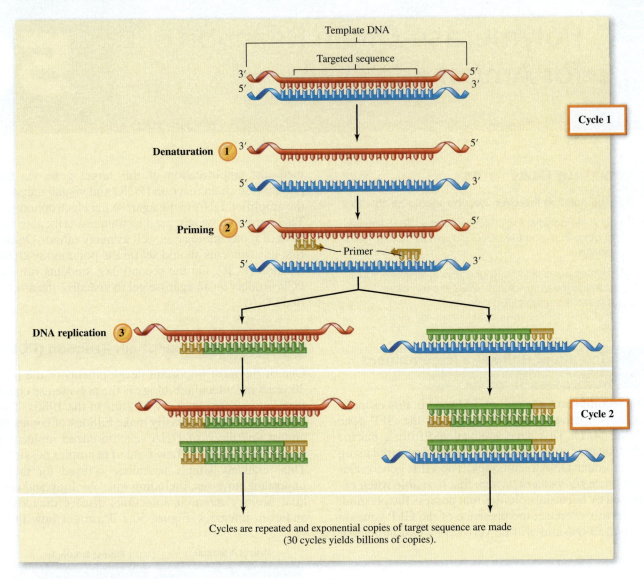

Figure 50.3 Amplification of DNA by polymerase chain reaction.

amplification can be used to detect specific sequences of bacterial DNA in patient samples to confirm infection. In forensics, the fact that only one copy of the template DNA is needed to amplify and obtain many copies means that one hair follicle or even one drop of dried blood is enough for a crime scene investigator to obtain the necessary genetic information via PCR.

The polymerase chain reaction, as shown in figure 50.3, involves the attachment of a pair of short complementary DNA primers to specific sites on template DNA and the use of a thermostable DNA polymerase to replicate the target DNA sequence between these two primers. Specifically, the technique involves denaturation, or separation of the two strands, of template DNA at 95°C, followed by annealing the primers at the appropriate temperature, and DNA replication at 72°C, the optimal temperature for the activity of

many thermostable DNA polymerases. The PCR cycle is repeated 30 to 40 times and, because each newly synthesized DNA molecule serves as a template for additional rounds of replication, this results in exponential amplification of the specific DNA sequence between the two primers. PCR is normally performed in a computer-controlled apparatus called a thermocycler, which is programmed to ensure that each step is conducted at the correct temperature for the correct amount of time.

The first thermostable DNA polymerase to be used in PCR was the *Taq* enzyme, which was isolated from *Thermus aquaticus*, a bacterium found to grow in the hot springs of Yellowstone National Park by Dr. Thomas D. Brock of Indiana University in 1965. Since then, many other thermostable DNA polymerases have been isolated from other thermophilic organisms and used for amplification of DNA in the lab. The heat-stable property of *Taq* and other thermostable polymerases is essential for *in vitro* amplification of DNA because they can withstand the high temperature (95°C) that is necessary to denature the DNA template between replication cycles.

When designing a PCR experiment, several parameters need to be considered: quantity of template DNA, quality of the primers, annealing temperature of primers to the template, and the time needed to replicate the DNA of interest. Of these considerations, proper design of the primers is the most important factor in achieving success in PCR. Each primer should hybridize to only one specific site on the template DNA, flanking both ends of the target DNA sequence. Most primers are designed using computer programs that take all of these points into account. In this exercise, primers will be used to amplify a specific portion of the GFP gene to confirm its presence in the pGLO plasmid. The length (kb) of the amplified target sequence is based on the exact primers chosen by your instructor.

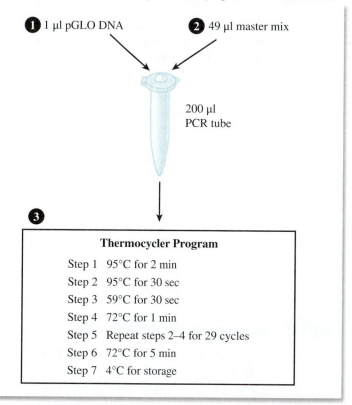

Figure 50.4 **Amplification of portion of GFP gene on pGLO plasmid.**

Materials

- thermocycler
- pGLO DNA (0.01 ng/μl)
- master mix (*Taq* polymerase, 10X buffer for *Taq*, primer 1, primer 2, and dNTP mix)
- thin-walled PCR tubes (200 μl volume)
- pipetters (P-20 and P-200) and pipette tips (sterile)
- gloves
- ice and a beaker to hold ice

Note: For PCR analysis, wear gloves throughout the experiment to avoid contaminating samples with bacteria from your hands or introducing nucleases from your hands that will degrade the template DNA or inactivate the polymerase enzyme.

Proceed through the following steps, referring to the flowchart in figure 50.4 as you work.

1. Prepare one thin-walled 200 μl PCR tube by labeling it with your group number and placing it on ice.
2. Using a pipetter (P-20), add 1 μl of purified pGLO DNA (0.01 ng/μl) to the PCR tube.
3. Use a pipetter (P-200) to add 49 μl of the PCR master mix to the PCR tube and return to the ice beaker (figure 50.5). This master mix contains several important components for the reaction. It includes the specific primers required to amplify the target sequence in the template DNA, the individual nucleotides that are used to build the new DNA strands, and the thermostable enzyme that catalyzes the building process.

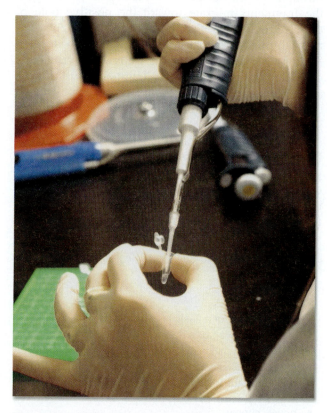

Figure 50.5 **Addition of master mix to a PCR tube.**
© McGraw-Hill Education/Auburn University Photographics Services

4. Place the PCR tube in a thermocycler and start the program according to the steps that follow or as indicated by your instructor.

Program:

Step 1:   95°C for 2 minutes
Step 2:   95°C for 30 seconds to denature
Step 3:   59°C for 30 seconds to anneal primers
Step 4:   72°C for 1 minute to replicate DNA
Step 5:   Repeat steps 2–4 for 29 additional cycles
Step 6:   72°C for 5 minutes to complete extensions
Step 7:   4°C for storage until ready to end the run

5. Once the PCR run is complete, store PCR tube at 4°C until you are ready to start part 2.

Part 2: Agarose Gel Electrophoresis

DNA can be separated based on fragment length using agarose or polyacrylamide gels. Agarose gels are used for separation of DNA fragments that are larger in size (more than 100 nucleotides). For gel electrophoresis, DNA samples are loaded into wells at one end of an agarose gel, and then an electrical current is applied causing the DNA fragments to migrate through the gel. Because DNA is negatively charged, the fragments migrate toward the positive pole of the electrophoresis chamber. Shorter fragments will move through the gel more quickly and will be located farther away from the loading wells when visualized after electrophoresis. Longer fragments will move a shorter distance through the gel and will be located closer to the wells after electrophoresis. Nucleic acids can be visualized on the gel by using fluorescent stains and looking at the completed gels under UV light. In this exercise, students will run an agarose gel to visualize the DNA molecules from their PCR experiment in part 1. This will provide confirmation that the GFP gene was present in the original pGLO DNA sample and was amplified through PCR.

Materials

- agarose
- 1× TBE
- 200 ml bottle or a flask
- microwave oven or a boiling water bath
- oven mitt
- GelRed™ nucleic acid gel stain
- gel apparatus (gel tray, a comb with 8-10 wells, gel chamber)
- power supply
- 1.7 ml microcentrifuge tubes

- 10× DNA sample loading dye
- P-20 pipetter and pipette tips
- 1 kb DNA ladder (Invitrogen, CA, or equivalent)
- UV safety glasses
- UV transilluminator
- photodocument system for UV (optional)
- gloves

Refer to the flowchart in figure 50.6 to complete the steps of the gel electrophoresis process.

1. Add 0.6 g of agarose in 50 ml of 1× TBE, pH 8.0 (Tris-Borate-EDTA buffer) in a 200 ml bottle to make a solution of 1.2% agarose (w/v). Bottle cap should be loose.

2. Place the bottle in a microwave oven and heat at 50% power for 5 minutes to melt the agarose. If the agarose is not fully melted, continue heating for 2–5 more minutes until it is melted, and allow the agarose to cool on the bench top until it is comfortable to touch (~55°C).

Caution

Be careful not to shake the hot liquid because it can boil over and cause severe injury! Wear an oven mitt or equivalent when removing agarose from the microwave oven.

3. Add 5 µl of GelRed™ nucleic acid gel stain to the melted agarose.

4. Assemble an agarose gel tray according to your instructor's directions.

5. When the agarose has cooled to the touch, slowly pour the slurry into the gel tray, preassembled with a comb, until a gel of approximately 5 mm in thickness is obtained.

6. Allow the agarose gel to solidify at room temperature. As agarose solidifies, it will turn from clear to opaque gray in color.

7. While the gel is solidifying, prepare your PCR sample for gel electrophoresis by transferring 9 µl from your PCR tube to a labeled sterile microcentrifuge tube.

8. Add 1 µl of the 10× loading dye to this tube and mix.

9. Once the gel is solid, place the gel in the running chamber and cover the gel completely with 1× TBE running buffer (figure 50.7). Make sure that the side of the gel with the wells is placed toward the negative (black) end of the electrophoresis chamber.

10. Using a pipetter (P-20), carefully load 3 µl of the provided 1 kb DNA ladder (0.1 µg/µl) to the first well (figure 50.8). Change the pipette tip and load 10 µl of your prepared sample into the well assigned to you by your instructor. If you

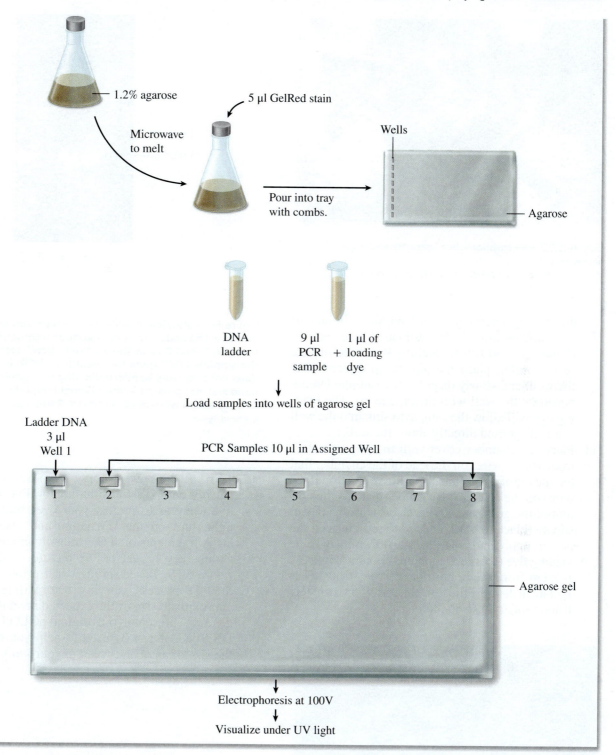

Figure 50.6   **Agarose gel electrophoresis.**

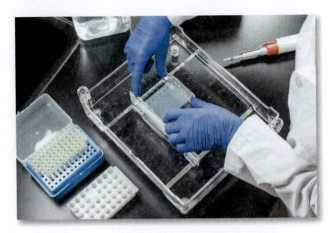

Figure 50.7 Electrophoresis apparatus and a gel assembly.
© McGraw-Hill Education. Area9/Julie Skotte, photographer

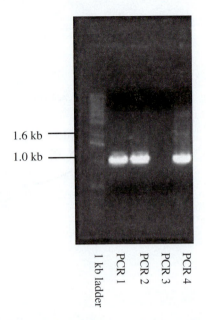

1.6 kb
1.0 kb

1 kb ladder
PCR 1
PCR 2
PCR 3
PCR 4

Figure 50.9 Possible results. Lane 1 represents the DNA ladder with bands showing at various fragment lengths. Lanes 2, 3, and 5 each show a bright band, representing the amplified GFP gene fragment from PCR. Lane 4 does not show any bands, indicating that either the GFP gene was not present in the original template pGLO DNA or errors were made in the PCR process.
© Harold Benson

are sharing a gel with other lab groups, be sure to keep a record of where you loaded your sample. (**Note:** To load DNA, carefully break the surface of the buffer, place the pipette tip directly over the well, and slowly dispense the sample. Do not penetrate the well with the pipette tip. The loading dye will allow the sample to sink into the well if it is dispensed directly above the well.)

11. Place the chamber cover with the negative electrode (black wire) on the side of the chamber that has the top of the gel and the positive electrode (red wire) on the side of the chamber that has the bottom of the gel. Connect the electrodes to a power supply (black to black and red to red), turn on the power supply, and run the gel at 100 V until the blue loading dye (bromophenol blue) is approximately 1–2 cm from the bottom of the gel. If the gel is 10 cm in length, the run time is approximately 1 hour under these conditions.

12. Turn off the power supply. Remove the gel and place it on a UV transilluminator. Put on protective eyewear to protect your eyes from the UV light, turn on the UV lamp, and visualize the DNA fragments on a UV transilluminator.

13. Compare your gel to the picture of the gel in figure 50.9. Based on the exact primers used in the PCR process, your instructor will tell you the approximate size of the expected band if the GFP gene was present in the original pGLO DNA. The DNA ladder can be used to help you determine the size of any bands seen in your sample lane.

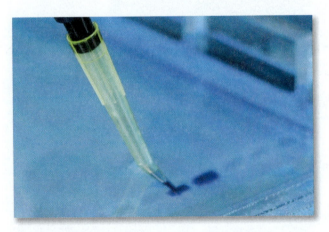

Figure 50.8 Adding DNA sample to a well of the agarose gel.
© McGraw-Hill Education. Area9/Julie Skotte, photographer

50 Polymerase Chain Reaction for Amplifying DNA

A. Results

1. Draw a picture of your gel in the space below, marking locations of all bands seen in each lane of the gel.

2. Did your PCR sample contain the expected amplified DNA fragment? If so, label the correct band in your drawing above.

    _____

3. If the band was present, what does this tell you about the original pGLO DNA sample used for PCR? If the band was not present, suggest some possible reasons for this.

    _____

    _____

    _____

B. Short-Answer Questions

1. For PCR, what property of DNA does temperature influence? Why do the "denaturation" and "annealing" steps proceed at different temperatures?

   _____

   _____

2. Before the development of thermocyclers, how do you think PCR cycling could have been accomplished? What would have been the drawbacks of such a procedure?

   _____

   _____

3. Why are the correct primer sequences essential for successful amplification?

   _____

   _____

4. Why is it important to wear gloves when setting up the PCR tubes?

   _____

   _____

5. Explain how PCR is used to detect the presence of pathogens in patient samples. Look up an example of where this technique is used routinely in diagnosis.

   _____

   _____

   _____

   _____

6. What causes the separation of DNA fragments during agarose gel electrophoresis?

   _____

   _____

7. What side of the electrophoresis chamber (red or black) should the wells be on when the gel is placed in the chamber? What would happen if you placed the agarose gel with the wells on the opposite side of the chamber?

   _____

   _____

Bacterial Transformation

Learning Outcomes

After completing this exercise, you should be able to

1. Understand important regions of the pGLO plasmid and gene expression in bacterial cells that possess this plasmid.

2. Introduce the pGLO plasmid to competent bacterial cells through transformation.

3. Test for the presence of the plasmid by growing cells on a medium containing ampicillin.

4. Observe the expression of the green fluorescent protein (GFP) gene when cells are grown on a medium enriched with arabinose.

Horizontal Gene Transfer Mechanisms in Bacteria

Acquisition of genetic material, in addition to mutation of genes and loss of unnecessary genes, is an important part of bacterial adaptation and evolution. In many cases, acquisition of new genes (i.e., antibiotic resistance) may mean the difference between survival and death of an organism in the environment. Thus, bacteria have developed various mechanisms, including transformation, transduction, and conjugation, to acquire external genetic information. **Bacterial transformation** refers to the uptake of naked DNA molecules by bacterial cells. Transduction involves transfer of genetic material carried by bacteriophages, or bacterial viruses. Conjugation involves transfer of genetic material from one living bacterium to another using an external appendage known as a pilus.

Bacterial transformation was first discovered in the pathogenic bacterium *Streptococcus pneumoniae*. When non-capsule-forming—and thus non-pathogenic—*S. pneumoniae* were exposed to killed capsule-forming pathogenic cells, the nonpathogenic cells became virulent because they acquired the genes for capsule formation from the dead virulent cells. This ability of a bacterium to take up naked DNA molecules is termed "competence." Some bacteria, including *S. pneumoniae* and *Bacillus subtilis*, are naturally competent and can take up naked DNA molecules from the environment. Many other bacteria, including *Escherichia coli*, are not naturally

competent, but they can be made competent via laboratory manipulation.

Recombinant DNA Technology

The ability to manipulate *E. coli* to be competent, and hence take up naked DNA, revolutionized the field of biotechnology and facilitated the cloning of human genes into *E. coli* for important research and medical purposes (figure 51.1). In order to transform bacteria that express specifically desired characteristics,

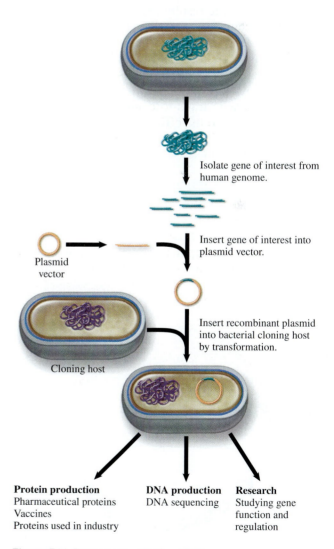

Isolate gene of interest from human genome.

Plasmid vector

Insert gene of interest into plasmid vector.

Cloning host

Insert recombinant plasmid into bacterial cloning host by transformation.

Protein production
Pharmaceutical proteins
Vaccines
Proteins used in industry

DNA production
DNA sequencing

Research
Studying gene function and regulation

Figure 51.1 **Process for cloning DNA.**

a vector is used to carry the gene(s) of interest into the competent cells. Plasmids make excellent cloning vectors for several reasons. They can replicate independent of the bacterial chromosome in the cell, and they are often present in multiple copies in one cell. Plasmids also often carry a gene, such as an antibiotic resistance gene, which can be used as a selectable marker when attempting to identify cells that have been transformed by the uptake of the plasmid.

In 1978, the City of Hope National Medical Center and Genentech, Inc. in California announced that they had successfully produced insulin using these recombinant DNA techniques. They isolated the human insulin gene, inserted it into a bacterial plasmid, and transformed bacteria with this recombinant plasmid in the laboratory. The transformed bacterial cells reproduced, and these clones were able to express the human insulin gene, producing insulin protein in large quantities. This insulin product was harvested and purified for therapeutic use in humans. In 1982, the FDA approved the first recombinant insulin drug, humulin, improving access to insulin treatment for diabetics.

In this exercise, you will replicate this general process used to create therapeutic drugs through recombinant DNA technology. You will attempt to transform cells with the **pGLO plasmid** (figure 51.2), a recombinant plasmid that will introduce two new observable genetic traits to competent *E. coli* cells. In order to improve the efficiency of transformation, cells are treated with calcium chloride and subjected to "heat shock" to increase competence.

The pGLO Plasmid

The pGLO plasmid (figure 51.2) has several important DNA segments. First of all, the **ori** site allows the plasmid to be replicated by DNA polymerase. The ***bla*** gene, β-lactamase gene, codes for an enzyme

that hydrolyzes antibiotics such as ampicillin. Bacterial cells that possess this plasmid should always be ampicillin-resistant because this gene is always expressed. The **GFP** gene is a "foreign" gene from a jellyfish (*Aequorea victoria*) that was genetically engineered into this plasmid. This gene codes for a green fluorescent protein, but its expression is controlled by an arabinose promoter, so bacterial cells that possess the pGLO plasmid will not always produce this protein. The ***araC*** gene on this plasmid codes for a protein that blocks this promoter if arabinose is not present in the environment of the cell. If cells possessing the plasmid are grown on a medium that does not contain arabinose, they cannot express the GFP gene. However, if arabinose is present, it binds to the *araC* protein, a conformational change occurs, and RNA polymerase can now bind to the promoter. Transcription and translation of the GFP gene can occur, and the cells will express the green fluorescent protein, which is visible under UV light.

Transformation of *E. coli* with pGLO

Refer to the flowchart in figure 51.3 as you work through the following steps of the transformation procedure.

Materials

- pGLO plasmid DNA
- plate culture of *E. coli* HB101
- ice in a large beaker
- water bath at 42°C
- transformation (50 mM CaCl$_2$, pH 6.1) solution
- LB broth
- 1 of each agar plate: LB, LB/amp, LB/amp/ara
- inoculating loop or glass spreader
- 2 microcentrifuge tubes (sterile)
- micropipetters (P-20 and P-200) and sterile tips
- sterile 1 ml pipettes and a pipette aid
- UV light source

1. Label two sterile microcentrifuge tubes as +pGLO and −pGLO, and place the tubes on ice.
2. Using a sterile 1 ml pipette, transfer 0.25 ml of transformation solution into each tube and return tubes to ice.
3. Use a sterile inoculating loop to transfer one colony of *E. coli* into each of the tubes.
4. Use a P-20 micropipetter (or a sterile inoculating loop) to add 10 μl of plasmid DNA to the +pGLO tube only. *Do not add this to the −pGLO tube.*
5. Incubate on ice for 10 minutes.
6. Transfer both test tubes to a 42°C water bath for 50 seconds. Immediately return both tubes to ice for 2 minutes.

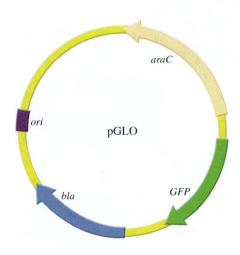

Figure 51.2 **The pGLO plasmid.**

araC

ori

pGLO

bla

GFP

7. Add 0.25 ml of LB broth to both tubes and incubate at room temperature for 10 minutes. This step is necessary for the plasmid to express the antibiotic resistance gene (ampicillin resistance, in this case).

8. During incubation, label the three agar plates as shown here.

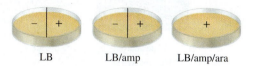

LB   LB/amp   LB/amp/ara

9. After 10 minutes of incubation, use a micro-pipetter to add samples to the agar surface of each labeled plate. First, add 100 µl of −pGLO to the − sides of the LB and LB/amp plates. Then, add 100 µl of +pGLO to the + sides of these plates. Finally, add 200 µl of +pGLO to the LB/amp/ara plate.

10. Use a sterile loop or glass spreader to spread the samples across their respectively labeled areas of each plate. Be sure to sterilize the spreading tool between samples.

11. Incubate the plates at 37°C.

12. Before looking at your incubated plates, make predictions on the type of growth expected for each area of the three plates by completing the table at the top of Laboratory Report 51.

13. Observe the incubated plates using a UV light source, and record the actual results in the Laboratory Report.

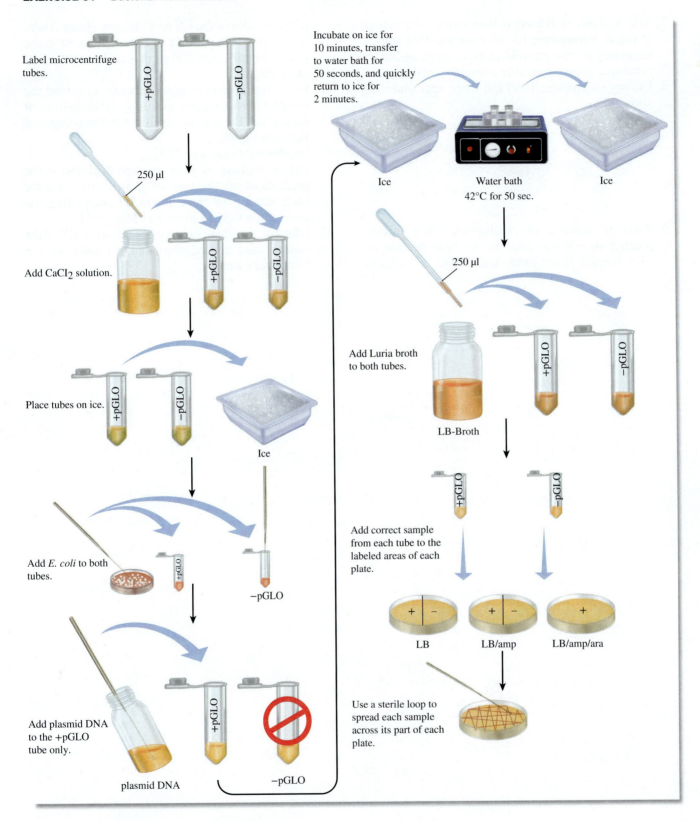

Figure 51.3 Steps in the transformation of *E. coli*.

Source: Adapted from Bio-Rad Laboratories, Inc. www.bio-rad.com/webroot/web/pdf/lse/literature/Bulletin_3052.pdf

51 Bacterial Transformation

A. Predictions

Draw the growth you expect to observe in each area of the three agar plates after incubation. Be sure to indicate specifics such as expected amount of growth and the presence or absence of green fluorescent protein.

LB

LB/amp

LB/amp/ara

B. Results

1. Observing the actual growth on your incubated plates, complete the table below. Describe the growth as a heavy lawn, isolated colonies, or absent. Indicate the presence or absence of GFP by entering yes or no.

	GROWTH	GFP
−pGLO on LB		
+pGLO on LB		
−pGLO on LB/amp		
+pGLO on LB/amp		
+pGLO on LB/amp/ara		

2. Did your results match your predictions? If not, explain why this may have occurred.

_____

_____

_____

C. Short-Answer Questions

1. What was the purpose of the LB plate in this exercise?

_____

_____

2. Which plate(s) contain transformed bacteria, and how do you know? Be specific about which side of each plate.

_____

_____

_____

3. Explain why there are differences in the amount of growth observed on the three plates.

_____

_____

_____

4. The GFP gene was expressed on which plate(s)? Explain.

_____

_____

5. What was the importance of the LB/amp plate in this exercise?

_____

_____

6. If growth had been present on the –pGLO side of the LB/amp plate, what might have been a reason for this unexpected result?

_____

_____

_____

7. Explain the purpose of the antibiotic resistance gene in this experiment. Why is this genetic trait an important part of the recombinant DNA technology process in the biotechnology industry?

_____

_____

_____

8. What are the advantages of using organisms such as bacteria and yeast as cloning hosts? What safety considerations must be made when choosing a cloning host to make proteins for human use?

_____

_____

_____

Medical Microbiology

Medical microbiology is the study of microorganisms which affect human health. Microorganisms can affect human health either positively or negatively. Although the microbiology laboratory is primarily focused on the negative ways microorganisms affect human health, it must also be appreciated that microorganisms are vital for human existence. The human body plays host to trillions of microbes that make up its own unique microbiota (microscopic life). These microbes help digest carbohydrates, make vitamins, compete for space so that pathogens cannot invade, and aid the immune system.

The microbial population within the human body is in a continual state of flux and is affected by a variety of factors (health,

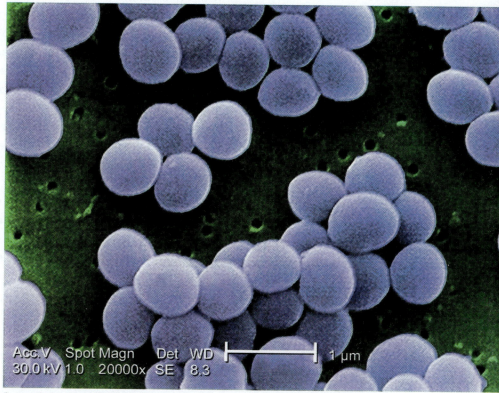

Source: Janice Haney/Centers for Disease Control

age, diet, hygiene). When the human body is exposed to an outside organism, three results are possible: (1) the organism can pass through the body, (2) the organism can colonize the body (either transiently or permanently) and become part of the microbiota, or (3) the organism can produce disease. Organisms that colonize humans don't interfere with normal body functions and may have a positive effect on the body (see above). In contrast, organisms that cause disease produce damage to the human host. It is important to differentiate between colonization and disease.

Disease-causing organisms can be divided into opportunistic pathogens and true pathogens. Opportunistic pathogens do not produce disease in their normal environment but can cause disease when they are given access to unprotected sites of the body or if a patient's immune system is not functioning properly. Opportunistic pathogens are usually typical members of the human microbiota. In Exercises 52 and 53, you will isolate and enrich organisms from within your own unique microbiota. True pathogens are always associated with human disease. Exercise 54 employs attenuated (less dangerous) strains of true pathogens.

In the microbiology laboratory, we protect ourselves from both kinds of pathogens by taking extensive safety measures to reduce our exposure to them. The safety measures taken are specific to the pathogen and are based on the primary portal of entry and communicability of the pathogen. Please review the

Basic Microbiology Laboratory Safety information pages at the beginning of this lab manual before conducting any of these exercises. All of these exercises call for lab coats and gloves, and they require special precautions for contaminated equipment and waste. Consult with your instructor to determine whether eye protection is recommended in your setting. Hands should be washed before gloves are put on and after they are removed. Eyes and mouth should not be touched while you are in the microbiology laboratory because they are common portals of entry for bacteria. Nothing should ever be placed into your mouth while you are in the microbiology laboratory. Gloves, swabs, petri dishes, and other disposable items used during these exercises should be placed in the biohazard bag in the laboratory. Supplies that are to be reused should be placed in a container designated for autoclaving. Additionally, many institutions require *Salmonella, Shigella,* and *Staphylococcus aureus* to have their own separate biohazard waste containers.

The following exercises focus on bacteria that are of medical importance and the methods used to isolate and identify these bacteria. Some tests used to characterize these bacteria have been covered in the section on the identification of unknown bacteria. However, other tests will be new and will apply specifically to a particular bacterium. Exercises 52–54 involve the methods used to characterize the staphylococci, the streptococci, and the procedures to differentiate and identify the gram-negative intestinal bacteria. Exercise 55 is a simulated epidemic that will demonstrate how the spread of disease through a population can be tracked.

The Staphylococci:
Isolation and Identification

Learning Outcomes

After completing this exercise, you should be able to

1. Enrich and isolate staphylococci from human sources and from fomites using selective media and culture techniques.

2. Identify unknown staphylococci that you have isolated using differential media and biochemical tests specific for these bacteria.

The name "staphylococcus" is derived from Greek, meaning "bunch of grapes." Staphylococci are gram-positive spherical bacteria that divide in more than one plane to form irregular clusters of cells (figure 52.1). In *Bergey's Manual of Determinative Bacteriology,* the staphylococci are currently grouped with other gram-positive cocci such as *Micrococcus, Enterococcus,* and *Streptococcus.* The staphylococci consist of 40 different species with 24 subspecies. The staphylococci are non-motile, non-spore-forming, and able to grow in media containing high salt concentrations. Most are considered facultative anaerobes. Although the staphylococci were originally isolated from pus in wounds, they were later demonstrated to be part of the normal microbiota of nasal membranes, hair follicles,

skin, and the perineum in healthy individuals. Infections by staphylococci are initiated when a breach of the skin or mucosa occurs, when a host's ability to resist infection occurs, or when a staphylococcal toxin is ingested.

The Centers for Disease Control and Prevention estimate that one-third of the U.S. population carries *Staphylococcus aureus*, and this bacterium is responsible for many serious infections. To further complicate matters, *S. aureus* has developed resistance to many antibiotics, including methicillin. MRSA, or methicillin-resistant *S. aureus*, is a major epidemiological problem in hospitals where it is responsible for some healthcare-acquired infections (HAIs). More recently, a community form of MRSA has been isolated from infections in individuals who have not been hospitalized. It has been isolated from school gymnasiums where it has caused infections in student athletes. It is estimated that about 1–2% of the U.S. population now carries MRSA. Although *S. aureus* species are considered to be the most virulent members of the genus, *Staphylococcus epidermidis, S. saprophyticus, S. haemolyticus,* and *S. lugdunensis* are also associated with human diseases.

S. aureus, the most clinically significant staphylococcal pathogen, can cause skin infections, wound infections, bone tissue infections, scalded skin syndrome, toxic shock syndrome, and food poisoning. It has a wide variety of virulence factors and many unique characteristics. The most notable virulence factor possessed by *S. aureus* is coagulase production. Virtually all strains of *S. aureus* are coagulase-positive and will cause serum to form a clot. The role of coagulase in the pathogenesis of disease is unclear, but coagulase may cause a clot to form around the staphylococcal infection, thus protecting the bacterium from host defenses. Another enzyme associated with *S. aureus* is DNase, a nuclease that digests DNA. *S. aureus* also produces a hemolysin called α-toxin that causes a wide, clear zone of beta-hemolysis on blood agar. This powerful toxin plays a significant role in virulence because it not only lyses red blood cells but also damages leukocytes, heart muscle, and renal tissue. Additionally, many strains of *S. aureus* produce a pigment that can act as a virulence factor. The pigment (staphyloxanthin) has antioxidant properties which prevent reactive oxygen (superoxide) produced by the host immune system from killing the bacteria. This pigment is

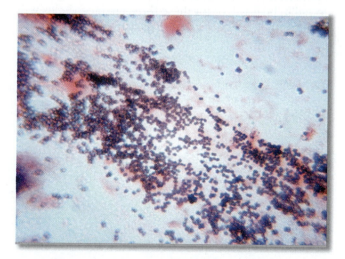

Figure 52.1 Gram stain of *Staphylococcus aureus*.
Source: Centers for Disease Control

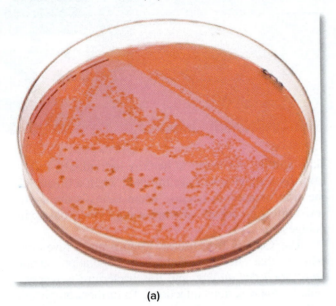

(a)

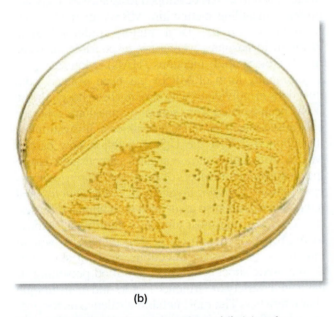

(b)

Figure 52.2 *Staphylococcus epidermidis* (a) and *Staphylococcus aureus* (b) growing on a mannitol salt agar plate. Note the color change that occurs due to acid production from the fermentation of mannitol by *S. aureus*.

© McGraw-Hill Education/Gabriel Guzman, photographer

responsible for the golden color of *S. aureus* when it is grown on blood agar and staphylococcus 110 plates. Finally, *S. aureus* ferments mannitol to produce acid. This metabolic characteristic can be observed when cultures of *S. aureus* are grown on mannitol salt agar (MSA). The production of acid lowers the pH of the medium, causing the phenol red indicator to turn from red to yellow (figure 52.2b).

The coagulase-negative staphylococci (CNS), *S. epidermidis* and *S. saprophyticus*, differ from

S. aureus in many ways. These species of staphylococci, as indicated by their name, do not produce coagulase. They also do not produce DNase or α-toxin. All people have CNS on their skin, and these species were at one time thought of as harmless commensals. However, their clinical significance has greatly increased, particularly in patients who have compromised immune systems or prosthetic or indwelling devices. *S. epidermidis* is the most common cause of hospital-acquired urinary tract infections. Infections involving *S. epidermidis* have also been documented for catheters, heart valves, and other prosthetic devices. *S. saprophyticus* is the second most common cause of urinary tract infections in sexually active young women. The CNS are unpigmented and appear opaque when grown on blood agar and staphylococcus 110 plates. *S. saprophyticus* is the only clinically important staphylococci species that is resistant to novobiocin. Some strains of *S. saprophyticus* are able to ferment mannitol to acid.

In this experiment, we will attempt to isolate and differentiate staphylococci species from (1) the nose, (2) a **fomite** (inanimate object), and (3) an "unknown control." We will follow the procedure illustrated in figure 52.3. If the nasal membranes and fomite fail to yield a positive isolate, the unknown control will yield a positive isolate provided all the inoculations and tests are performed correctly. The organisms collected will first be cultured in media containing 7.5% sodium chloride. The high salt concentration will inhibit the growth of most bacteria, while allowing species of staphylococci to grow. Once the cultures have been enriched for staphylococci, the exercise will focus on identifying the isolated bacteria as *S. aureus*, *S. epidermidis*, *S. saprophyticus*, or an unidentified CNS. The characteristics we will look for in our isolates will be (1) beta-type hemolysis (α-toxin), (2) mannitol fermentation, and (3) coagulase production. Organisms found to be positive for these three characteristics will be presumed to be *S. aureus* (please see table 52.1).

Table 52.1   Differentiation of Three Species of Staphylococci

	S. aureus	S. epidermidis	S. saprophyticus
α-toxin	+	−	−
Mannitol fermentation	+	−	(+)
Coagulase	+	−	−
DNase	+	−	−
Novobiocin	S	S	R

Note: S = sensitive; R = resistant; + = positive; (+) = mostly positive

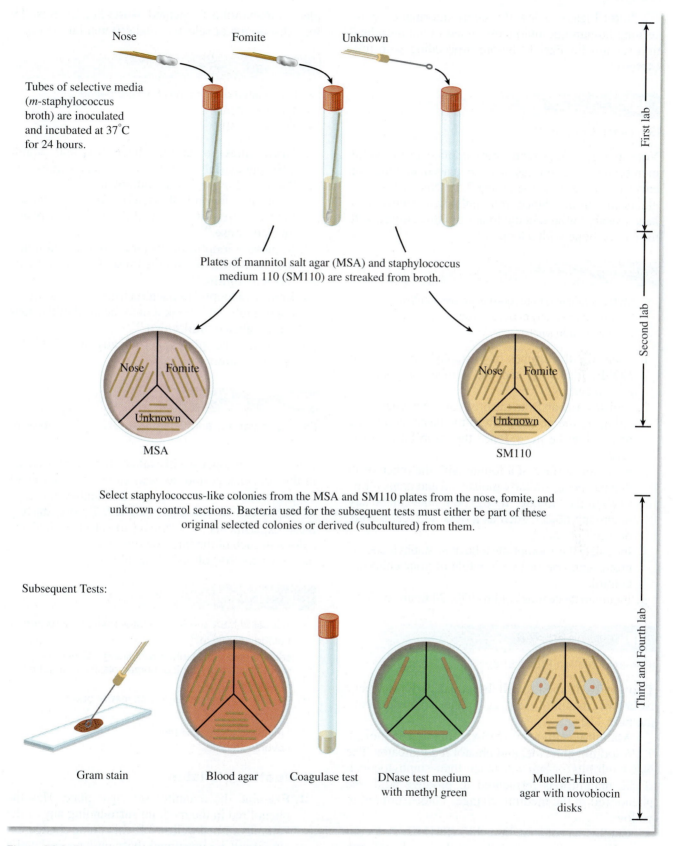

Tubes of selective media (*m*-staphylococcus broth) are inoculated and incubated at 37°C for 24 hours.

First lab

Plates of mannitol salt agar (MSA) and staphylococcus medium 110 (SM110) are streaked from broth.

Second lab

MSA

SM110

Select staphylococcus-like colonies from the MSA and SM110 plates from the nose, fomite, and unknown control sections. Bacteria used for the subsequent tests must either be part of these original selected colonies or derived (subcultured) from them.

Subsequent Tests:

Third and Fourth lab

Gram stain   Blood agar   Coagulase test   DNase test medium with methyl green   Mueller-Hinton agar with novobiocin disks

Figure 52.3 **Procedure for presumptive identification of staphylococci.**

Note: Please review the safety information concerning human microbiota discussed in the introduction section for Part 12 before proceeding with this exercise.

⏰ First Period

(Specimen Collection)

Note in figure 52.3 that swabs applied to the nasal membranes and fomites will be placed in tubes of enrichment medium containing 7.5% NaCl (*m*-staphylococcus broth). Since your unknown control will lack a swab, initial inoculations from this culture will have to be done with a loop.

Materials

- 1 tube containing numbered unknown control
- 3 tubes of *m*-staphylococcus broth
- 2 sterile cotton swabs

1. Label the three tubes of *m*-staphylococcus broth NOSE, FOMITE, and the number of your unknown control.
2. Moisten a swab from the broth tube designated for culturing the nose and swab the nasal membrane just inside the nose. Place the swab back in the tube.
3. Swab the surface of a fomite with the other swab that has been similarly moistened and deposit this swab in the "fomite" tube. The fomite can be any inanimate object, such as a desk surface, a coin, or a cell phone.
4. Inoculate the appropriate tube of *m*-staphylococcus broth with one or two loopfuls of your unknown control.
5. Incubate these tubes of broth for 24 hours at 37°C.

⏰ Second Period

(Primary Isolation Procedure)

Two kinds of media will be streaked for primary isolation: mannitol salt agar and staphylococcus medium 110.

Mannitol salt agar (MSA) contains mannitol, 7.5% sodium chloride, and phenol red indicator. The NaCl inhibits organisms other than staphylococci. If the mannitol is fermented to produce acid, the phenol red in the medium changes color from red to yellow.

Staphylococcus medium 110 (SM110) also contains NaCl and mannitol, but it lacks phenol red. The advantage over MSA is that this medium favors colony pigmentation by different strains of *S. aureus.* Since SM110 lacks phenol red, no color change takes

place if mannitol is fermented. **Note:** See Exercise 18 for a discussion of selective and differential media.

Materials

- 3 incubated culture tubes from last period
- 1 plate of MSA
- 1 plate of SM110

1. Draw lines to divide the MSA and Staph 110 plates into thirds. Label each third as "nose," "fomite," and "unknown."
2. Remove a loopful of bacteria from the "nose" tube and streak it onto the third of the agar plates labeled **"nose."**
3. Remove a loopful of bacteria from the "fomite" tube and streak it onto the third of the agar plates labeled **"fomite."**
4. Remove a loopful of bacteria from the "unknown" control tube and streak it onto the third of the agar plates labeled **"unknown."**
5. Incubate the plates aerobically at 37°C for 24 to 36 hours.

⏰ Third Period

(Plate Evaluations and Coagulase/DNase/Novobiocin Tests)

During this period, we will evaluate the plates streaked in the previous period, as well as set up a series of experiments that will help you to identify the type of staphylococci you have isolated. Before starting these experiments, it is crucial to select an isolated colony in each of the three sections (nose, fomite, and unknown control) of one of the plates.

Materials

- incubated MSA and SM110 plates from previous period
- 1 blood agar plate
- capped serological tubes containing 0.5 ml of 1:4 saline dilution of rabbit or human plasma (one tube for each isolate)
- 1 plate of DNase test agar with methyl green
- Gram staining kit
- 1 Mueller-Hinton agar plate
- novobiocin disks

Evaluation of Plates

1. Examine the mannitol salt agar plate. Has the phenol red in the medium surrounding any of the colonies turned yellow? If this color change exists, it can be presumed that you have isolated a strain of *S. aureus* or *S. saprophyticus.*
2. Examine the plate of SM110. The presence of growth here indicates that the organisms are

salt-tolerant. Note color of the colonies. *S. aureus* colonies will appear yellow or orange, while CNS colonies will appear colorless.

3. Record your observations of these plates in Laboratory Report 52.

Blood Agar Inoculations

1. Divide one blood agar plate (BAP) into thirds. Label the sections "nose," "fomite," and "unknown control."

2. Select staphylococcus-like colonies from the MSA and SM110 plates from the nose, fomite, and unknown control sections. If you do not have an isolated colony from each section, consult with your instructor about any modifications you may need to make as you proceed through this lab (for example, if you only have two cultures, divide the BAP in half as opposed to thirds).

3. Streak out part of the selected "nose" colony onto the third of the BAP labeled "nose." Streak out part of the selected "fomite" colony onto the third of the BAP labeled "fomite." Streak out part of the selected "unknown" colony onto the third of the BAP labeled "unknown."

4. Incubate the blood agar plates at 37°C for 18 to 24 hours. *Don't leave plates in incubator longer than 24 hours.* Overincubation will cause blood degeneration.

[At this point in the lab, you can either proceed to the following tests using the remaining portion of your selected colonies, or incubate your BAP for 24 hours and then proceed with the following tests. Incubating your BAP for 24 hours before proceeding will provide you with a larger amount of pure culture to use in the coagulase, DNase, and novobiocin tests, all of which require a heavy inoculum of bacteria.]

Coagulase Tests

The fact that 97% of the strains of *S. aureus* are coagulase-positive and that the other two species are *always* coagulase-negative makes the coagulase test an excellent definitive test for confirming identification of *S. aureus*.

The coagulase test involves inoculating a small tube of plasma with several loopfuls of the organism and incubating it at 37°C for several hours. If the plasma coagulates (forms a clot), the organism is coagulase-positive. Coagulation may occur in 30 minutes or several hours later. *Any degree of coagulation, from a loose clot suspended in plasma to a solid immovable clot, is considered to be a positive result, even if it takes 24 hours to occur.*

It should be emphasized that this test is valid only for gram-positive, staphylococcus-like

Figure 52.4 Coagulase test: Positive test with negative test shown in inset.
© McGraw-Hill Education/Lisa Burgess, photographer

bacteria because some gram-negative rods, such as *Pseudomonas*, can cause clotting. The mechanism of clotting, however, in such organisms is not due to coagulase.

Proceed as follows:

1. Label the plasma tubes "nose," "fomite," or "unknown control," depending on which of your plates have staph-like colonies.

2. With a wire loop, inoculate the appropriate tube of plasma with organisms from the colonies on SM110 or MSA. Success is more rapid with a heavy inoculation. If positive colonies are present on both nose and fomite sides, be sure to inoculate a separate tube for each sample.

3. Securely cap tubes and place in a 37°C water bath.

4. Check for solidification of the plasma every 30 minutes for the remainder of the period (figure 52.4). Note that solidification may be complete or semisolid.

   Any cultures that are negative at the end of the period will be left in the water bath. At 24 hours, your instructor will remove them from the water bath and place them in the refrigerator, so that you can evaluate them in the next laboratory period.

5. Record your results in Laboratory Report 52.

DNase Test

The fact that coagulase-positive bacteria are also able to hydrolyze DNA makes the DNase test another reliable means of confirming *S. aureus* identification. The following procedure can be used to determine if a staphylococcus can hydrolyze DNA.

1. Heavily streak test organisms on a plate of DNase test agar. One plate can be used for several test cultures by making short streaks about 1 inch long.

2. Incubate for 18 to 24 hours at 37°C.

Gram-Stained Slides

While your tubes of plasma are incubating in the water bath, prepare Gram-stained slides from the same colonies that were used for the blood agar plates and coagulase tests.

Examine the slides under an oil immersion lens, and draw the organisms in the appropriate areas of Laboratory Report 52.

Novobiocin Susceptibility Test

The coagulase-negative organisms, *S. saprophyticus* and *S. epidermidis*, will be indistinguishable if *S. saprophyticus* does not ferment mannitol. Novobiocin resistance can be used to identify *S. saprophyticus*.

1. Divide one Mueller-Hinton agar plate into thirds. Label the sections "nose," "fomite," and "unknown control."
2. Using a loop, inoculate each third of the plate with the appropriate test culture. Be sure to inoculate the entire surface of the agar for each section.
3. Aseptically transfer a novobiocin disk (either using forceps or a disk dispenser) to the center of each sector of the plate.
4. Incubate the plate at 37°C for 24 hours.

 **Fourth Period**

(Confirmation)

During this period, we will make a final assessment of all tests and perform any other confirmatory tests that might be available to us.

Materials

- coagulase tubes from previous period
- incubated blood agar plate from previous period
- incubated DNase test agar plate from previous period

1. Examine any coagulase tubes that were carried over from the last laboratory period that were negative at the end of that period.
2. Examine the colonies on your blood agar plate. Look for clear (beta-type) hemolysis around the colonies. The presence of α-toxin is a definitive characteristic of *S. aureus*.
3. Look for zones where the dye has cleared around the streak on the DNase agar plate (figure 52.5).
4. Examine the colonies on your Mueller-Hinton agar plate with novobiocin disks. Susceptibility to novobiocin is indicated by an area of no growth surrounding the novobiocin disk. For a culture to be considered susceptible to novobiocin, the zone of inhibition surrounding the disk must be greater than 16 mm in diameter (figure 52.6).

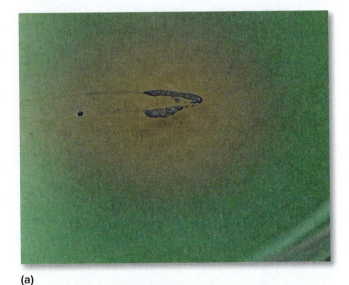

(a)

(b)

Figure 52.5 (a) DNase test on DNase methyl green agar: clearing indicates DNA breakdown. (b) Negative DNase test: no clearing around the streak.
© McGraw-Hill Education/Lisa Burgess, photographer

Figure 52.6 **Novobiocin test showing sensitivity for *Staphylococcus epidermidis*.**
© McGraw-Hill Education/Lisa Burgess, photographer

5. Record your results for all tests in the chart in Laboratory Report 52.

Further Testing

Consult with your instructor for any additional tests that may be performed to identify staphylococci. The latex agglutination slide test (Exercise 57) can be used for this purpose.

Laboratory Report

Collect data from your classmates to complete the chart in Laboratory Report 52 and answer all the questions.

Student: _____

Date: _____ Section: _____

52 The Staphylococci: Isolation and Identification

A. Results

1. **Tabulation**

   After examining your mannitol salt agar and staphylococcus medium 110 plates, record the presence (+) or absence (−) of staphylococcus growth in the appropriate columns. After performing coagulase tests on the various isolates, record the results also as (+) or (−) in the appropriate columns. Collect data from your classmates to complete the chart.

STUDENT INITIALS	FOMITE				UNKNOWN			
	Item	Staph Colonies		Coagulase	Unknown #	Staph Colonies		Coagulase
		MSA	SM110			MSA	SM110	

CULTURE	TOTAL TESTED	TOTAL COAGULASE POSITIVE	PERCENTAGE COAGULASE POSITIVE	TOTAL COAGULASE NEGATIVE	PERCENTAGE COAGULASE NEGATIVE
Fomite					
Unknown					

2. **Microscopy**

   Provide drawings here of your various isolates as seen under oil immersion (Gram staining).

UNKNOWN CONTROL	NOSE	FOMITE

3. **Record of Culture and Test Results**

	GROWTH ON MSA (+/−)	FERMENTATION OF MANNITOL (+/−)	ALPHA-HEMOLYSIS ON BAP (+/−)	COAGULATION OF PLASMA (+/−)	DNASE ZONE OF CLEARING (+/−)	NOVOBIOCIN (S/R)
Nose						
Fomite						
Unknown						

B. Short-Answer Questions

1. Describe the selective and differential properties of mannitol salt agar (MSA) for the isolation and identification of staphylococci.

   _____

   _____

2. Describe the differential property of blood agar for the isolation and identification of staphylococci.

   _____

   _____

3. Why is the coagulase test considered to be the definitive test for *S. aureus*?

   _____

   _____

4. What is the role of coagulase in the pathogenesis of *S. aureus*?

   _____

   _____

5. What is the role of α-toxin in the pathogenesis of *S. aureus*?

    _____

    _____

6. What are healthcare-associated infections?

    _____

    _____

7. Why are the staphylococci among the leading causes of healthcare-associated infections?

    _____

    _____

8. Why are staphylococcal infections becoming increasingly difficult to treat?

    _____

    _____

9. Why might hospital patients be tested for nasal carriage of *S. aureus*?

    _____

    _____

10. Describe results from a coagulase, DNase, and novobiocin test that would suggest a mixed culture was used for the tests, as opposed to a pure culture.

    _____

    _____

11. Why is MRSA not confined to transmission only in hospitals?

    _____

    _____

The Streptococci and Enterococci:
Isolation and Identification

Learning Outcomes

After completing this exercise, you should be able to

1. Isolate streptococci from a mixed culture or from the human throat using selective media and culturing techniques.

2. Identify unknown streptococci that you have isolated using selective media and biochemical tests specific for the streptococci.

The streptococci and enterococci differ from the staphylococci discussed in Exercise 52 in two significant characteristics: (1) Most isolates occur in chains rather than in clusters (figure 53.1), and (2) they lack the enzyme catalase, which degrades hydrogen peroxide to form water and oxygen.

The streptococci and enterococci comprise a large and varied group of gram-positive cocci. They are facultative anaerobes and generally considered nonmotile. They can occur singly or in pairs, but they are best known for their characteristic formation of long chains (figure 53.1). At one time, streptococci and

enterococci because of their similarities were placed in the same genus. However, molecular genetic analysis has determined that they are different enough to be placed in separate families, *Streptococcaceae and Enterococcaceae.*

There are many ways to group and identify the medically important species of these two families. Initial identification of these bacteria is often based on their hemolytic pattern when grown on blood agar. Some species of streptococci and enterococci produce exotoxins that completely destroy red blood cells in blood agar. Complete lysis of red blood cells around a colony is known as **beta-hemolysis** and results in a clear zone surrounding the colonies. Other species of streptococci and enterococci partially break down the hemoglobin inside red blood cells on a blood agar plate, producing a greenish discoloration around the colonies known as **alpha-hemolysis.** Species of bacteria that do not exhibit any hemolysis of blood display **gamma-hemolysis**, that is, they have no effect on the red blood cells in a blood agar plate. The three kinds of hemolysis on blood agar are shown in figure 53.2. After patterns of hemolysis have been identified, species of streptococci and enterococci can be further differentiated based on their cell wall carbohydrates. A method developed by Rebecca Lancefield in the 1930s uses an alphabetic system (A, B, C, etc.) to designate different groups of bacteria. Serological tests are used to differentiate antigenic differences in cell wall carbohydrates that occur in these bacteria. Along with hemolytic patterns and serologic grouping, physiologic and biochemical characteristics are also used to identify streptococcal and enterococcal isolates.

Although *Streptococcus* and *Enterococcus* genera contain a large number of species, only a small number of them are human pathogens.

Beta-Hemolytic Groups

The Lancefield serological groups that fall into the beta-hemolytic category are groups A, B, and C.

Group A Streptococci

Streptococcus pyogenes, the main representative of group A streptococci, is by far the most serious streptococcal pathogen. Humans are the primary

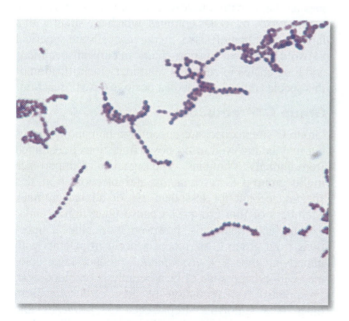

Figure 53.1 Gram stain of *Streptococcus*.
Source: Centers for Disease Control

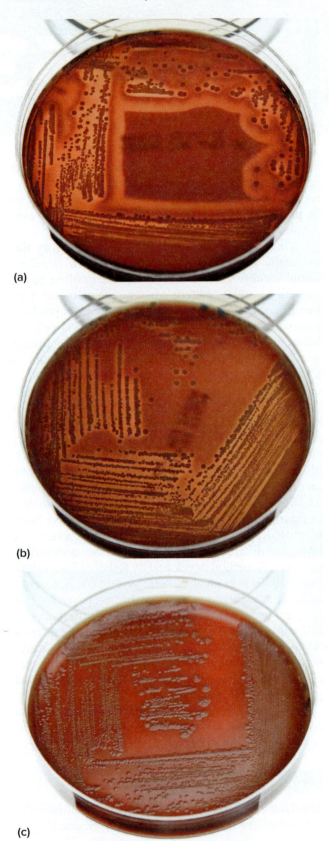

(a)

(b)

(c)

Figure 53.2 The three kinds of hemolysis produced by streptococci growing on blood agar plates: (a) beta-hemolysis, (b) alpha-hemolysis, and (c) gamma-hemolysis.

© McGraw-Hill Education/Gabriel Guzman, photographer

reservoir of *S. pyogenes*. Although the pharynx is the most likely place to find this species, it may be isolated from the skin and rectum. Infections from *S. pyogenes* range from pharyngitis and skin infections to scarlet fever, rheumatic fever, and acute glomerulonephritis. *S. pyogenes* also causes childbirth fever (puerperal sepsis), a serious infection that occurs in women after giving birth. When grown on blood agar, colonies are small (0.5 mm diameter), transparent to opaque, and domed. *S. pyogenes* produce hemolysins which rapidly injure cells and tissues. These hemolysins result in complete lysis of red blood cells around each *S. pyogenes* colony when grown on a blood agar plate, producing a clear zone usually two to four times the diameter of the colony (figure 53.2a). These bacteria are spherical cocci, arranged in short chains in clinical specimens and longer when grown in broth. In order to differentiate *S. pyogenes* from other streptococci and enterococci, isolates are tested for resistance to bacitracin. If a bacterial isolate is beta-hemolytic and sensitive to bacitracin, it is presumed to be *S. pyogenes*.

Group B Streptococci

S. agalactiae is the only recognized species of Lancefield group B. Like *S. pyogenes*, this pathogen may be found in the pharynx, skin, and rectum; however, it is more likely to be found in the genital and intestinal tracts of healthy adults and infants. This organism is an important cause of a serious neonatal infection involving sepsis and meningitis. Colonization of the maternal genital tract is associated with colonization of infants and risk of disease. In the adult population, *S. agalactiae* infections consist of abscesses, endocarditis, septicemia, bone and soft tissue infections, and pneumonia. *S. agalactiae* colonies are large, with a narrow zone of beta-hemolysis, in contrast to *S. pyogenes* colonies, which are small with a large zone of hemolysis. *S. agalactiae* cells are spherical to ovoid and occur in short chains in clinical specimens and long chains in culture. Preliminary identification of this species relies heavily on a positive CAMP reaction.

Group C Streptococci

Group C streptococci are uncommon human pathogens but may be involved in zoonoses (infections transmitted from animals to humans). The organism of importance in this group is *S. dysgalactiae*, and infections from this species account for less than 1% of all bacteremias. Members of this group can cause pharyngitis, endocarditis, and meningitis; however, most clinical infections from group C streptococci occur in patients with underlying illness. *S. dysgalactiae* produce large colonies with a large zone of beta-hemolysis on blood agar. Presumptive differentiation of *S. dysgalactiae* from other beta-hemolytic streptococci (*S. pyogenes* and *S. agalactiae*) is based primarily on resistance to bacitracin and a negative CAMP test.

Table 53.1 Physiological Tests for Streptococci and Enterococci Differentiation

	Bergey's Group	Lancefield Group	Hemolysis	Bacitracin Susceptibility	CAMP Reaction	SXT Sensitivity	Bile Esculin Hydrolysis	Tolerance to 6.5% NaCL	Optochin Susceptibility
S. pyogenes	Pyogenic	A	β	+	–	R	–	–	–
S. agalactiae	Pyogenic	B	β	–	+	R	–	±	–
S. pneumoniae	Pyogenic	none	α	–	–	–	–		+
S. dysgalactiae	Pyogenic	C	β	–	–	S	–	–	–
E. faecalis	Enterococci	D	γ or α¹	–	–	R	+	+	–
E. faecium	Enterococci	D	γ or α	–	–	R	+	+	–
S. bovis	Other	D	γ or α	–	–	R/S	+	–	–
S. mitis²	Oral (Viridans)	none	α¹	–	–	S	–	–	–
S. salivarius²	Oral (Viridans)	none	α¹	–	–	S	–	–	–
S. mutans²	Oral (Viridans)	none	none	–	–	S	–	–	–

Note: R = resistant; S = sensitive; blank = not significant.
¹Weakly alpha.
²Oral streptococci commonly isolated from throat swabs. Differentiation is based on additional biochemical tests not performed in this exercise.

Alpha-Hemolytic Groups

The grouping of streptococci and enterococci on the basis of alpha-hemolysis is not as clear-cut as it is for beta-hemolytic groups. Note in table 53.1 that some groups exhibit weak alpha-hemolysis, and a few alpha-hemolytic types may exhibit variable hemolysis, highlighting the notion that hemolysis can be a misleading characteristic in identification.

Streptococcus pneumoniae

Although *S. pneumoniae* does not possess a Lancefield antigen, it is a significant human pathogen. *S. pneumoniae* is the causative agent of bacterial pneumonia, and can cause meningitis and otitis media (ear infections) in children. It normally colonizes the pharynx, but in certain situations it can spread to the lungs, sinuses, or middle ear. Virulent strains of *S. pneumoniae* are covered with a polysaccharide capsule (avirulent lab strains do not produce the capsule). When this organism is grown on blood agar (figure 53.2b), its colonies appear smooth, mucoid, and surrounded by a zone of greenish discoloration (alpha-hemolysis). In culture these cells usually grow as diplococci, but they can also occur singly or in short chains. Presumptive identification of *S. pneumoniae* can be made with a positive optochin susceptibility test.

Viridans Streptococci Group

The viridans streptococci are a heterogeneous cluster of alpha-hemolytic and nonhemolytic streptococci. These organisms colonize the oral cavity, pharynx, gastrointestinal tract, and genitourinary tract. Most of the viridans are opportunists, and they are usually regarded as having low pathogenicity. Two species in the viridans group are thought to be the primary cause of dental caries (cavities). Although these organisms have few virulence factors and are constituents of the normal flora, introduction of these species into tissues through dental or surgical means can cause severe infections. The most serious complication of all viridans infections is subacute endocarditis. When grown on blood agar, viridans colonies appear very small, gray to whitish gray, and opaque. In culture they appear rod-like and grow in chains. The viridans group can be differentiated from the pneumococci and enterococci by a negative result in the bile esculin hydrolysis test, the salt-tolerance test, and the optochin susceptibility test.

Group D Enterococci

The enterococci of serological group D are part of the *Enterococcaceae* family and are considered variably hemolytic. There are two principal species of this enterococcal group, *E. faecalis* and *E. faecium*. They are predominantly inhabitants of the gastrointestinal tract; however, they have also been isolated from the genitourinary tract and oral cavity. Although the enterococci do not possess many virulence factors, they are important pathogens in hospitalized patients where they can cause urinary tract infections, bacteremia, and endocarditis. When grown on blood agar, enterococci form large colonies that can appear nonhemolytic (figure 53.2c), alpha-hemolytic, or rarely beta-hemolytic. In culture, they grow as diplococci in short chains. The enterococci can grow under extreme conditions, and this phenotype can be exploited to help differentiate them from various streptococcal species. Isolates that are able to grow in the presence of 6.5% NaCl and are able to hydrolyze bile esculin are presumed to be enterococci in this exercise.

Group D Nonenterococci

S. bovis is the only clinically relevant nonenterococcal species of group D. It is found in the intestinal tracts of humans, as well as cows, sheep, and other ruminants. Although it is found in many animals, *S. bovis* is a human pathogen and has been implicated as a causative agent of endocarditis and meningitis, and is associated with malignancies of the gastrointestinal tract. Colonies of *S. bovis* appear large, mucoid (many strains have a capsule), and are either nonhemolytic or alpha-hemolytic. In culture *S. bovis* grows in pairs and short chains. Key reactions for this group are a positive bile esculin test and negative salt broth test.

The purpose of this exercise is twofold: (1) to learn the standard procedures for isolating streptococci and enterococci and (2) to learn how to differentiate the medically important species of these families. Figure 53.3 illustrates the overall procedure to be followed in the exercise. To broaden the application of the tests for identifying streptococci and enterococci, your instructor may supply you with unknown cultures to be identified along with your pharyngeal isolates. Keep in mind as you complete this exercise that these tests result in a presumptive identification of your isolates. Commercial kits are available, such as the RapID Streptoccoci panel, which can be used to confirm the identification of your isolates.

Note: Please review the safety information concerning human microbiota discussed in the introduction to Part 12 before proceeding with this exercise.

⏱ First Period

(Making a Streak-Stab Agar Plate)

During this period, a plate of blood agar is swabbed and streaked in a special way to determine the type of hemolytic bacteria that are present in the pharynx and in an unknown mixture.

Since swabbing one's own throat properly can be difficult, it will be necessary for you to work with your laboratory partner to swab each other's throats. Once your throat has been swabbed, you will use the swab and a loop to inoculate a blood agar plate according to a special procedure shown in figure 53.4.

Materials

- 1 tongue depressor
- 1 sterile cotton swab
- inoculating loop
- 2 blood agar plates
- unknown mixture

1. With the subject's head tilted back and the tongue held down with the tongue depressor, rub the back surface of the pharynx up and down with the sterile swab.

   Also, *look for white patches* in the tonsillar area. Avoid touching the cheeks and tongue.
2. Since streptococcal hemolysis is most accurately analyzed when the colonies develop anaerobically beneath the surface of the agar, it will be necessary to use a streak-stab technique as shown in figure 53.4. The essential steps are as follows:

   - Roll the swab over an area approximating one-fifth of the surface. The entire surface of the swab should contact the agar.
   - With a wire loop, streak out three areas, as shown, to thin out the organisms.
   - Stab the loop into the agar to the bottom of the plate at an angle perpendicular to the surface to make a clean cut without ragged edges.
   - Be sure to make the stabs in an unstreaked area so that streptococcal hemolysis will be easier to interpret with a microscope.

Caution

Dispose of swabs and tongue depressors in a beaker of disinfectant or biohazard bag.

3. Repeat the inoculation procedure for the unknown mixture.
4. Incubate the plates aerobically at 37°C for 24 hours. Do not incubate the plates longer than 24 hours.

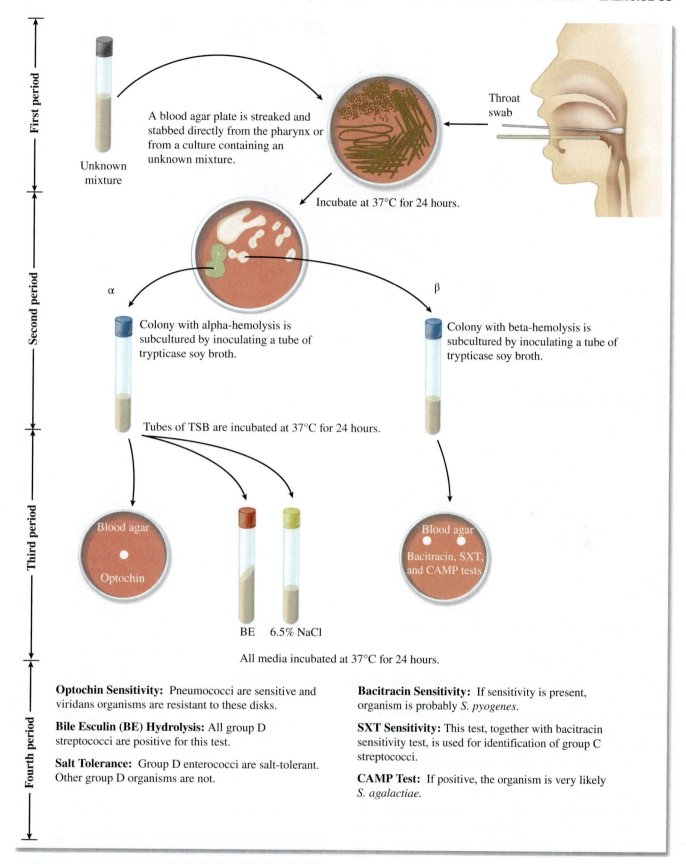

First period

A blood agar plate is streaked and stabbed directly from the pharynx or from a culture containing an unknown mixture.

Throat swab

Unknown mixture

Incubate at 37°C for 24 hours.

Second period

α

Colony with alpha-hemolysis is subcultured by inoculating a tube of trypticase soy broth.

β

Colony with beta-hemolysis is subcultured by inoculating a tube of trypticase soy broth.

Tubes of TSB are incubated at 37°C for 24 hours.

Third period

Blood agar

Optochin

BE   6.5% NaCl

Blood agar

Bacitracin, SXT, and CAMP tests

All media incubated at 37°C for 24 hours.

Fourth period

Optochin Sensitivity: Pneumococci are sensitive and viridans organisms are resistant to these disks.

Bile Esculin (BE) Hydrolysis: All group D streptococci are positive for this test.

Salt Tolerance: Group D enterococci are salt-tolerant. Other group D organisms are not.

Bacitracin Sensitivity: If sensitivity is present, organism is probably *S. pyogenes*.

SXT Sensitivity: This test, together with bacitracin sensitivity test, is used for identification of group C streptococci.

CAMP Test: If positive, the organism is very likely *S. agalactiae*.

Figure 53.3 **Media inoculations for the presumptive identification of streptococci and enterococci.**

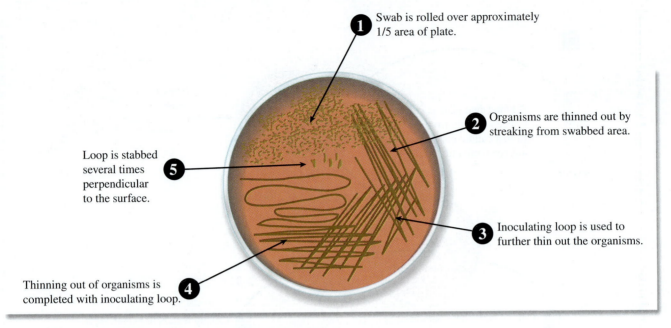

① Swab is rolled over approximately 1/5 area of plate.

② Organisms are thinned out by streaking from swabbed area.

Loop is stabbed several times perpendicular to the surface. ⑤

③ Inoculating loop is used to further thin out the organisms.

Thinning out of organisms is completed with inoculating loop. ④

Figure 53.4 Streak-stab procedure for blood agar inoculations.

⏱ Second Period

(Analysis and Subculturing)

During this period, two objectives must be accomplished: first, the type of hemolysis must be correctly determined, and second, well-isolated colonies must be selected for making subcultures. The importance of proper subculturing cannot be overemphasized; without a pure culture, future tests are certain to fail. Proceed as follows:

Materials

- incubated blood agar plates from previous period
- tubes of trypticase soy broth (TSB), one for each different type of colony
- dissecting microscope

1. Look for isolated colonies that have hemolysis surrounding them.
2. Do any of the stabs appear to exhibit hemolysis? Examine these hemolytic zones near the stabs under 60× magnification with a dissecting microscope or with the scanning objective (4×) of a light microscope.
3. Hemolysis patterns are divided into beta- and alpha-hemolysis, and alpha-hemolytic patterns can be furthered subdivided into either alpha or alpha-prime (consult figure 53.5). Complete lysis of red blood cells around a colony (a clear area around the colony) is beta-hemolysis. Partial lysis of red blood cells around a colony results in greenish discoloration of the area around the colony and

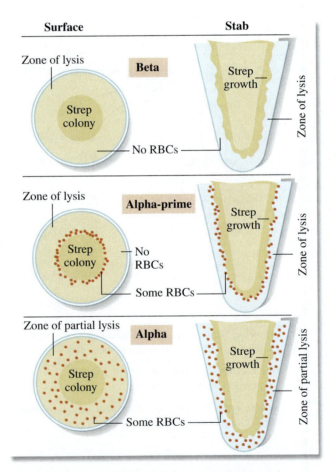

Figure 53.5 Comparison of the types of hemolysis seen on blood agar plates. The illustrations on the left indicate the appearance of red blood cells surrounding a bacterial colony on the surface of a blood agar plate. The illustrations on the right indicate the appearance of red blood cells surrounding stabs into the blood agar.

is called alpha-hemolysis. A small area of intact red blood cells around a colony surrounded by a wider zone of complete hemolysis is **alpha-prime-hemolysis.** Some streptococci from the viridans group display alpha-prime-hemolysis. Alpha-prime-hemolysis is sometimes mistaken for beta-hemolysis, which is why it is recommended the hemolytic colonies/stabs be viewed under either the dissecting scope or the lowest power of a light microscope. Colonies that do not affect the red blood cells surrounding them are said to exhibit nonhemolysis or **gamma-hemolysis.**

4. Record your observations in Laboratory Report 53.
5. Select well-isolated colonies that exhibit hemolysis (alpha, beta) for inoculating tubes of TSB. Be sure to label the tubes ALPHA or BETA. Whether or not the organism is alpha or beta is crucial in identification.

    Since the chances of isolating beta-hemolytic streptococci from the pharynx are usually quite slim, notify your instructor if you think you have isolated one.
6. Incubate the tubes at 37°C for 24 hours.

🕐 Third Period

(Inoculations for Physiological Tests)

Presumptive identification of the various groups of streptococci is based on the physiological tests in table 53.1. Note that groups A, B, and C are all beta-hemolytic. The remainder are usually alpha-hemolytic or nonhemolytic.

Since each of the physiological tests is specific for differentiating only two or three groups, it is not desirable to do all the tests on all unknowns. For economy and precision, carefully follow figure 53.3 to select which tests you will perform on an isolate or unknown based on the type of hemolysis it exhibits.

Before any inoculations are made, however, it is desirable to do a purity check on each TSB culture from the previous period. To accomplish this, it will be necessary to make a Gram-stained slide of each of the cultures.

Gram Stained Slides (Purity Check)

Materials

- TSB cultures from previous period
- Gram staining kit

1. Make a Gram stained slide from your isolates and examine them under an oil immersion lens. Do they appear to be pure cultures?
2. Draw the organisms in the appropriate circles in Laboratory Report 53.

Beta-Type Inoculations

Use the following procedure to perform tests on each isolate that has beta-type hemolysis:

Materials

for each isolate:
- 1 blood agar plate
- 1 bacitracin differential disk (0.04 units)
- 1 SXT sensitivity disk
- 1 broth culture of *S. aureus*
- dispenser or forceps for transferring disks

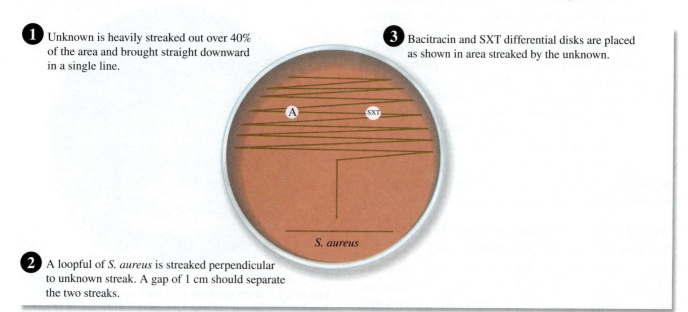

① Unknown is heavily streaked out over 40% of the area and brought straight downward in a single line.

③ Bacitracin and SXT differential disks are placed as shown in area streaked by the unknown.

② A loopful of *S. aureus* is streaked perpendicular to unknown streak. A gap of 1 cm should separate the two streaks.

Figure 53.6 **Blood agar inoculation technique for the CAMP, bacitracin, and SXT tests.**

1. Label a blood agar plate with the proper identification information of each isolate and unknown to be tested.
2. Follow the procedure outlined in figure 53.6 to inoculate each blood agar plate with the isolate (or unknown) and *S. aureus.*

   Note that a streak of the unknown is brought down perpendicular to the *S. aureus* streak, keeping the two organisms about 1 cm apart.
3. With forceps or dispenser, place one bacitracin differential disk and one SXT disk on the heavily streaked area at points shown in figure 53.6. Press down lightly on each disk.
4. Incubate the blood agar plates at 37°C, aerobically, for 24 hours.

Alpha-Type Inoculations

As shown in figure 53.3, three inoculations will be made for each isolate or unknown that is alpha-hemolytic.

Materials

- 1 blood agar plate (for up to 4 unknowns)
- 1 6.5% sodium chloride broth tube per unknown
- 1 bile esculin (BE) slant per unknown
- 1 optochin disk per unknown
- candle jar setup or CO_2 incubator

1. Mark the bottom of a blood agar plate to divide it into halves, thirds, or quarters, depending on the number of alpha-hemolytic organisms to be tested. Label each space with the code number of each test organism.
2. Completely streak over each area of the blood agar plate with the appropriate test organism, and place one optochin disk in the center of each area. Press down slightly on each disk to secure it to the medium.
3. Inoculate one tube of 6.5% sodium chloride broth and streak the surface of one bile esculin slant.
4. Incubate all media at 35–37°C as follows:
   Blood agar plates: 24 hours in a candle jar
   6.5% NaCl broths: 24, 48, and 72 hours
   Bile esculin slants: 48 hours

   Note: While the blood agar plates should be incubated in a candle jar or CO_2 incubator, the remaining cultures can be incubated aerobically.

Inoculations of Colonies That Do Not Exhibit Hemolysis (Gamma-Type)

Refer to table 53.1. Isolates of this nature are probably members of group D or viridans. Perform an SXT sensitivity test, a bile esculin hydrolysis test, and a salt-tolerance test as described under Alpha-Type Inoculations.

Note: Optochin susceptibility can be evaluated as well if an optochin disk is placed on the streaked area of the SXT sensitivity BAP, in the same manner as the SXT disk is placed.

⏱ Fourth Period

(Evaluation of Physiological Tests)

Once all of the inoculated media have been incubated for 24 hours, begin to examine the plates and tubes and add test reagents to some of the cultures. Some of the tests will also have to be checked at 48 and 72 hours.

After the appropriate incubation period, assemble all the plates and tubes from the last period, and examine the blood agar plates first that were double-streaked with the unknowns and *S. aureus.* Note that the CAMP reaction, bacitracin susceptibility test, and SXT sensitivity test can be read from these plates. Proceed as follows:

CAMP Reaction (β test)

If you have an unknown that produces an enlarged arrowhead-shaped hemolytic zone at the juncture where the unknown meets the *Staphylococcus aureus* streak, as seen in figure 53.7, the organism is *Streptococcus agalactiae.* This phenomenon is due to what is called the *CAMP factor*, named for the developers of the test, Christie, Atkins, and Munch-Peterson. The CAMP factor produced by *Streptococcus agalactiae* acts synergistically with staphylococcal hemolysins, causing an enhanced breakdown of red blood cells and therefore producing the arrowhead zone of clearing.

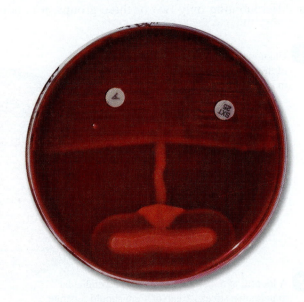

Figure 53.7 Note positive SXT disk on right, negative bacitracin disk on left, and positive CAMP reaction (arrowhead). Organism: *S. agalactiae.*
© Harold Benson

The only problem that can arise from this test is that if the plate is incubated anaerobically, a positive CAMP reaction can occur on *S. pyogenes* inoculated plates.

Record the CAMP reactions for each of your isolates or unknowns in Laboratory Report 53.

Bacitracin Susceptibility (β test)

Any size zone of inhibition seen around the bacitracin disks should be considered to be a positive test result. Note in table 53.1 that *S. pyogenes* is positive for this characteristic.

This test has two limitations: (1) the disks must be of the *differential type*, not sensitivity type, and (2) the test should not be applied to alpha-hemolytic streptococci. Reasons: Sensitivity disks have too high a concentration of the antibiotic, and many alpha-hemolytic streptococci are sensitive to these disks.

Record the results of this test in the table under number 2 of Laboratory Report 53.

SXT Sensitivity Test (β test)

The disks used in this test contain 1.25 mg of trimethoprim and 27.75 mg of sulfamethoxazole (SXT). The purpose of this test is to distinguish groups A and B from other beta-hemolytic streptococci. Note in table 53.1 that both groups A and B are uniformly resistant to SXT.

If a beta-hemolytic streptococcus proves to be bacitracin resistant and SXT susceptible, it is classified as being a **non–group A or B beta-hemolytic streptococcus.** This means that the organism is probably a species within group C. *Keep in mind that an occasional group A streptococcal strain is susceptible to both bacitracin and SXT disks.* One must always remember that exceptions to most tests do occur; that is why this identification procedure leads us only to *presumptive* conclusions.

Record any zone of inhibition (resistance) as positive for this test.

Note: A few strains of *E. faecalis* are beta-hemolytic. A beta-hemolytic isolate of *E. faecalis* would have a negative CAMP reaction and would appear resistant to bacitracin and SXT. Although this exercise does not outline performing a bile esculin hydrolysis test or salt-tolerance test on a beta-hemolytic colony, these physiological tests would help confirm identification of a beta-hemolytic *E. faecalis* strain.

Bile Esculin (BE) Hydrolysis (α test)

This is the best physiological test that we have for the identification of group D streptococci. Both enterococcal and non-enterococcal species of group D are able to hydrolyze esculin in the agar slant, causing the slant to blacken.

Figure 53.8 **Positive bile esculin hydrolysis on left; negative on right.**
© Harold Benson

A positive BE test tells us that we have a group D streptococcus; differentiation of the two types of group D streptococci (*Enterococcus* and *S. bovis*) depends on the salt-tolerance test.

Examine the BE agar slants, looking for **blackening of the slant,** as illustrated in figure 53.8. If less than half of the slant is blackened, or if no blackening occurs within 24 to 48 hours, the test is negative.

Salt-Tolerance (6.5% NaCl) (Group D) (α test)

All enterococci of group D produce heavy growth in 6.5% NaCl broth. As indicated in table 53.1, *S. bovis* does not grow in this medium. This test, then, provides a good method for differentiating the organisms of group D.

A positive result shows up as turbidity within 72 hours. A color change of **purple to yellow** may also be present. If the tube is negative at 24 hours, incubate it and check it again at 48 and 72 hours. *If the organism is salt tolerant and BE positive, it is considered to be an enterococcus.* Parenthetically, it should be added here that approximately 80% of group B streptococci will grow in this medium.

Optochin Susceptibility (α test)

Optochin susceptibility is used for differentiation of the alpha-hemolytic viridans streptococci from the pneumococci. The pneumococci are sensitive to these disks; the viridans organisms are resistant.

Materials

- blood agar plates with optochin disks
- plastic metric ruler

1. Measure the diameters of zones of inhibition that surround each optochin disk, evaluating whether the zones are large enough to be considered positive. The standards are as follows:
   - For 6 mm diameter disks, the zone must be at least 14 mm diameter to be considered positive.
   - For 10 mm diameter disks, the zone must be at least 16 mm diameter to be considered positive.
2. Record your results in Laboratory Report 53.

Final Confirmation

All the laboratory procedures performed so far lead us to presumptive identification. To confirm these conclusions, it is necessary to perform serological tests on each of the unknowns. If commercial kits are available for such tests, they should be used to complete the identification procedures.

Laboratory Report

Complete Laboratory Report 53.

53 The Streptococci and Enterococci: Isolation and Identification

A. Results

1. Microscopy
   Provide drawings here of your isolates and unknowns as seen under oil immersion (Gram staining).

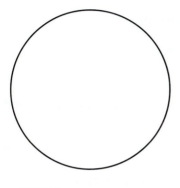 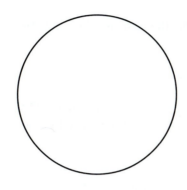 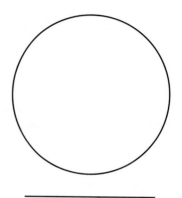

_____ _____ _____

2. **Record of Test Results**
   Record here all information pertaining to the identification of pharyngeal isolates and unknowns.

SOURCE OF UNKNOWN	Hemolysis	Bacitracin Susceptibility	CAMP Reaction	SXT Sensitivity	Bile Esculin Hydrolysis	Tolerance to 6.5% NaCl	Optochin Susceptibility

3. **Final Determination**
   Record here the identities of your various isolates and unknowns:

   Pharyngeal isolates: _____

   _____

   Unknowns: _____

   _____

   _____

391

B. Short-Answer Questions

1. When bacteria from a throat swab are streaked on blood agar, why is the agar stabbed several times with the loop?

    _____

    _____

2. Differentiate between alpha- and beta-hemolysis.

    _____

    _____

3. What compound in the cell wall is the basis for the Lancefield classification?

    _____

    _____

4. In the CAMP reaction, which organism produces the CAMP factor? What substance does the CAMP factor react with to cause enhanced breakdown of red blood cells?

    _____

5. Humans may carry both staphylococci and streptococci as normal microbiota. How might you easily differentiate between the two genera?

    _____

6. Name two tests that are useful in differentiating *S. pyogenes* and *S. agalactiae*.

    _____

    _____

7. Name two tests that are useful for the differentiation of pneumococci and oral viridans streptococci.

    _____

    _____

8. What test can be performed to differentiate the enterococci from other group D streptococci?

    _____

    _____

9. What test can be performed to differentiate between group A and group C streptococci?

    _____

    _____

10. Describe the appearance of an *S. agalactiae* colony grown on blood agar. Describe how that colony would differ in appearance from a colony of *S. pyogenes*.

    _____

    _____

11. Vaginal swabs are taken from pregnant women in their third trimester. Which streptococcal species is the focus of the investigation, and why is this test conducted?

     _____

     _____

12. Which streptococci are implicated in the development of dental caries? What is the mechanism of their formation?

     _____

     _____

13. Why is *S. pneumoniae* not able to be classified by the Lancefield system?

     _____

     _____

MICROBIOLOGY

Gram-Negative Intestinal Pathogens

Learning Outcomes

After completing this exercise, you should be able to

1. Enrich for *Salmonella* and *Shigella* using differential and selective media.

2. Differentiate the coliforms and other lactose fermenters from the non-lactose fermenters, *Salmonella, Shigella,* and *Proteus,* using specific biochemical tests and differential media.

The *Enterobacteriaceae* are a large family of diverse gram-negative rods (figure 54.1). They are ubiquitous and can be found in soil, water, vegetation, and the intestinal tracts of most organisms. All members of this family are facultative anaerobes. This heterogeneous collection of organisms is responsible for a variety of human diseases, including bacteremias, urinary tract infections, and numerous intestinal infections. Organisms in this group are divided into potential human pathogens and true intestinal pathogens. The potential human pathogens are part of the normal commensal microbiota of the gastrointestinal tract and can cause opportunistic infections if they are not confined to their natural environment or if a person's immune response is compromised. The true pathogens are not normally present as commensal microbiota in the gastrointestinal tract of humans, and they are always associated with human disease.

E. coli is an important opportunistic pathogen. Normally, *E. coli* is found in large numbers as a resident of the colon; however, it can grow outside its normal body site and cause urinary tract infections, sepsis, wound infections, and meningitis. Most infections involving *E. coli* are endogenous, meaning the resident *E. coli* of the commensal microbiota established the infection when it grew outside of its natural site. *E. coli* can also cause clinical disease in immune compromised patients. Additionally, this species can acquire virulence factors encoded on plasmids or in bacteriophage DNA (lysogenic conversion), causing some strains to have enhanced virulence. *E. coli* 0157:H7 is a particularly virulent strain that has been associated with various kinds of contaminated food and has caused many deaths. This strain of *E. coli* produces a toxin that damages blood vessels and causes

very severe diarrhea. From a nomenclature standpoint, *E. coli* are defined as **coliforms** because they are gram-negative organisms that can ferment lactose. Other biochemical characteristics used in the identification of *E. coli* include its motility, its ability to produce indole, and its inability to use citrate as a carbon source. Additional potential human pathogens that are also coliforms include *Enterobacter, Citrobacter, Klebsiella,* and *Serratia.* These organisms can cause infections in individuals whose defenses are compromised, but they rarely cause infections in immune competent individuals.

Proteus are opportunistic pathogens that are normally found in the intestinal tract and considered harmless, but they can cause urinary tract infections, wound infections, and septicemia. They are motile and produce many different types of fimbriae, which account for their ability to adhere to the epithelium of the urinary tract and cause urinary tract infections. Most species of *Proteus* produce large quantities of urease, which then raises the urine pH and can cause the formation of crystals. Organisms of the genus *Proteus* are not considered coliforms because they are unable to ferment lactose; however, they are able to ferment glucose.

The true intestinal pathogens of the *Enterobacteriaceae* family are *Salmonella, Shigella,* and *Yersinia.* These organisms differ from the potential human pathogens by not being part of the normal microbiota of humans and by having well-developed virulence factors. *Salmonella* infections involve gastroenteritis, septicemia, and typhoid fever. Some strains of *Salmonella,* such as the one that causes typhoid fever, can exist in patients for more than a year. This carrier state following *Salmonella* infection represents an important source of human infection. These organisms are also transmitted through poultry and dairy products. *Salmonella* is one of the major causes of food-borne illnesses and is often spread in improperly prepared poultry. When grown on differential or selective media, *Salmonella* produce non-lactose-fermenting colonies with black centers if the media contain indicators for hydrogen sulfide production. These pathogens have flagella (are motile) and are negative for indole and urease.

Shigella intestinal pathogens cause dysentery; however, different species vary in the severity for the disease. Humans are the only known reservoir

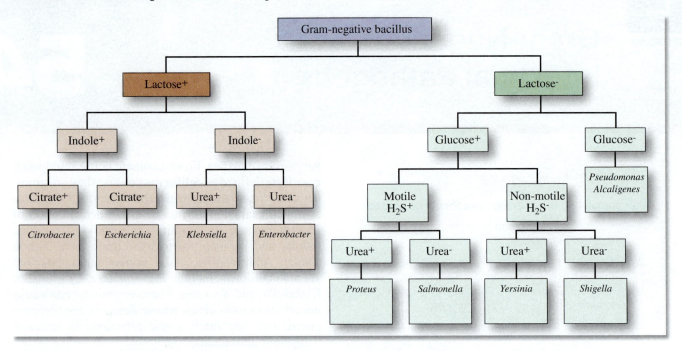

Figure 54.1 Separation outline of gram-negative bacilli, including oxidase-negative *Enterobacteriaceae* and oxidase-positive, non-glucose-fermenting, gram-negative bacilli.

of *Shigella*, and a small number of patients develop asymptomatic colonization of the colon, thus creating a reservoir for infection. *Shigella* species do not ferment lactose (with the exception of *Shigella sonnei*, which slowly ferments lactose) or hydrolyze urea. Unlike *Salmonella*, *Shigella* are non-motile and do not produce hydrogen sulfide. When outbreaks of *Salmonella* and *Shigella* occur, serological typing (Exercise 56) is useful in tracing epidemics caused by a particular serotype of the respective organism. Serotypes (also called serovars) are strains of bacteria within a species that are biochemically similar but differ in their antigenic composition.

The final clinically relevant primary intestinal pathogen is *Yersinia*. *Yersinia pestis*, the most famous human pathogen within this genus, causes the highly fatal systemic disease known as the bubonic plague. The bubonic plague killed approximately one-third of the population of Europe during the 1300s. Humans are accidental hosts of *Yersinia*, and infections occur through ingestion of contaminated animals or the handling of contaminated animal tissues. It is also transmitted by infected fleas when they bite to take a blood meal. Similar to *Shigella* and *Salmonella*, *Yersinia* are also non-lactose fermenters when cultured. *Yersinia* do not produce hydrogen sulfide and are generally considered non-motile and urease positive.

In this exercise you will be presented with a mixture of organisms that represents a "simulated" GI tract sample. In actual practice, stool samples are used and are plated onto isolation media and grown

in enrichment broth. In this experiment you will be given a mixed culture containing a coliform, *Proteus*, and either *Salmonella* or *Shigella*. The pathogens will be attenuated, but their presence will naturally demand utmost caution and handling. Your goal will be to isolate the primary (true) pathogen from the mixed culture and to make a genus identification of the pathogen. Figure 54.1 summarizes the biochemical characteristics that will be used for genus identification.

Note: Please review the safety information concerning human microbiota discussed in the introduction to Part 12 before proceeding with this exercise.

🕐 First Period

(Isolation)

There are several excellent selective differential media that have been developed for the isolation of these pathogens. Various inhibiting agents such as brilliant green, bismuth sulfite, sodium desoxycholate, and sodium citrate are included in them.

Widely used media for the isolation of intestinal pathogens include MacConkey agar, Hektoen Enteric agar (HE), and Eosin Methylene Blue agar (EMB). These media may contain bile salts and/or sodium desoxycholate to inhibit gram-positive bacteria. To inhibit coliforms and other nonenterics, they may contain citrate. All of them contain lactose and a dye so that if an organism is a lactose fermenter, its colony will take on a color characteristic of the dye present.

Proceed as follows to inoculate selective media with your unknown mixture. Your instructor will indicate which media should be used. Figure 54.4 outlines the entire process that will be carried out over several lab periods.

Materials

- unknown culture (mixture of a coliform, *Proteus*, and *Salmonella* or *Shigella*)
- plates of different selective media: MacConkey, Hektoen Enteric (HE), or Eosin Methylene Blue (EMB) agar

1. Label each plate with your name and unknown number.
2. With a loop, streak each plate with your unknown to obtain isolated colonies.
3. Incubate the plates at 37°C for 24 hours.

⏲ Second Period

(Fermentation Tests)

As stated above, the fermentation characteristic that separates the *Salmonella* and *Shigella* pathogens from the coliforms is their *inability to ferment lactose*. Once you have isolated colonies on differential media that resemble *Salmonella* and *Shigella* (non-lactose fermenters), the next step will be to determine whether the isolates can ferment glucose. Kligler's iron agar is often used for this purpose. It contains two sugars, glucose and lactose, as well as phenol red to indicate when acid is produced by fermentation, and iron salts for the detection of H_2S. The glucose concentration of Kligler's iron agar is only 10% of the lactose concentration. Non-lactose fermenters, such as *Shigella* and *Salmonella,* initially produce a yellow slant due to acid produced by the fermentation of glucose, which lowers the pH, causing the agar to turn yellow. When the glucose supply (remember the agar is only 10% glucose) is depleted by non-lactose fermenters they begin to break down amino acids in the medium, producing ammonia and raising the pH, causing part of the agar to turn red. At the end of 24 hours of incubation, a non-lactose fermenter has a red slant and a yellow butt (bottom of the tube). The bottom of the tube is yellow due to the production of acid during the fermentation of glucose. After the glucose is exhausted, amino acids are degraded, producing ammonia that causes the pH indicator to turn red at the slanted surface. Lactose fermenters produce yellow slants and butts. Because these organisms can ferment lactose, and because there is 10 times more lactose in the media than glucose, this sugar is not exhausted over a 24-hour time period, and the pH is kept low (yellow) due to the acids produced during fermentation. If the test bacteria reduce sulfur, hydrogen sulfide

will be produced, reacting with iron compounds in the medium to cause a black precipitate of metal sulfides.

Proceed as follows to inoculate three KIA slants from colonies on the selective media that look like either *Salmonella* or *Shigella.* The reason for using three slants is that you may have difficulty distinguishing *Proteus* from the *Salmonella* or *Shigella* pathogens. By inoculating three tubes from different colonies, you will be increasing your chance of success.

Materials

- 3 agar slants (Kligler's iron)
- streak plates from first period

1. Label the three slants with your name and the number of your unknown.
2. Look for isolated colonies that look like *Salmonella* or *Shigella* organisms. The characteristics to look for on each medium are as follows:

- **MacConkey agar**—*Salmonella, Shigella,* and other non-lactose-fermenting species produce smooth, colorless colonies. Coliforms that ferment lactose produce reddish, mucoid, or dark-centered colonies (figure 54.2).
- **Hektoen Enteric (HE) agar**—*Salmonella* and *Shigella* colonies are greenish-blue. Some species of *Salmonella* will have greenish-blue colonies with black centers due to H_2S production. Coliform colonies are salmon to orange and may have a bile precipitate.
- **Eosin Methylene Blue (EMB) agar**—Bacteria that ferment lactose produce blue-black colonies,

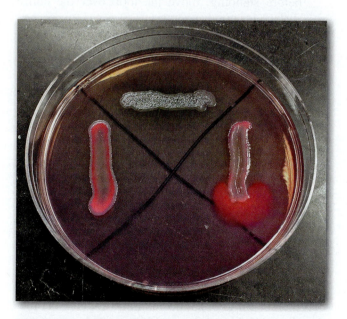

Figure 54.2 MacConkey agar results: (top) non-lactose fermenter, (left and right) lactose fermenters.
© McGraw-Hill Education/Lisa Burgess, photographer

397

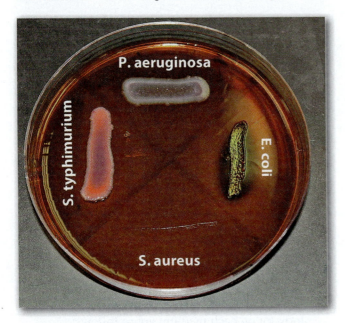

Figure 54.3 EMB agar: (top and left) non-lactose fermenters, (right) lactose fermenter with metallic sheen, (bottom) inhibition of gram-positive bacteria.
© McGraw-Hill Education/Lisa Burgess, photographer

and strong lactose fermenters (such as *E. coli*) have a characteristic green, metallic sheen. Non-lactose fermenters such as *Shigella* and *Salmonella* produce colorless or pink colonies (figure 54.3).

3. With an inoculating loop and needle, inoculate the three agar slants from separate *Salmonella* or *Shigella* appearing colonies. Use the streak-stab technique. When streaking the surface of the slant before stabbing, move the loop over the entire surface for good coverage.

4. Loosen the caps on the slants and incubate at 37°C for 18 to 24 hours. The timing of the incubation of your slant should be carefully monitored. If your slant is read too early (prior to 24 hours) you may erroneously conclude that the organism is capable of fermenting both glucose and lactose. The glucose supply in the tube will last approximately 12 hours, thus tubes containing non-lactose fermenters will have a yellow slant and butt in the beginning of the incubation. If your slant is older than 24 hours you may observe a condition known as **alkaline reversion,** which produces a false negative result for both glucose and lactose. This phenomenon occurs when all of the lactose in the tube has been depleted (usually around 48 hours). Once the lactose is completely metabolized, the lactose fermenters then begin to break down amino acids in the medium. The breakdown of amino acids forms ammonia, and ammonia increases the pH of the agar, causing the color to change back to red.

⏱ Third Period

(Slant Evaluations and Final Inoculations)

During this period, you will inoculate tubes of SIM medium and urea broth with organisms from the slants of the previous period. Examination of the separation outline in figure 54.1 shows that the final step in the differentiation of the *Salmonella* or *Shigella* pathogens is to determine whether a non-lactose fermenter can do three things: (1) exhibit motility, (2) produce hydrogen sulfide, and (3) produce urease. You will also make a Gram-stained slide to perform a purity check.

Materials

- incubated Kligler's iron agar slants from previous period
- 1 tube of SIM medium for each positive KIA slant
- 1 urea slant for each positive KIA slant
- Gram staining kit

1. Examine the slants from the previous period and **select those tubes that have a yellow butt with a red slant.** These tubes contain organisms that ferment only glucose (non-lactose fermenters). If you used Kligler's iron agar, a black precipitate in the medium will indicate that the organism produces H_2S (refer to figure 38.1).

   Note that slants in figure 54.4 that are completely yellow can ferment lactose as well as glucose. Tubes that are completely red are either nonfermenters or examples of alkaline reversion. Ignore those tubes.

2. Make Gram-stained slides from these slants and confirm the presence of gram-negative rods.

3. With a loop, inoculate one urea slant from each KIA slant that has a yellow butt and red slant (non-lactose fermenter).

4. With an inoculating needle, stab one tube of SIM medium from each of the same agar slants. Stab in the center to two-thirds of the depth of the medium.

5. Incubate these tubes at 37°C for 18 to 24 hours.

6. Refrigerate the positive Kligler's iron slants for future use as a stock culture, if needed.

⏱ Fourth Period

(Final Evaluation)

During this last period, the tubes of SIM medium and urea slants from the last period will be evaluated. Serotyping can also be performed, if desired.

Materials

- incubated urea slant tubes and SIM medium
- Kovacs' reagent
- 5 ml pipettes

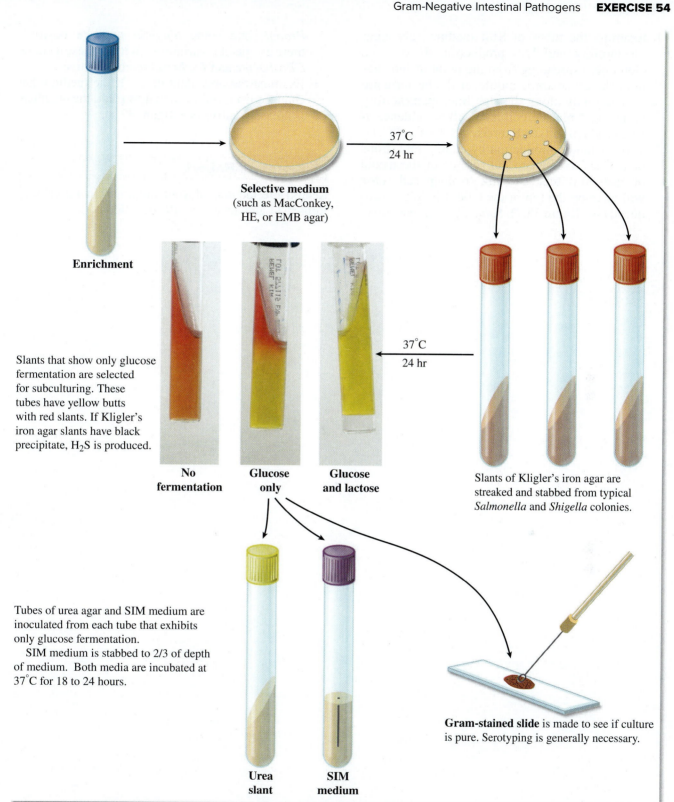

Enrichment

Selective medium
(such as MacConkey,
HE, or EMB agar)

37°C
24 hr

Slants that show only glucose
fermentation are selected
for subculturing. These
tubes have yellow butts
with red slants. If Kligler's
iron agar slants have black
precipitate, H$_2$S is produced.

**No
fermentation**

**Glucose
only**

**Glucose
and lactose**

37°C
24 hr

Slants of Kligler's iron agar are
streaked and stabbed from typical
Salmonella and *Shigella* colonies.

Tubes of urea agar and SIM medium are
inoculated from each tube that exhibits
only glucose fermentation.
SIM medium is stabbed to 2/3 of depth
of medium. Both media are incubated at
37°C for 18 to 24 hours.

Gram-stained slide is made to see if culture
is pure. Serotyping is generally necessary.

**Urea
slant**

**SIM
medium**

Figure 54.4 **Isolation and presumptive identification of *Salmonella* and *Shigella*.**
© McGraw-Hill Education/ Auburn University Photographic Services

1. Examine the tubes of SIM medium, checking for motility and H$_2$S production. If you see cloudiness spreading from the point of inoculation, the organism is motile. If the bacteria are growing only along the stab line, then motility is absent. A black precipitate will be evidence of H$_2$S production (see Exercise 38, figure 38.2).
2. Test for indole production by adding four drops of Kovacs' reagent to the surface of the media of each SIM tube. A **pink to deep red color** will form on the top of the tube if indole is produced (see figure 38.2). *Salmonella* are negative.

Proteus and some *Shigella* may be positive; there is species variability within these genera. *Citrobacter* and *Escherichia* are positive.

3. Examine the urea slant tubes. If the medium has changed from yellow to **bright pink,** the organism is urease-positive (see figure 37.4).

Laboratory Report

Record the identity of your unknown in Laboratory Report 54 and answer all the questions.

54 Gram-Negative Intestinal Pathogens

A. Results

1. Record of test results

UNKNOWN NUMBER	FERMENTS LACTOSE	FERMENTS GLUCOSE	H₂S PRODUCED	MOTILITY	INDOLE	UREASE

2. Microscopy

   Provide a drawing of your unknown as seen under oil immersion (Gram staining).

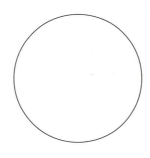

2. What was the genus of your unknown?

   _____   _____

   Genus   Unknown No.

3. What problems, if any, did you encounter?

   _____

   _____

   _____

   _____

   _____

   _____

4. Now that you know the genus of your unknown, what steps would you follow to determine the species?

   _____

   _____

   _____

B. Short-Answer Questions

1. Name three enteric pathogens of primary medical importance.

   _____

2. The ability of *Salmonella* to produce H_2S is one characteristic that helps differentiate it from *Shigella.* List the three opportunities you had in this exercise to determine whether or not your unknown produced H_2S.

   _____

   _____

3. What selective agents are added to media to preferentially grow enterobacteria for study? What type of growth is inhibited?

   _____

4. What characteristic separates *Salmonella* and *Shigella* from most of the other enterobacteria? What media can be used for this differentiation?

   _____

   _____

5. What two characteristics separate *Salmonella* from *Shigella?* What media can be used for this differentiation?

   _____

   _____

6. Which coliform bacteria are the most difficult to distinguish from the *Salmonella* or *Shigella* pathogens? What is the primary characteristic used to differentiate them?

   _____

   _____

7. How can acid production by glucose and lactose fermentation be differentiated in the same tube?

   _____

8. What is alkaline reversion? Explain why this condition gives a false negative result.

   _____

   _____

9. In this lab exercise, were the results of the indole test necessary to differentiate between *Salmonella* and *Shigella*? Explain why or why not.

   _____

   _____

10. What food is a common source of *Salmonella* infections?

    _____

    _____

A Synthetic Epidemic

Learning Outcomes

After completing this exercise, you should be able to

1. Define terminology related to the field of epidemiology.

2. Differentiate between common source and propagated epidemics.

3. Perform a simple test, using detectable reagents as a "microbe," to demonstrate how an infectious agent can be passed from person to person.

4. Describe how epidemiology is used to trace the source and spread of communicable diseases.

5. Explain how herd immunity can reduce the spread of communicable diseases through a population and protect susceptible individuals from infection.

A disease caused by microorganisms that enter the body and multiply in the tissues at the expense of the host is said to be an **infectious disease.** Infectious diseases that are transmissible from one person to another are considered to be **communicable.** The transfer of communicable infectious agents between individuals can be accomplished by direct contact, such as in handshaking, kissing, and sexual intercourse, or these agents can be spread indirectly through food, water, objects, animals, and so on.

Epidemiology is the study of how, when, where, what, and who are involved in the spread and distribution of diseases in human populations. An epidemiologist is, in a sense, a medical detective who searches out the sources of infection so that the transmission cycle can be broken.

Whether an epidemic actually exists is determined by comparing the number of new cases with previous records. If the number of newly reported cases in a given period of time in a specific area is excessive, an **epidemic** is considered to be in progress. Notable epidemics in the United States today include chlamydia and pertussis. If the disease spreads to one or more continents, a **pandemic** is occurring. An example of a pandemic disease is HIV/AIDS. According to the World Health Organization, over 36.9 million people are living with HIV worldwide. Tuberculosis, caused by *Mycobacterium tuberculosis,* is also considered a pandemic, with an estimated one-third of the world's

population now infected with the causative bacterium. An infectious disease that exhibits a steady frequency over a long period of time in a particular region is considered **endemic.** In tropical regions of the world, malaria is endemic.

Epidemics fall into two categories: common source epidemics and host-to-host epidemics (see figure 55.1). Common source epidemics occur rapidly and have many new cases immediately after the initial case. This type of epidemic usually involves a contaminated fomite (inanimate object) or contaminated food or water. After the infected source has been identified and removed in a common source epidemic, the number of new cases of disease drops rapidly, and the epidemic quickly subsides. From August to October 2011, 147 people across the United States were infected with *Listeria monocytogenes* after eating contaminated cantaloupes from a Colorado farm. The identification of this source led to recalls of the fruit and a quick end to the epidemic.

Propagated, or host-to-host, epidemics grow much more slowly and are also slower to subside. These epidemics involve transmission of the infectious agent through direct contact with the infected individual, a carrier, or a vector. A **vector** is a living

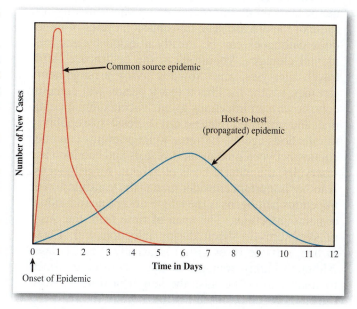

Figure 55.1 **Comparison of common source and host-to-host epidemics.**

carrier, such as an insect or rodent, which transmits an infectious agent. Control of propagated epidemics can involve education, vaccination, administration of antivirals or antibiotics, improved hygiene, and voluntary quarantine. The 2009 outbreak of H1N1 influenza was an example of a host-to-host epidemic and was eventually declared a pandemic as it spread from person to person across continents. Many of these control measures were implemented to slow the spread of the infection.

Human history has been shaped by epidemics. Population growth has been curbed, wars have been lost and won, and migrations have occurred all due to past outbreaks of disease. In the Middle Ages, "The Black Death," or bubonic plague caused by *Yersinia pestis*, killed one-third of the European population, redistributing wealth and property. In 1918, the flu pandemic affected one-third of the world population. The rapid spread of the influenza virus during this epidemic is attributed to the living conditions and travel of soldiers during World War I. Smallpox was brought to Mexico by European explorers, and the spread of this disease through the native population of Mexico is thought to be the reason they were defeated by the Spaniards. AIDS has wreaked havoc on the sub-Saharan African population and has slowed economic growth throughout Africa. Continual changes in human behavior and the threat of bioterrorism make pandemics a concern for the future. Today, it is common to travel across continents and expand our presence into environments that historically have been untouched. This creates opportunities for the spread of infection throughout the world as well as interactions with microbes and animals that may transmit diseases to humans.

The ability of infectious diseases to become epidemic and pandemic depends on the transmission cycle of the disease. The transmission cycles of infectious pathogens can be greatly affected by environmental changes. Vocations in the healthcare field, travel, lifestyles, and crowded living arrangements can increase the likelihood that a person will come in contact with an infectious agent. Vaccination, quarantine, and improved hygiene can decrease susceptibility to infectious agents. Extreme weather events can alter contact between humans and vectors and can increase reservoirs, the natural hosts or habitats for pathogens. Another important consideration in the transmission of infectious pathogens is the length of time an individual remains infectious. The span of time from the onset of symptoms to death during infection with Ebola (a hemorrhagic fever virus) is usually 2 to 21 days. Although highly contagious, Ebola is often confined to small regions because the length of time between the onset of symptoms and death is short, thus restricting the amount of time an infected individual can act as a carrier/host. In the case of HIV/AIDS, patients

are infectious for the remainder of their lifetimes. The length of time they can act as carriers is much longer, thus greatly enhancing the number of individuals to whom they can transmit the disease. Finally, some infections, such as chlamydia and hepatitis B, can be asymptomatic. Carriers can unknowingly transmit the infection to others, resulting in significant spread of the causative agent through the population.

Another principle that affects the spread of an epidemic through a population is referred to as community, or herd, immunity. When a certain proportion of a population is vaccinated or already immune to an infectious agent, an outbreak is unlikely in the population. Even individuals who cannot be vaccinated, such as infants, pregnant women, and immunocompromised individuals, are somewhat protected from infection because the spread of the agent is limited in the community (figure 55.2).

The National Institutes of Health has developed an online simulation that you can use to explore the impact of these variables on the transmission of a disease through a population. Visit this activity at the following URL: science.education.nih.gov/supplements/nih1/diseases/activities/activity4.html.

The microbiology laboratory plays a crucial role in preventing an outbreak from becoming an epidemic. Laboratory support involves, but is not limited to, culturing of infectious agents and environmental sites, isolate identification, and serological typing. Results from the microbiology lab help epidemiologists determine the first incident in a given outbreak, known as the **index case.** Identification of the index case aids in determining the original source of infection in an outbreak. Additionally, the microbiology lab is responsible for reporting suspicion or identification of infectious diseases to health departments, the Centers for Disease Control and Prevention (CDC), and the World Health Organization.

In this experiment, we will have an opportunity to approximate, in several ways, the work of the epidemiologist. Each member of the class will take part in the spread of a "synthetic infection" with simple reagents used to simulate the spread of an imaginary disease.

Two different experiments will be conducted: Procedures A and B. In Procedure A, students transfer a detectable agent by handshaking. The agent used is visible under UV light. Procedure A represents an epidemic in which the infectious agent is transferred from person to person by physical contact, and class data will be analyzed to determine the index case. In Procedure B, students transfer a detectable agent by the exchange of fluid in a test tube. The detectable agent in Procedure B is NaOH, which causes a change in pH that can be detected by the addition of a pH indicator. Procedure B represents an epidemic in which

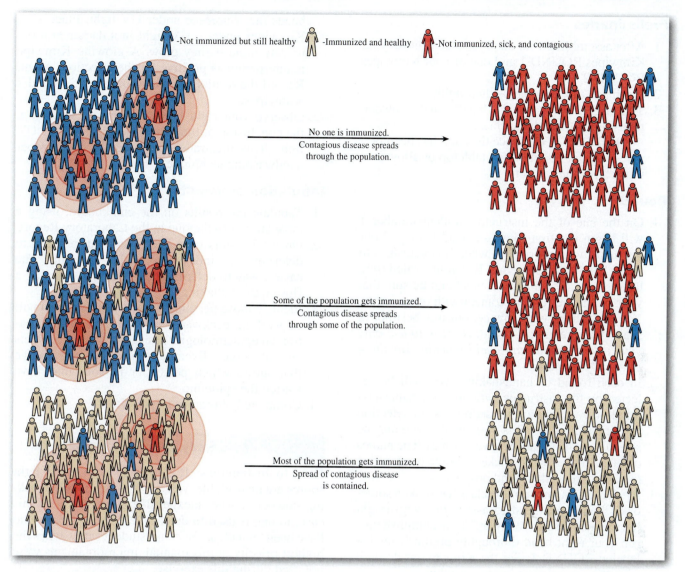

Figure 55.2 Community or herd immunity.
Courtesy: National Institute of Allergy and Infectious Diseases.

the infectious agent is transferred by the exchange of body fluid (saliva, sweat, urine, secretions). This procedure will be slightly modified and repeated three times to demonstrate the concept of community (herd) immunity.

Procedure A

In this experiment, each student will be given a numbered container of white unknown powder. Only one member in the class will be given a detectable agent that is to be considered the infectious agent and is visible under UV light. The other members will be issued a transmissible agent that is considered noninfectious. After each student has spread the powder on his or her hands, all members of the class will engage in two rounds of handshaking, directed by the instructor. A record of the handshaking contacts will be recorded

on a chart similar to the one in the Laboratory Report. After each round of handshaking, the hands will be rubbed on a Kimwipe or tissue that will later be placed under UV light to determine the presence or absence of the infectious agent.

Once all the data are compiled, an attempt will be made to determine two things: (1) the original source of the infection, and (2) who the carriers are. The type of data analysis used in this experiment is similar to the procedure that an epidemiologist would employ. Proceed as follows:

Materials

- 1 numbered petri dish containing an unknown white powder (either powder that is detectable using a UV light, such as GloGerm, or powder that is undetectable under UV light, such as baking soda)
- 2 Kimwipes or tissues

Preliminaries

1. After assembling your materials, label one of your Kimwipes ROUND 1 and one of your Kimwipes ROUND 2.
2. Wash and dry your hands thoroughly.
3. Thoroughly coat your right hand with the powder, focusing on the palm surface.
   IMPORTANT: Once the hand has been prepared, do not rest it on the tabletop or allow it to touch any other object.

Round 1

1. On the cue of the instructor, student number 1 will begin the first round of handshaking. Your instructor will inform you when it is your turn to shake hands with someone. Use your treated right hand to make firm hand contact and be sure that your palms are fully in contact with one another. You may shake with anyone, but it is best not to shake your neighbor's hand. *Be sure to use only your treated hand, and avoid touching anything else with that hand.*
2. In each round of handshaking, you will be selected by the instructor *only once* for handshaking; however, due to the randomness of selection by the handshakers, it is possible that you may be selected as the "shakee" several times. The names of each "shaker" and "shakee" should be recorded on the board as the round progresses.
3. After every member of the class has shaken someone's hand, you need to assess just who might have picked up the "microbe." To accomplish this, rub your right hand thoroughly on the Kimwipe labeled ROUND 1. Set the Kimwipe aside until after the second round of handshaking.
   IMPORTANT: Don't allow your hand to touch any other object. A second round of handshaking follows.

Round 2

1. On cue of the instructor, student number 1 will again select a person at random to shake hands with, proceeding as in Round 1 until everyone has had a turn to initiate a handshake. Avoid contact with any other objects. Be sure to keep a record of each handshake on the board.
2. Once the second handshaking episode is finished, rub the fingers and palm of the contaminated hand on the Kimwipe labeled Round 2. Set the Kimwipe aside and thoroughly wash hands.

Chemical Identification

1. The powder that is considered the infectious agent in this lab exercise is comprised of synthetic beads that fluoresce under UV light. Place your Kimwipes under a UV light in a darkened room to determine if they glow. A glowing Kimwipe is interpreted as positive for the infectious agent. Record the results of your Round 1 and Round 2 Kimwipes.
2. Observe your right hand, which was washed at the conclusion of the handshaking, under the UV light. This illustrates the importance of proper handwashing technique.

Tabulation of Results

1. Tabulate the results on the chalkboard, using a table similar to the one in the Laboratory Report.
2. Once all results have been recorded, proceed to determine the index case in this epidemic. The easiest way to determine this is to put together a flowchart of shaking.
3. Identify those persons that test positive. You will be working backward with the kind of information an epidemiologist has to work with (contacts and infections). Eventually, a pattern will emerge that shows which person or persons may have started the epidemic.
4. Complete Laboratory Report 55.

Procedure A: Results and Analysis

Note to instructors: If GloGerm and a UV light source are unavailable, you can use corn starch or baking powder as safe, inexpensive alternatives. In this case, iodine is used to determine those students who have been "infected." Because student safety takes the highest priority in this manual, microorganisms were not used during this exercise.

Procedure B

In this experiment, you will be investigating the benefit of community immunity for a population. In each round, students will be given a test tube containing water (representing a susceptible individual), 0.1 M NaOH (representing an infected individual), or a pH buffer solution (representing a vaccinated individual). After receiving their tubes, students will exchange a portion of the fluid in the tubes with one another as directed by the instructor. During each round, each student will exchange fluid with three different students. Once all of the exchanges have been completed for each round, a pH indicator will be added to all tubes to determine the presence or absence of the "infectious agent."

Proceed as follows for each round, recording class data in the Laboratory Report.

Materials

- sterile gloves
- 1 numbered test tube with lid containing liquid (either distilled water, a pH buffer solution such as a sodium phosphate buffer pH = 6.8, or 0.1 M NaOH)
- 1 dropper
- phenolphthalein solution, dissolved in alcohol and diluted in water (pH indicator) by instructor

Caution

Sodium hydroxide (NaOH) and phenolphthalein can irritate the eyes and skin. Wear gloves, and alert your instructor if any spills occur.

Round 1: 100% Susceptible Population

For this round, one student will unknowingly be infected with the agent and the rest of the students will be susceptible to infection (sample fluid of water).

1. On cue from the instructor, each student will participate in the first fluid exchange by selecting another student at random to swap fluid with. Each participant will use a dropper to trade a few drops of fluid.
2. After the class has conducted the first exchange, your instructor will announce that each student needs to find a second random student to exchange fluid with. Complete the fluid exchange with this second classmate.
3. Repeat with a third exchange when indicated by your instructor.

4. Add one drop of phenolphthalein to your test tube. If the fluid turns pink, you are positive for the infectious agent.
5. Record the class results in Laboratory Report 55, Procedure B: Results and Analysis.

Round 2: 50% Susceptible Population

For this second round, you will receive a new test tube sample of fluid. One student will unknowingly be infected with the agent, 50% of the students will be susceptible to infection (sample fluid of water), and 50% will represent vaccinated individuals (sample fluid of pH buffer).

1. Repeat the same procedure as above with your new fluid sample, exchanging with three individuals as your instructor leads.
2. Use the indicator to test for infection, and record the class results in Laboratory Report 55, Procedure B: Results and Analysis.

Round 3: 10% Susceptible Population

For this final round, repeat the preceding experiment starting with one infected individual, 90% vaccinated students (sample fluid of pH buffer), and 10% susceptible students (sample fluid of water). Record the class results in Laboratory Report 55, Procedure B: Results and Analysis.

Laboratory Report

Complete Laboratory Report 55.

55 A Synthetic Epidemic

A. Procedure A: Results and Analysis

Record the class information from the board into the table below. The SHAKER is the person designated by the instructor to shake hands with another class member. The SHAKEE is the individual chosen by the shaker. A Kimwipe that glows under UV light is positive; a Kimwipe that does not glow when exposed to UV light is negative.

SHAKER Round 1	RESULT + or −	SHAKEE Round 1	RESULT + or −	SHAKER Round 2	RESULT + or −	SHAKEE Round 2	RESULT + or −
1.				1.			
2.				2.			
3.				3.			
4.				4.			
5.				5.			
6.				6.			
7.				7.			
8.				8.			
9.				9.			
10.				10.			
11.				11.			
12.				12.			
13.				13.			
14.				14.			
15.				15.			
16.				16.			
17.				17.			
18.				18.			
19.				19.			
20.				20.			

1. Was it possible to determine an index case? If so, who was it? If not, explain why._____

2. What knowledge can be gained from determining the index case?_____

3. How many carriers resulted after Round 1?_____

4. How many carriers resulted after Round 2?_____

5. What percentage of the population was initially infected? What percentage of the population was infected at the end of the experiment?_____

6. What was the mode of transmission in this epidemic?_____

7. Describe an environment that would have a high rate of disease transmission._____

8. If this were a real infectious agent, such as a cold virus or influenza, list some factors that would affect transmission._____

9. How would it have been possible to stop this infection cycle?_____

10. Describe how the incubation period of a disease affects the spread of the disease._____

11. What factors are compared to confirm that an epidemic is occurring?_____

B. Procedure B: Results and Analysis

Record the class data for each round of this experiment in the table below. Calculate the percentage of susceptible individuals who contracted the disease during each round of this procedure. To calculate, use the following equation:

$$\frac{\text{Number of New Infected Individuals} \times 100}{\text{Initial Number of Susceptible Individuals}}$$

ROUND	INITIAL NUMBER OF INFECTED INDIVIDUALS	INITIAL NUMBER OF SUSCEPTIBLE INDIVIDUALS	NUMBER OF NEW INFECTED INDIVIDUALS (FINAL NUMBER OF INFECTED INDIVIDUALS MINUS 1 INITIAL)	PERCENTAGE OF SUSCEPTIBLE INDIVIDUALS THAT CONTRACTED INFECTION
1	1	100% of class = _____ individuals		
2	1	50% of class = _____ individuals		
3	1	10% of class = _____ individuals		

1. Explain how vaccines work to protect individuals from infection with pathogenic organisms.

   _____

   _____

2. Based on your results, describe how the number of vaccinated individuals affects the spread of a disease through a population.

   _____

   _____

3. Describe community (herd) immunity and its benefits to public health.

   _____

   _____

Immunology and Serology

When the human body becomes infected by a microbial agent such as a bacterium or virus, the invading microbe will encounter many different kinds of cells and host factors that will eventually destroy the microbe and prevent further damage or death to the host. These cells and factors comprise the **immune system,** which is composed of organs such as the liver, the spleen, and the lymph glands. The immune system recognizes the microbe as "foreign," and cells associated with the immune system engulf and remove the invading microbe, thus preventing its multiplication and further spread.

Many of the immune cells are derived from undifferentiated cells in the bone marrow called **stem cells.** Stem cells processed by the thymus gland, a butterfly-shaped gland at the base of the neck, undergo development into **T-lymphocytes (or T-cells).** Other stem cells processed by various lymphatic tissue, such as bone marrow in humans, will develop into

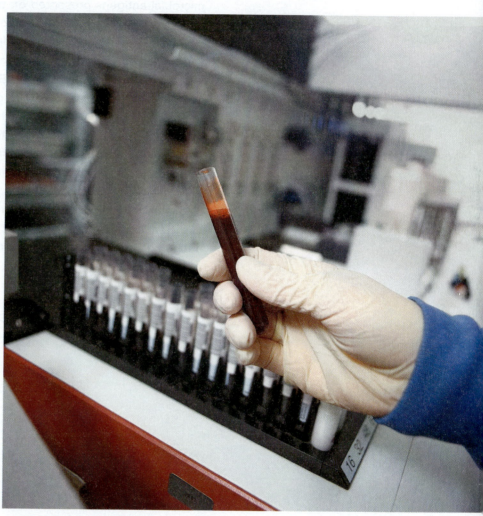

© liquidlibrary/PictureQuest RF

B-lymphocytes (or B-cells). Both of these types of lymphocytes have specific roles in various aspects of the immune response.

An invading microorganism is initially encountered by white blood cells such as **macrophages** and **neutrophils.** The microbe is engulfed and destroyed by these cells. However, components of the microbe or virus are preserved and "presented" on the surfaces of phagocytic cells such as the macrophages. These components are usually composed of proteins and carbohydrates and are referred to as **antigens.** Antigens are substances such as proteins, carbohydrates, nucleic acids, and even some lipids that are foreign to the host and therefore elicit the immune response. A unique property of the immune system is that it can differentiate the substances that comprise the host from those associated with the invader. That is, the immune system of an animal or human normally does not respond to its own protein and carbohydrate makeup. A breakdown in this recognition of self can result in an immune response to its own antigenic makeup and the development of an **autoimmune disease** such as some forms of arthritis.

Microbial antigens that are presented on the surfaces of macrophages stimulate a variety of other immune cells. For example, some T-lymphocytes are stimulated to seek out and destroy any cells infected with the microbe or virus. These cytotoxic T-lymphocytes recognize infected cells by the microbial antigens presented on their surfaces. Destruction of these infected cells ensures that the microbe or virus cannot reproduce to infect new cells. Other T-cells, called helper cells, are also stimulated, and they send chemical messages in the form of leukotrienes and cytokines that interact with **B-lymphocytes.** The B-lymphocytes differentiate into plasma cells that synthesize host proteins called **antibodies,** which specifically react with antigens and immobilize them so that phagocytic cells can engulf the complexes of antigens and antibodies that are formed by this interaction. Antibodies are usually specific for a portion of an antigen molecule called an **epitope,** which is a surface feature of the antigen. Hence, there is a tremendous diversity in the numbers of antibodies directed against a single protein molecule. Antibody molecules are Y-shaped, and the prongs of the antibody molecule are what specifically attach to the epitope of the antigen. Because the antibody molecule has two prongs (binding sites),

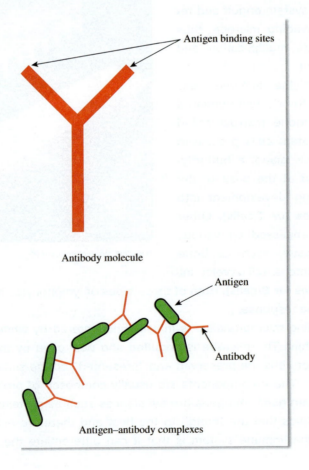

Antigen binding sites

Antibody molecule

Antigen

Antibody

Antigen–antibody complexes

it can simultaneously bind to the same epitope on different antigens, thereby forming a lattice complex of antigens and antibody molecules. When these complexes are destroyed by phagocytic cells, it ensures that no free antigen is present to cause further damage to the host.

A host produces different classes of antibodies, also called immuno-globulins. Immunoglobulin D (IgD) is associated with the surface of B-cells and aids in the recognition of antigens "presented" by phagocytic cells. IgA occurs on the surface of cells lining the respiratory and intestinal tracts, and it is thought to play a protective role in the respiratory and digestive systems. IgE is found on the surface of mast cells and binds to antigens called allergens. When this occurs, the chemical histamine is released by mast cells, and it is histamine that is responsible for many of the symptoms associated with allergic reactions. IgM and IgG are found primarily in the serum, which is the clear liquid in which blood cells are suspended. These antibodies are important in immobilizing free antigens for destruction by phagocytic cells. IgM is made in the primary response, that is, the first time an antigen is encountered by a host, whereas IgG is made in the secondary or amnestic (memory) response to an antigen. The following is a list of the various classes of antibodies and their functions.

ANTIBODY CLASSES AND THEIR FUNCTIONS

ANTIBODY CLASS	FUNCTION
IgM	Primary immune response
IgG	Secondary immune response
IgE	Release of histamine from mast cells
IgA	Protection of respiratory and intestinal tracts
IgD	Recognition of antigens found on B-cells

When antibodies react with an antigen, the result will differ depending on the nature of the antigen. If an antigen is particulate, such as a cell or a virus, antibodies specific for the antigen will cause it to **agglutinate,** or form visible clumps. However, if the antigen is soluble, as is the case for many proteins and polysaccharides, the antibodies will cause the antigen to **precipitate.** These types of reactions are useful for diagnosing disease in a patient. During the course of a disease, the amount of antibodies in a patient's serum, or the **titer** of the antibodies, will increase. The presence of an increasing titer of an antibody can therefore be used as an indicator of the course of a disease. Blood is collected from a patient, and diluted serum is reacted with the antigen. The lowest dilution that still demon-strates agglutination or precipitation is the serum titer. An increasing titer over time indicates an active infection.

In the following exercises, you will study antigen–antibody reactions and how they are used to monitor and detect disease.

Slide Agglutination Test:
Serological Typing

Organisms of different species differ not only in their morphology and physiology but also in the various components that make up their molecular structure. Macromolecules such as proteins, polysaccharides, nucleic acids, and phospholipids define the molecular structure of a cell. Some of the macromolecules of a bacterial cell can act as **antigens** because when introduced into an animal, they stimulate the immune system to form antibodies that are specific for these substances. The production of antibodies occurs because the antigens are foreign to the animal and different from the animal's own unique makeup. The antigenic structure of each species of bacteria is unique to that species, and much like a fingerprint of a human, its antigenic makeup can be used to identify the bacterium. The antigens that comprise lipopolysaccharide (O-antigens), capsules (K-antigens), and flagellar antigens (H-antigens) of the *Enterobacteriaceae* can be used to differentiate these bacteria into **serotypes.** Serotypes are more specific than species because they can be identical physiologically but differ significantly in their antigenic makeup. For example, *Escherichia coli* O157:H7, a serotype of *E. coli,* is a serious pathogen that can be transmitted in contaminated food to cause disease and death. It can be differentiated by its unique antigenic makeup from strains of *E. coli* that normally inhabit the intestinal tract of humans.

When a human is challenged by a microbe, **B-cells** differentiate into plasma cells that are responsible for producing immunoglobulins or antibodies that are specific for the antigens of the invading microorganism. The immunoglobulins occur in abundance in circulating blood in an immunized individual. If the blood cells are separated from the serum, the clear liquid that remains contains the immunoglobulins or antibodies and is called **antiserum.**

The antigens of a microorganism can be determined by a procedure called **serological typing** or serotyping. It consists of adding a suspension of microorganism to antiserum that contains antibodies specific for antigens associated with the microorganism. The antibodies in the antiserum will react with specific antigens on the bacterial cell and cause the cells to **agglutinate,** or form visible clumps. Serotyping is particularly useful in the identification of *Salmonella* and *Shigella,* which cause infections in humans such as typhoid fever and bacillary dysentery. For example, *Salmonella* can be differentiated into more than 2500 different serotypes based on antigenic differences associated with the cell. Serotyping of *Salmonella* or *Shigella* is useful in tracing epidemics caused by a particular strain or serotype of the respective organism. In the identification of these two genera, biochemical tests are first used to identify the organism as either *Salmonella* or *Shigella* (Exercise 54), followed by serotyping to identify specific strains.

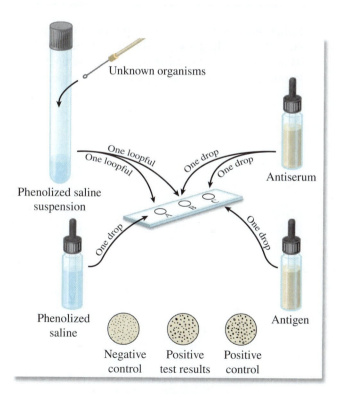

Figure 56.1 Slide agglutination technique.

In this exercise, you will be issued two unknown organisms, one of which is a *Salmonella*. By following the procedure shown in figure 56.1, you will determine which one of the unknowns is *Salmonella*. Note that you will use two test controls. A **negative test control** will be set up in depression A on the slide to see what the absence of agglutination looks like. The negative control is a mixture of antigen and saline (antibody is lacking). A **positive test control** will be performed in depression C with standardized antigen and antiserum to give you a typical reaction of agglutination.

Materials

- 2 numbered unknowns per student (slant cultures of a *Salmonella* and a coliform)
- *Salmonella* O-antigen, group B (Difco #2840-56)
- *Salmonella* O antiserum, poly A-I (Difco #2264-47)
- depression slides or spot plates
- dropping bottle of phenolized saline solution (0.85% sodium chloride, 0.5% phenol)
- 2 serological tubes per student
- 1 ml pipettes

Caution

Keep in mind that *Salmonella typhimurium* is a BSL2 pathogen and can cause gastroenteritis. Please review all safety procedures described in the Basic Microbiology Laboratory Safety information at the beginning of this manual.

1. Label three depressions on a spot plate or depression slide **A, B,** and **C,** as shown in figure 56.1.
2. Make a phenolized saline suspension of each unknown in separate serological tubes by suspending one or more loopfuls of organisms in 1 ml of phenolized saline. Mix the organisms sufficiently to ensure complete dispersion of clumps of bacteria. The mixture should be very turbid.
3. Transfer 1 loopful (0.05 ml) from the phenolized saline suspension of one tube to depressions A and B.
4. To depressions B and C, add 1 drop of *Salmonella* O polyvalent antiserum. To depression A, add 1 drop of phenolized saline, and to depression C, add 1 drop of *Salmonella* O-antigen, group B.
5. Mix the organisms in each depression with a clean wooden swab stick. Use a new stick for each depression.
6. Compare the three mixtures. Agglutination should occur in depression C (positive control), but not in depression A (negative control). If agglutination occurs in depression B, the organism is *Salmonella*.
7. Repeat this process on another slide for the other organism.

Caution

Deposit all slides and serological tubes in a container of disinfectant provided by the instructor.

Laboratory Report

Record your results in the first portion of Laboratory Report 56–57.

Slide Agglutination Test for *S. aureus*

EXERCISE

57

Learning Outcomes

After completing this exercise, you should be able to

1. Understand how antibodies can be adsorbed to latex spheres and then used to detect soluble antigens by agglutination reactions.
2. Perform serological tests using antibodies adsorbed to latex spheres to detect the coagulase or protein A of *Staphylococcus*.

When antibodies react with soluble antigens such as proteins or polysaccharides, the result is a precipitation reaction in which a visible precipitate is formed. The reaction, however, is best seen when a precise ratio of antibody to antigen occurs, called the **equivalence point.** At the equivalence point, a visible precipitate forms between the antigen and antibody. If excess antibody or antigen is present, soluble complexes can form, and no visible precipitate will form. Hence, unless the precise ratio of antigen to antibody is achieved in a precipitin reaction, a result cannot be readily seen. However, soluble antigens such as proteins can also be detected using a modified procedure involving an agglutination reaction. In these tests, antibodies that have been produced against soluble antigens are adsorbed, or chemically linked, to polystyrene latex particles (figure 57.1). The polystyrene particles act as carriers of the antibodies, and when they react with their soluble antigen an agglutination reaction occurs between the antibodies carried on the polystyrene particles and the antigen (figure 57.2).

Many manufacturers have devised such tests to detect specific pathogens or their products. In this exercise, you will use reagents manufactured by Difco Laboratories to determine if a suspected staphylococcus is producing the enzyme **coagulase** and/or **protein A.** Both protein A and coagulase are thought to be important virulence factors for *Staphylococcus aureus.* Coagulase may cause clot formation in capillaries at infection sites, thus preventing leukocytes, which are critical in the inflammatory response, from reaching areas of the infection. Protein A is a cell wall component of *S. aureus* that reacts with host immunoglobulins and coats the surface of the bacterial cell with these molecules. This immunoglobulin coat protects the bacterial cell by inhibiting phagocytosis by white blood cells and destruction of the bacterial cell. In the test for coagulase/protein A, the test reagent is a suspension of yellow latex particles to which antibodies against coagulase and protein A have been adsorbed. If the test strain of staphylococcus is producing coagulase and/or protein A, these antigens will

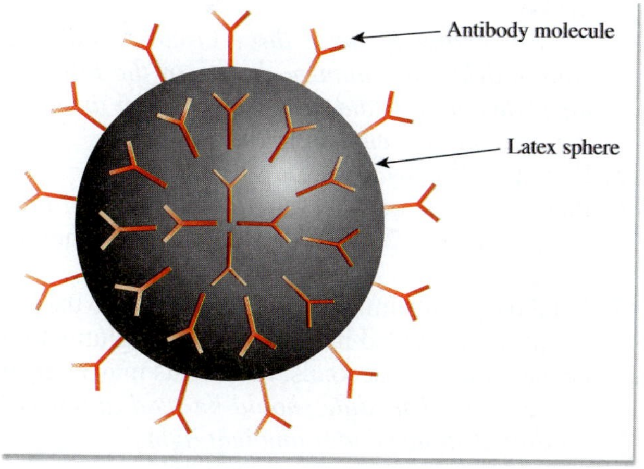

Figure 57.1 **Antibody molecules adsorbed to a latex sphere.**

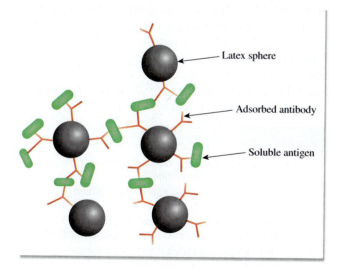

Figure 57.2 **Reaction between antibodies adsorbed to latex spheres and soluble antigens resulting in agglutination.**

Figure 57.3 **Slide agglutination test (direct method) for the presence of coagulase and/or protein A.**

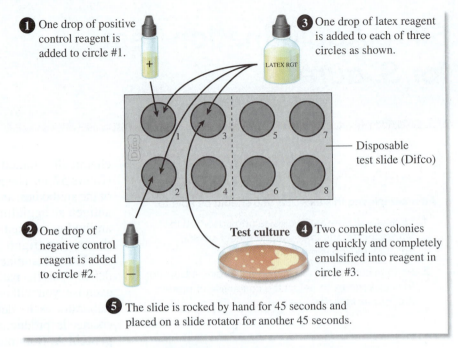

① One drop of positive control reagent is added to circle #1.

③ One drop of latex reagent is added to each of three circles as shown.

Disposable test slide (Difco)

② One drop of negative control reagent is added to circle #2.

Test culture

④ Two complete colonies are quickly and completely emulsified into reagent in circle #3.

⑤ The slide is rocked by hand for 45 seconds and placed on a slide rotator for another 45 seconds.

react with the latex particles and antibodies to cause agglutination of the latex spheres. No agglutination means that the test organism does not produce the virulence factors. This test is specific enough that it can be used instead of culturing.

Reagents are included in the test kit to perform both positive and negative controls. The test is performed on disposable cards with eight black circles. As indicated in figure 57.3, three circles are used to test one unknown. The remaining circles are used for testing five additional unknowns at the same time. The black background facilitates the observation of the agglutination reaction by providing good contrast for the reaction.

There are two versions of this test: direct and indirect. The procedure for the direct method is illustrated in figure 57.3. The indirect method differs in that saline is used to suspend the organism being tested.

It should be pointed out that the reliability correlation between this test for coagulase and the tube test (figure 52.4) is very high. Studies reveal that a reliability correlation of over 97% exists. Proceed as follows to perform this test.

Materials

- plate culture of staphylococcus-like organism (trypticase soy agar plus blood)
- Difco Staph Latex Test kit #3850-32-7, which consists of bottles of:
    Bacto Staph Latex Reagent, Positive Control, Negative Control, and Normal Saline Reagent
    disposable test slides (black circle cards)
    mixing sticks (minimum of 3)
- slide rotator

Direct Method

If the direct method is used, as illustrated in figure 57.3, follow this procedure:

1. Place 1 drop of Bacto Staph Positive Control reagent onto circle #1.
2. Place 1 drop of Bacto Staph Negative Control reagent on circle #2.
3. Place 1 drop of Bacto Staph Latex Reagent onto circles #1, #2, and #3.
4. Using a sterile inoculating needle or loop, quickly and completely emulsify *two isolated colonies* from the culture to be tested into the drop of Staph Latex Reagent in circle #3.

    Also, emulsify the Staph Latex Reagent in the positive and negative controls in circles #1 and #2 using separate mixing sticks supplied in the kit.

    All mixing in these three circles should be done quickly to minimize drying of the latex on the slide and to avoid extended reaction times for the first cultures emulsified.
5. Rock the slide by hand for 45 seconds.
6. Place the slide on a slide rotator capable of providing 110 to 120 rpm and rotate it for another 45 seconds.
7. Read the results immediately, according to the descriptions provided in table 57.1. If agglutination occurs before 45 seconds, the results may be read at that time. *The slide should be read at normal reading distance under ambient light.*

Indirect Method

The only differences between the direct and indirect methods pertain to the amount of inoculum and the use of saline to emulsify the unknown being tested. Proceed as follows:

1. Place 1 drop of Bacto Staph Positive Control reagent onto test circle #1.
2. Place 1 drop of Bacto Staph Negative Control onto circle #2.
3. Place 1 drop of Bacto Normal Saline Reagent onto circle #3.
4. Using a sterile inoculating needle or loop, completely emulsify *four isolated colonies* from the culture to be tested into the circle containing the drop of saline (circle #3).
5. Add 1 drop of Bacto Staph Latex Reagent to each of the three circles.
6. Quickly mix the contents of each circle, using individual mixing sticks.
7. Rock the slide by hand for 45 seconds.
8. Place the slide on a slide rotator capable of providing 110 to 120 rpm and rotate it for another 45 seconds.
9. Read the results immediately according to the descriptions provided in table 57.1. If agglutination occurs before 45 seconds, the results may be read at that time (figure 57.4). *The slide should*

Table 57.1

POSITIVE REACTIONS	
4 +	Large to small clumps of aggregated yellow latex beads; clear background
3 +	Large to small clumps of aggregated yellow latex beads; slightly cloudy background
2 +	Medium to small but clearly visible clumps of aggregated yellow latex beads; moderately cloudy background
1 +	Fine clumps of aggregated yellow latex beads; cloudy background
NEGATIVE REACTIONS	
–	Smooth, cloudy suspension; particulate grainy appearance that cannot be identified as agglutination
–	Smooth, cloudy suspension; free of agglutination or particles

be read at normal reading distance under ambient light.

Laboratory Report

Record your results in the last portion of Laboratory Report 56–57.

(a)

(b)

Figure 57.4   Slide agglutination test (a) positive; (b) negative.
© McGraw-Hill Education/ Lisa Burgess, photographer

56 Slide Agglutination Test: Serological Typing

A. Results

1. Describe the appearance of the mixtures in each of the wells.

   _____

   _____

2. Which unknown number proved to be *Salmonella*?

   _____

   _____

B. Short-Answer Questions

1. What types of compounds in bacterial cells can serve as antigens?

   _____

   _____

2. What are immunoglobulins?

   _____

   _____

3. What is a serotype of an organism?

   _____

   _____

4. What is agglutination?

   _____

   _____

5. What types of controls are used for the slide agglutination?

   _____

   _____

6. Why are serotypes more specific than species?

   _____

   _____

A. Results

1. Describe the appearance of the mixtures in each of the wells.

    _____

    _____

2. Was there a positive result for *S. aureus*? If so, what was the degree of the positive reaction?

    _____

    _____

B. Short-Answer Questions

1. What two *S. aureus* antigens are being detected with the use of this test kit?

    _____

    _____

2. What definitive test for *S. aureus* is highly correlated with this agglutination test?

    _____

    _____

3. What advantages does the agglutination test have over the definitive *S. aureus* test?

    _____

    _____

4. What factor can cause an antigen–antibody precipitate not to form?

    _____

    _____

Slide Agglutination Test for *Streptococcus*

Learning Outcomes

After completing this exercise, you should be able to

1. Understand the importance of carbohydrate antigens in the Lancefield classification of streptococci.

2. Identify an unknown streptococcus using carbohydrate antigens extracted from cells and reacting them with latex beads coated with antibodies developed against the carbohydrate antigens.

The Lancefield classification of streptococci divides this group of bacteria into immunological groups based on carbohydrate antigens associated with the cells (see Exercise 53). These groups are designated A through V. The majority of pathogenic streptococci occur in groups A, B, C, D, F, and G, and these bacteria usually also display beta-hemolysis when grown on blood agar. The presence of these antigenic carbohydrate groups in streptococcal cells has been used to develop an agglutination test for streptococci belonging to these Lancefield groups. In the test, the antigenic carbohydrates are first enzymatically extracted from cells and then reacted with latex beads which have been coated with antibodies developed against the specific carbohydrate antigens. If the carbohydrate antigens are present on the bacterial cells and then extracted by the enzymatic treatment, they will cause the latex beads to agglutinate when they react with the adsorbed antibodies.

In practice, suspected streptococci are isolated from clinical samples onto blood agar plates and grown for 18 to 24 hours at 37°C. Several colonies (2–5) showing beta-hemolysis are selected and treated with an extraction enzyme that cleaves the antigens from the cells. The extracts are then mixed with the antibody-coated latex beads and observed for agglutination (figure 58.1). In the following exercise, you will use this procedure to test for pathogenic streptococci.

Materials

- cultures of *Streptococcus pyogenes, S. agalactiae, S. bovis, Enterococcus faecalis,* and *E. faecium* grown on blood agar plates at 37°C for 24 hours.

- unknown cultures of the above organisms (per group)
- Oxoid Diagnostic Streptococcal grouping kit, which consists of:
  - latex grouping reagents A, B, C, D, F, and G
  - polyvalent positive control
  - extraction enzyme
  - disposable reaction cards
- test tubes (serological type)
- pipettes (1 ml)
- Pasteur pipettes
- pipette aids
- wooden sticks

Caution

Keep in mind that many streptococci are BSL2 pathogens. Please review all safety procedures described in the Basic Microbiology Laboratory Safety information at the beginning of this manual.

Procedure

Students will work in pairs. Each pair will test one known culture and an unknown culture assigned by the instructor.

ORGANISMS TO BE TESTED				
S. PYOGENES	S. AGALACTIAE	S. BOVIS	E. FAECALIS	E. FAECIUM
Group No. 1	2	3	4	5
6	7	8	9	10
11	12	13	14	15

1. Reconstitute the extraction enzyme with distilled water using the amount indicated on the label of the reagent bottle.
2. Label one test tube with the known organism that is to be tested and the second test tube with your unknown organism. Dispense 0.4 ml of extraction enzyme to each test tube.
3. Using a loop, select 2–5 colonies (2–3 mm) of your known organism and transfer them to the appropriate extraction enzyme tube.
4. Repeat this procedure for your unknown culture.

Reprinted by permission of Thermo Fisher Scientific.

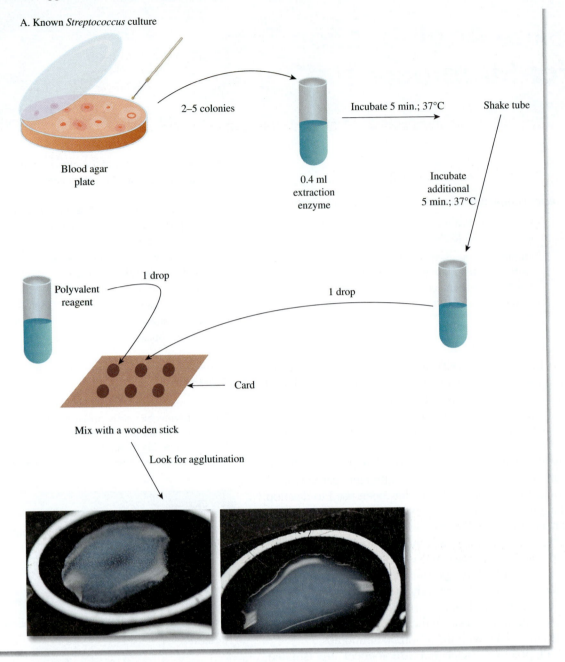

Figure 58.1 Procedure for performing the slide agglutination test for known and unknown *Streptococcus* colonies.
© McGraw-Hill Education/Lisa Burgess, photographer

5. Incubate the enzyme extraction tubes at 37°C for 5 minutes. After 5 minutes, shake each tube vigorously and continue incubation of the tubes for an additional 5 minutes.

6. Carefully mix the antibody-latex suspension corresponding to the group to which your known belongs (i.e., A, B, C, D, F, or G). Dispense one drop of the suspension to one of the rings on the reaction card.

7. Mix and dispense the polyvalent reagent to a separate ring on the card. This will be a positive control.

8. Using a Pasteur pipette, add one drop of extraction enzyme from your known culture to the ring on the card containing the latex suspension corresponding to the group of your known culture. Add a second drop of the extraction enzyme for your known culture to the ring with the polyvalent reagent.

9. Using a wooden stick, spread the extraction enzyme and latex suspension over the entire surface of the ring. Using a separate wooden stick, repeat the procedure for the polyvalent reagent and known enzyme extract.

B. Unknown *Streptococcus* culture

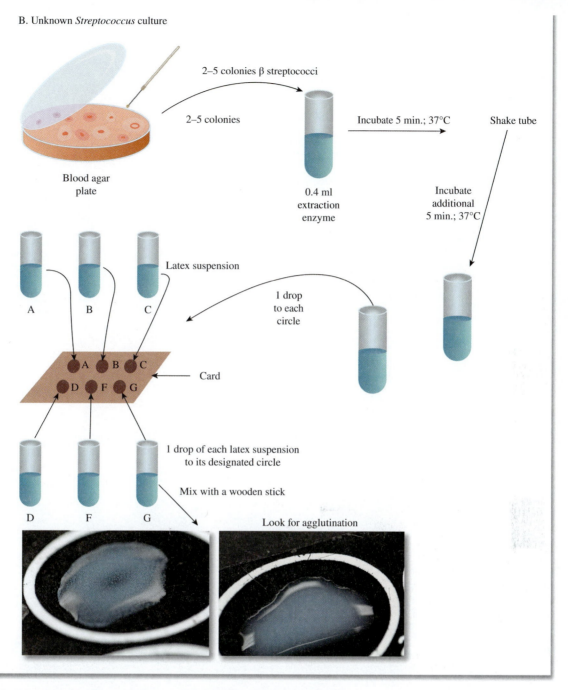

2–5 colonies β streptococci

2–5 colonies

Blood agar plate

0.4 ml extraction enzyme

Incubate 5 min.; 37°C

Shake tube

Incubate additional 5 min.; 37°C

Latex suspension

A   B   C

1 drop to each circle

A B C
D F G

Card

1 drop of each latex suspension to its designated circle

Mix with a wooden stick

D   F   G

Look for agglutination

Figure 58.1   Positive agglutination (left photo); no agglutination (right photo)
© McGraw-Hill Education/Lisa Burgess, photographer

10. Gently rock the cards back and forth. Agglutination will usually occur in 30 seconds.
11. To test your unknown culture, dispense one drop of each of the individual latex suspensions, A, B, C, D, F, and G to six separate circles on a reaction card.
12. Add one drop of the extraction enzyme from your unknown organism to each of the circles with the various latex suspensions.

13. Using a separate stick, mix the extraction enzyme and latex reagents.
14. Gently rock the card back and forth. Record which latex suspension caused the agglutination of the enzyme extract from your unknown organism.

58 Slide Agglutination Test for *Streptococcus*

A. Results

Appearance of known culture
with latex suspension

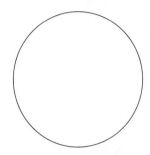

Appearance of known culture
with polyvalent suspension

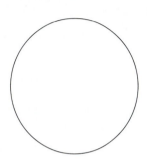

Unknown Organism

Latex Mixture   A   B   C   D   F   G

(Record a + for agglutination and a − for none.)

B. Short-Answer Questions

1. The Lancefield classification of streptococci is based on what property of these cells?

   _____

   _____

2. Why doesn't this test work for the identification of *Streptococcus pneumoniae*?

   _____

   _____

3. Streptococcal pathogens belonging to the groups tested also display what other important characteristic?

   _____

   _____

Enzyme-Linked Immunosorbent Assay (ELISA)

Learning Outcomes

After completing this exercise, you should be able to

1. Summarize the basics of HIV infection and the importance of early detection.

2. Understand the basis of enzyme-linked immunosorbent assays.

3. Perform an indirect ELISA to detect anti-HIV antibodies in patient serum.

Acquired immune deficiency syndrome (AIDS) is a viral immune system disease, resulting in the impaired ability of an infected patient to combat other infectious agents. AIDS is caused by human immunodeficiency virus (HIV), a retrovirus that targets and destroys immune cells. Helper T cells and macrophages are the main targets of HIV, and over time, HIV-infected patients have reduced counts of these cell types. Both innate and acquired immunity are affected by this cell loss, and the immunocompromised person is at risk of opportunistic bacterial, viral, and fungal infections as well as certain cancers.

HIV infection is a major global health concern, affecting millions across all inhabited continents. Recent World Health Organization statistics estimate 36.9 million people are living with HIV around the world. Thirty-nine million have died of AIDS since the pandemic began in the early 1980s. In the United States, the Centers for Disease Control estimates over 1.2 million individuals are infected with HIV, and 50,000 new cases develop each year. Even though the number of new HIV infections and AIDS deaths are decreasing, HIV remains among the top 10 causes of death worldwide.

A successful vaccine for HIV has been elusive despite considerable efforts over the four decades since this infectious agent was discovered. Instead, prevention education and advanced methods of diagnostic testing have been the keys to slowing the spread of HIV. Identifying HIV-positive individuals increases their access to healthcare and antiretroviral therapy, which improves patient outcomes and overall public health. Early detection and treatment can lessen the severity of the disease, slow its progression, and reduce the rates of transmission to others.

Several tests are available for the diagnosis of HIV infection. Currently, the CDC recommends that everyone be tested at least once in their lifetime and that those with risk factors be tested annually. The initial screening test for HIV is a fourth-generation immunoassay aimed at detecting antibodies to HIV-1 and HIV-2 (two strains of HIV found in different parts of the world) as well as HIV p24, an antigenic component of HIV's core. If this immunoassay is positive, follow-up tests such as the Western blot, an indirect immunofluorescence assay, or a nucleic acid test are used to confirm HIV infection.

ELISA

Immunological testing involves *specific* interactions between antigens and antibodies and is highly *sensitive,* detecting even minute amounts in a patient sample. **Immunoassays** are *in vitro* tests that use labeled antibodies to visualize these antigen–antibody interactions as part of the diagnosis of infections such as strep throat, syphilis, West Nile virus, HIV, hepatitis B and C, and measles. Antibodies may be labeled with fluorescent dyes, radioactive isotopes, or enzymes. *Direct* immunoassays use the labeled antibodies to detect unknown antigens in a clinical specimen, whereas *indirect* immunoassays detect specific antibodies in a patient's serum.

Enzyme-linked immunosorbant assays (ELISA) use enzyme-labeled antibodies to detect either unknown antigens or antibodies in a patient sample. Specifically, how does an indirect ELISA work? First, known antigens are bound to the wells of a clear microtiter plate. Then, test serum from a patient is added to the wells. If the patient has detectable levels of antibodies to that known antigen, they will bind to the antigen in the well. Next, a labeled antibody that can specifically bind to those test antibodies is added to the well. This secondary antibody is linked to an enzyme. If the patient's serum contains the antibodies to the known antigen, these secondary antibodies will bind to them and remain in the well after washing. Finally, when the enzyme's substrate is added to that well, the enzyme catalyzes a color reaction that is easily visible. If the patient's serum does not contain the antibodies to the antigen, the secondary antibody with the enzyme is removed from the well by washing, and no visible reaction occurs.

In this exercise, you will use an indirect ELISA to detect the presence of anti-HIV IgG antibodies in hypothetical patient serum samples, with a positive result indicating likely infection with HIV. Figure 59.1 illustrates and explains each step of this process.

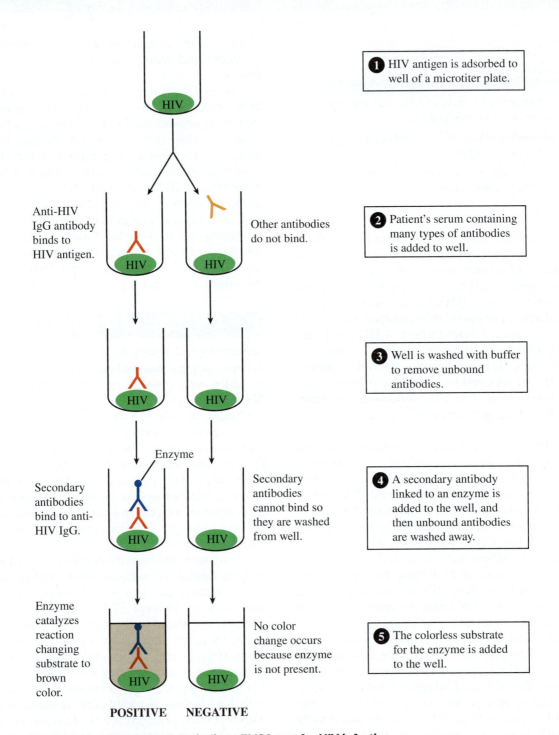

Figure 59.1   The steps of the indirect ELISA test for HIV infection.

Data from Edvotek, *AIDS Kit 1: Simulation of HIV-1 Detection,* www.edvotek.com/site/pdf/271.pdf

Materials

- microtiter plate (4 rows of three wells)
- micropipetter and sterile tips
- phosphate buffered saline (PBS) in small beaker
- small waste beaker
- biohazard bag for used tips
- reagents in labeled microcentrifuge tubes: HIV antigen, positive (+) control, patient serum #1, patient serum #2, secondary antibody, substrate

Carefully follow the steps below to minimize the risk of contamination between test wells. In a real testing lab, micropipette tips would be changed consistently throughout the process; however, in this simulated ELISA test, the use of new tips at specific intervals is indicated to minimize cost and materials but still provide accurate results.

Note

Gloves should be worn throughout this exercise, replicating the test process as it would be conducted in the medical laboratory.

1. Label your microtiter plate with a Sharpie, as shown below. Also, label the plate with your table number. Each student at the table should assume responsibility for one row of the ELISA throughout the exercise.

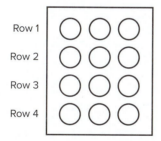

2. Using one micropipette tip for all wells, add 0.1 ml (100 µl) of HIV antigen to all 12 wells. Discard the tip into the biohazard receptacle.
3. Incubate the plate for 5 minutes at room temperature.
4. Using one micropipette tip for all wells, remove all the liquid in each well and discard into waste beaker. Discard your tip when finished with this step.
5. Using one micropipette tip for all wells, wash each well with 0.1 ml of PBS. This means that you will add the liquid and remove it immediately. Discard the PBS wash into the waste beaker, and discard the tip when finished with this step.

6. Using a new tip for each *row*, add the following reagents into each:

   Row 1: Add 0.1 ml of PBS (negative control).
   Row 2: Add 0.1 ml of + (positive control).
   Row 3: Add 0.1 ml of PS1 (patient serum #1).
   Row 4: Add 0.1 ml of PS2 (patient serum #2).

7. Incubate at 37°C for 15 minutes.
8. Using a new tip for each *row*, remove the liquid from each well and discard into the waste beaker.
9. Using a new tip for each *well*, wash all of the wells with PBS buffer as in step 5.
10. Using a new tip for each *row*, add 0.1 ml of the anti-IgG with peroxidase (secondary antibody) to all 12 wells.
11. Incubate at 37°C for 15 minutes.
12. Using a new tip for each *row*, remove the liquid from each well and discard into the waste beaker.
13. Using a new tip for each *well*, wash all of the wells with PBS buffer.
14. Using a new tip for each *row*, add 0.1 ml of substrate to all 12 wells.
15. Incubate at 37°C for 5 minutes.
16. Remove the plate for analysis, and record your observations in the Laboratory Report. Figure 59.2 shows an example of a microtiter plate with both positive (brown) and negative (colorless) wells.

Laboratory Report

Complete Laboratory Report 59.

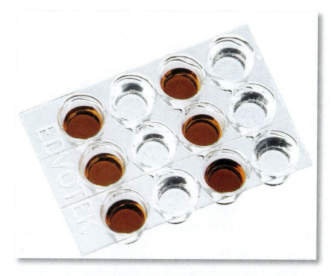

Figure 59.2   Sample results of indirect ELISA test for HIV. Brown wells indicate anti-HIV IgG antibodies were present in the sample, whereas colorless wells indicate these test antibodies were not present.

© Edvotek (www.edvotek.com)

59 Enzyme-Linked Immunosorbent Assay (ELISA)

A. Results

1. Color the diagram below to indicate the results of your ELISA.

```
        ┌─────────────────┐
Row 1   │  ○   ○   ○      │
        │                 │
Row 2   │  ○   ○   ○      │
        │                 │
Row 3   │  ○   ○   ○      │
        │                 │
Row 4   │  ○   ○   ○      │
        └─────────────────┘
```

2. Based on your results, which patient(s) are positive for the presence of anti-HIV IgG antibodies?

   _____

   _____

3. Describe the purpose and importance of rows 1 and 2.

   _____

   _____

4. How would you interpret a test plate where row 2 was brown across all wells and row 3 was brown in only two of the three wells?

   _____

   _____

B. Short-Answer Questions

1. Define *specificity* and *sensitivity* as these terms relate to immunological testing.

   _____

   _____

   _____

2. In your own words, explain why this procedure is referred to as an enzyme-linked immunosorbent assay.

   _____

   _____

   _____

3. If you forgot to wash the wells after you added the secondary antibody, what would you expect to see in each of the four rows of your ELISA?

_____

_____

4. What component must have been present in the positive control solution (antigen, anti-HIV IgG antibodies, or secondary antibody)? Explain how you know this is true.

_____

_____

_____

5. Explain the limitations of the ELISA for HIV testing. What might cause a false positive result? What might cause a false negative result?

_____

_____

_____

6. What other infections or conditions can be detected by an ELISA?

_____

_____

_____

Blood Grouping

Exercises 56 through 58 illustrate three uses of agglutination tests as related to (1) the identification of serological types (*Salmonella*), (2) staphylococcal species identification (*S. aureus*), and (3) streptococcal species identification. The typing of blood is another example of a medical procedure that relies on this useful phenomenon.

The procedure for blood typing was developed by Karl Landsteiner around 1900. He is credited with having discovered that human blood types can be separated into four groups on the basis of two antigens that are present on the surface of red blood cells. These antigens are designated as A and B. The four groups (types) are A, B, AB, and O, which are named based on the antigens that are present on each type (figure 60.1). The last group, type O, which is characterized by the absence of A or B antigens, is the most common type in the United States (45% of the population). Type A is next in frequency, found in 39% of the population. The frequencies of types B and AB are 12% and 4%, respectively. Type O has been called the "universal donor" because cells lack the A and B antigens. Type AB has been called the "universal recipient" because individuals with this blood type do not have anti-A or anti-B antibodies in their serum. However, blood is always specifically typed to avoid errors.

Blood typing is performed with antisera containing high titers of anti-A and anti-B antibodies. The test is usually performed by either slide or tube methods. In both instances, a drop of each kind of antiserum is added to separate samples of a saline suspension of red blood cells. If agglutination occurs only in the suspension to which the anti-A serum was added, the blood is type A. If agglutination occurs only in the anti-B mixture, the blood is type B. Agglutination in both samples indicates that the blood is type AB. The absence of agglutination indicates that the blood is type O. Figure 60.2 illustrates an example of the results for type A blood. Notice that agglutination occurs in the A well of this tray but not in the B well.

Between 1900 and 1940, a great deal of research was done to uncover the presence of other antigens in human red blood cells. Finally, in 1940, Landsteiner and Wiener reported that rabbit sera containing antibodies

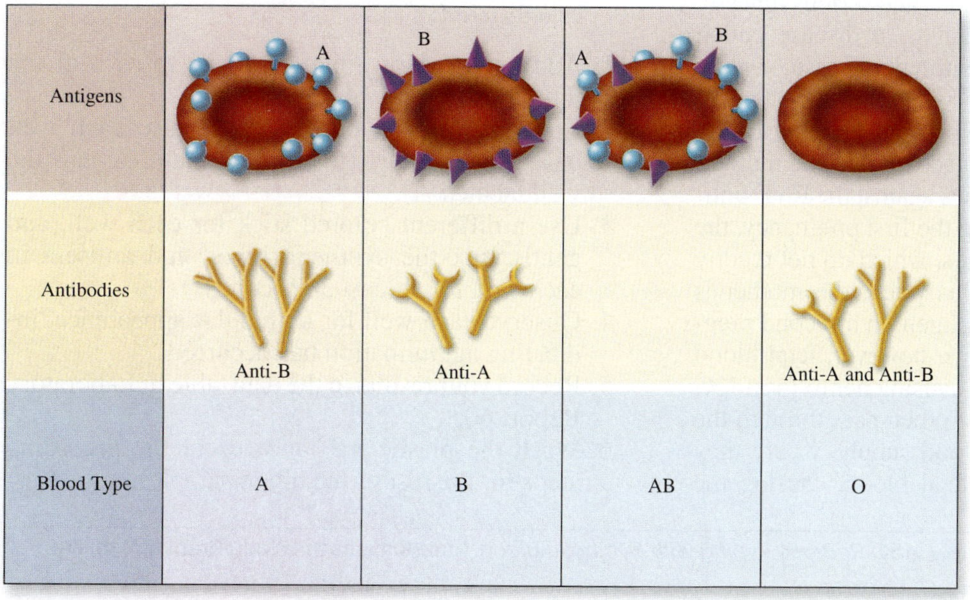

Figure 60.1 **Understanding ABO blood groups.**
© McGraw-Hill Education

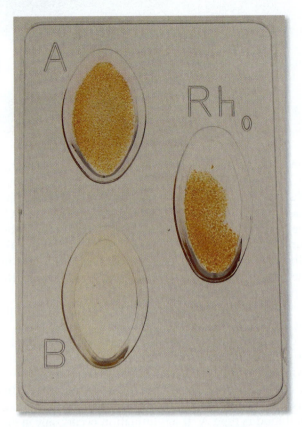

Figure 60.2   Sample blood typing test, indicating type A+ blood.
© Innovating Science

against the red blood cells of the rhesus monkey would agglutinate the red blood cells of 5% of humans. This antigen in humans, which was first designated as the **Rh factor** (for the rhesus monkey), was later found to exist as six antigens: C, c, D, d, E, and e. Of these six antigens, the D factor is responsible for the Rh-positive condition and is found in 85% of Caucasians, 94% of African-Americans, and 99% of Asians. Rh typing is also important to prevent incompatibilities and serious illness.

The Rh factor is important in a disease called **erythroblastosis fetalis,** or hemolytic disease of newborns. The condition occurs when an Rh-negative mother is pregnant with an Rh-positive fetus. Blood cells from the fetus cross the placenta into the mother, causing her to produce antibodies against the Rh D antigen on fetal blood cells. During the first pregnancy, the mother produces IgM antibodies, which do not readily pass back into the fetus. At this point, the mother is sensitized to the Rh-positive antigen. In a second pregnancy with an Rh-positive fetus, however, fetal blood cells will pass into the mother, causing her to form IgG antibodies. These antibodies can then pass through the placenta back into the fetal blood supply, where they react with the D antigen on fetal blood, causing the

blood cells to lyse and break up. This causes jaundice and a severe anemia in the fetus, sometimes requiring transfusion of the child at birth. To prevent this condition, the mother can be given an Rh immune globulin (RhoGAM) shot at week 28 and week 34 during every pregnancy and immediately after giving birth. Rho-GAM reacts with the D antigen on fetal blood cells and prevents the mother from forming any antibodies to the D antigen. It is also important to note that this condition can result in cases where there is an ABO incompatibility between the mother and the fetus, but it is usually rarer than the Rh-induced condition.

Typing blood for the Rh factor can also be performed by both tube and slide methods, but there are certain differences in the two techniques. First of all, the antibodies in the typing sera are of the incomplete albumin variety, which *will not agglutinate human red cells when they are diluted with saline.* Therefore, it is necessary to use whole blood or dilute the cells with plasma. Another difference is that the test *must be performed at higher temperatures:* 37°C for tube test; 45°C for the slide test.

Note

In the exercise, you will be working with synthetic blood that simulates human blood in typing reactions for the major blood groups A, B, O, and the Rh factor. When reacted with the "antisera" provided for typing, the same type of agglutination that would be seen with authentic blood is observed.

ABO and Rh Blood Typing

Materials

- blood typing kit containing 4 synthetic blood samples and A, B, and Rh antisera
- plastic trays with circular wells
- colored mixing sticks

1. Add one drop of a blood sample to each of the three wells on the plastic tray.
2. Add one drop of correct antisera to each well based on its label (for example, anti-A into the well labeled A).
3. Use a different colored stick for each well, and gently mix the synthetic blood and antisera in each well for at least 30 seconds.
4. Observe each well for a granular appearance, indicating agglutination has occurred.
5. Record your results in the data table in Laboratory Report 60.
6. Wash the plastic tray and repeat the preceding steps for the rest of the blood samples.

Data from Carolina Biological Supply, *ABO-RhBlood Typing with Synthetic Blood.* http://biolabs.tmcc.edu/PrepLabWebsite /BloodLab.pdf

60   Blood Grouping

A. Results

Record your results for all four synthetic blood samples in the following table. Put a + in each column for the wells that showed agglutination and a – for the wells that did not show agglutination. Based on these results, determine the blood type for each sample.

	BLOOD SAMPLE 1	BLOOD SAMPLE 2	BLOOD SAMPLE 3	BLOOD SAMPLE 4
A				
B				
Rh				
Blood type				

1. Which of the blood samples represents a universal recipient? Explain why this person could receive blood from any donor.

   _____

   _____

2. Which of the blood samples represents the most common blood type in the United States?

   _____

3. What type of antibodies would you expect to find in the serum of blood sample 2?

   _____

4. Why did the type O blood cells not react with either the anti-A or the anti-B antiserum?

   _____

   _____

B. Short-Answer Questions

1. Describe what is occurring in the wells with agglutination.

   _____

   _____

2. What happens when an individual receives a transfusion with an incompatible blood type?

   _____

   _____

3. Explain the difference between anti-A/anti-B antibodies and anti-Rh antibodies in terms of their presence in the plasma of individuals.

_____

_____

4. Describe the disease associated with an Rh-negative mother and an Rh-positive baby. How is it prevented in the United States?

_____

_____

5. At many blood donation centers, O-negative is the preferred donor for whole blood, whereas AB is the preferred donor for plasma. Explain why this is true.

_____

_____

_____

6. If you already know your blood type, fill out the following chart to review your understanding of the antigens and antibodies present in different blood types. If you do not know your blood type, fill out the following chart and drawing using a hypothetical blood type.

YOUR BLOOD TYPE	ANTIGENS ON YOUR RBCS	ANTIBODIES IN YOUR PLASMA

Draw what you would expect to see if you tested your own blood using the same test kit you used for the synthetic samples.

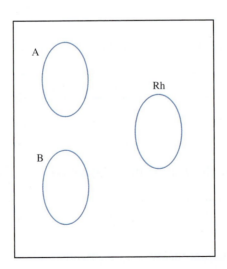

Table I   International Atomic Weights

ELEMENT	SYMBOL	ATOMIC NUMBER	ATOMIC WEIGHT
Aluminum	Al	13	26.97
Antimony	Sb	51	121.76
Arsenic	As	33	74.91
Barium	Ba	56	137.36
Beryllium	Be	4	9.013
Bismuth	Bi	83	209.00
Boron	B	5	10.82
Bromine	Br	35	79.916
Cadmium	Cd	48	112.41
Calcium	Ca	20	40.08
Carbon	C	6	12.010
Chlorine	Cl	17	35.457
Chromium	Cr	24	52.01
Cobalt	Co	27	58.94
Copper	Cu	29	63.54
Fluorine	F	9	19.00
Gold	Au	79	197.2
Hydrogen	H	1	1.0080
Iodine	I	53	126.92
Iron	Fe	26	55.85
Lead	Pb	82	207.21
Magnesium	Mg	12	24.32
Manganese	Mn	25	54.93
Mercury	Hg	80	200.61
Nickel	Ni	28	58.69
Nitrogen	N	7	14.008
Oxygen	O	8	16.0000
Palladium	Pd	46	106.7
Phosphorus	P	15	30.98
Platinum	Pt	78	195.23
Potassium	K	19	39.096
Radium	Ra	88	226.05
Selenium	Se	34	78.96
Silicon	Si	14	28.06
Silver	Ag	47	107.880
Sodium	Na	11	22.997
Strontium	Sr	38	87.63
Sulfur	S	16	32.066
Tin	Sn	50	118.70
Titanium	Ti	22	47.90
Tungsten	W	74	183.92
Uranium	U	92	238.07
Vanadium	V	23	50.95
Zinc	Zn	30	65.38
Zirconium	Zr	40	91.22

Table II Antibiotic Susceptibility Test Discs

ANTIBIOTIC AGENT	CONCENTRATION	INDIVIDUAL/10 PACK
Amikacin	30 µg	231596/231597
Amoxicillin/Clavulanic Acid	30 µg	231628/23629
Ampicillin	10 µg	230705/231264
Ampicillin/Subactam	10/10 µg	231659/231660
Azlocillin	75 µg	231624/231625
Bacitracin	2 units	230717/231267
Carbenicillin	100 µg	231235/231555
Cefaclor	30 µg	231652/231653
Cefazolin	30 µg	231592/231593
Cefixime	5 µg	231663/NA
Cefoperazone	75 µg	231612/231613
Cefotaxime	30 µg	231606/231607
Cefotetan	30 µg	231655/231656
Cefoxitin	30 µg	231590/231591
Ceftazidime	30 µg	231632/231633
Ceftriaxone	30 µg	231634/231635
Cefuroxime	30 µg	231620/231621
Cephalothin	30 µg	230725/231271
Chloramphenicol	30 µg	230733/231274
Clindamycin	2 µg	231213/231275
Doxycycline	30 µg	230777/231286
Erythromycin	15 µg	230793/231290
Gentamicin	10 µg	231227/231299
Imipenem	10 µg	231644/231645
Kanamycin	30 µg	230825/230829
Mezlocillin	75 µg	231614/231615
Minocycline	30 µg	231250/231251

Courtesy and © Becton, Dickinson and Company

Table II   Antibiotic Susceptibility Test Discs   (continued)

ANTIBIOTIC AGENT	CONCENTRATION	INDIVIDUAL/10 PACK
Moxalactam	30 µg	231610/231611
Nafcillin	1 µg	230866/231309
Nalidixic Acid	30 µg	230870/230874
Netilimicin	30 µg	231602/231603
Nitrofurantoin	100 µg	230801/231292
Penicillin	2 units	230914/231320
Piperacillin	100 µg	231608/231609
Rifampin	5 µg	231541/231544
Streptomycin	10 µg	230942/231328
Sulfisoxazole	0.25 mg	230813/231296
Tetracycline	5 µg	230994/231343
Ticarcillin	75 µg	231618/231619
Tobramycin	10 µg	231568/213569
Trimethoprim	5 µg	231600/231601
Vancomycin	30 µg	231034/231353

Table III   Indicators of Hydrogen Ion Concentration

Many of the following indicators are used in the media of certain exercises in this manual. This table indicates the pH range of each indicator and the color changes that occur. To determine the exact pH within a particular range, one should use a set of standard colorimetric tubes that are available from the prep room. Consult with your lab instructor.

INDICATOR	FULL ACID COLOR	FULL ALKALINE COLOR	pH RANGE
Cresol Red	red	yellow	0.2–1.8
Metacresol Purple (acid range)	red	yellow	1.2–2.8
Thymol Blue	red	yellow	1.2–2.8
Bromphenol Blue	yellow	blue	3.0–4.6
Bromcresol Green	yellow	blue	3.8–5.4
Chlorcresol Green	yellow	blue	4.0–5.6
Methyl Red	red	yellow	4.4–6.4
Chlorphenol Red	yellow	red	4.8–6.4
Bromcresol Purple	yellow	purple	5.2–6.8
Bromtheymol Blue	yellow	blue	6.0–7.6
Neutral Red	red	amber	6.8–8.0
Phenol Red	yellow	red	6.8–8.4
Cresol Red	yellow	red	7.2–8.8
Metacresol Purple (alkaline range)	yellow	purple	7.4–9.0
Thymol Blue (alkaline range)	yellow	blue	8.0–9.6
Cresolphthalein	colorless	red	8.2–9.8
Phenolphthalein	colorless	red	8.3–10.0

Appendix B—Indicators, Stains, Reagents

Indicators

All the indicators used in this manual can be made by (1) dissolving a measured amount of the indicator in 95% ethanol, (2) adding a measured amount of water, and (3) filtering with filter paper. The following chart provides the correct amounts of indicator, alcohol, and water for various indicator solutions.

INDICATOR SOLUTION	INDICATOR (GM)	95% ETHANOL (ML)	DISTILLED H$_2$O (ML)
Bromcresol Green	0.4	500	500
Bromcresol Purple	0.4	500	500
Bromthymol Blue	0.4	500	500
Cresol Red	0.4	500	500
Methyl Red	0.4	500	500
Phenolphthalien	1.0	50	50
Phenol Red	0.2	500	500
Thymol Blue	0.4	500	500

Stains and Reagents

Acid-alcohol (for Kinyoun stain)
    Ethanol (95%) . . . . .97 ml
    Concentrated HCl .3 ml

Alcohol, 70% (from 95%)
    Alcohol, 95% . . . . 368.0 ml
    Distilled water . . . . 132.0 ml

Barritt's Reagent (Voges-Proskauer test)
    Solution A: 6 g alpha-naphthol in 100 ml 95% ethyl alcohol.
    Solution B: 16 g potassium hydroxide in 100 ml water.
    Note that no creatine is used in these reagents as is used in O'Meara's reagent for the VP test.

Carbolfuchsin Stain (Kinyoun stain)
    Basic fuchsin . . . . 4 gm
    Phenol . . . . .8 ml

    Alcohol (95%) . . . . .20 ml
    Distilled/deionized water . . . . .100 ml
    Dissolve the basic fuchsin in the alcohol, and add the water while slowly shaking. Melt the phenol in a 56°C water bath and carefully add 8 ml to the stain.

Note: To facilitate staining of acid-fast bacteria, 1 drop of Tergitol No. 7 (Sigma Chemical Co.) can be added to 30–40 ml of the Kinyoun carbolfuchsin stain.

Crystal Violet Stain (Hucker modification)
    Solution A: Dissolve 2.0 g of crystal violet (85% dye content) in 20 ml of 95% ethyl alcohol.
    Solution B: Dissolve 0.8 g ammonium oxalate in 80.0 ml distilled water.
    Mix solutions A and B.

Diphenylamine Reagent (nitrate test)
    Dissolve 0.7 g diphenylamine in a mixture of 60 ml of concentrated sulfuric acid and 28.8 ml of distilled water.

Cool and add slowly 11.3 ml of concentrated hydrochloric acid. After the solution has stood for 12 hours, some of the base separates, showing that the reagent is saturated.

Gram's Iodine (Lugol's)

Dissolve 2.0 g of potassium iodide in 300 ml of distilled water and then add 1.0 g iodine crystals.

Iodine, 5% Aqueous Solution (Ex. 32)

Dissolve 4 g of potassium iodide in 300 ml of distilled water and then add 2.0 g iodine crystals.

Kovacs' Reagent (indole test)

n-amyl alcohol . . . . 75.0 ml
Hydrochloric acid (conc.) . . . . 25.0 ml
ρ-dimethylamine-benzaldehyde . . . . 5.0 g

Lactophenol Cotton Blue Stain

Phenol crystals . . . . 20 g
Lactic acid . . . . 20 ml
Glycerol . . . . 40 ml
Cotton blue . . . . 0.05 g

Dissolve the phenol crystals in the other ingredients by heating the mixture gently under a hot water tap.

Malachite Green Solution (spore stain)

Dissolve 5.0 g malachite green oxalate in 100 ml distilled water.

McFarland Nephelometer Barium Sulfate Standards (Ex. 42)

Prepare 1% aqueous barium chloride and 1% aqueous sulfuric acid solutions.

Add the amounts indicated in table 1 to clean, dry ampoules. Ampoules should have the same diameter as the test tube to be used in subsequent density determinations.

Seal the ampoules and label them.

Methylene Blue (Loeffler's)

Solution A: Dissolve 0.3 g of methylene blue (90% dye content) in 30.0 ml ethyl alcohol (95%).
Solution B: Dissolve 0.01 g potassium hydroxide in 100.0 ml distilled water. Mix solutions A and B.

Naphthol, alpha

5% alpha-naphthol in 95% ethyl alcohol
Caution: Avoid all contact with human tissues. Alpha-naphthol is considered to be carcinogenic.

Table 1   Amounts for Standards

TUBE	BARIUM CHLORIDE 1% (ML)	SULFURIC ACID 1% (ML)	CORRESPONDING APPROX. DENSITY OF BACTERIA (MILLION/ML)
1	0.1	9.9	300
2	0.2	9.8	600
3	0.3	9.7	900
4	0.4	9.6	1200
5	0.5	9.5	1500
6	0.6	9.4	1800
7	0.7	9.3	2100
8	0.8	9.2	2400
9	0.9	9.1	2700
10	1.0	9.0	3000

Nessler's Reagent (ammonia test)

Dissolve about 50 g of potassium iodide in 35 ml of cold ammonia-free distilled water. Add a saturated solution of mercuric chloride until a slight precipitate persists. Add 400 ml of a 50% solution of potassium hydroxide. Dilute to 1 liter, allow to settle, and decant the supernatant for use.

Nigrosin Solution (Dorner's)

Nigrosin, water soluble . . . . 10 g
Distilled water . . . . 100 ml

Boil for 30 minutes. Add as a preservative 0.5 ml formaldehyde (40%). Filter twice through double filter paper and store under aseptic conditions.

Nitrate Test Reagent

(see Diphenylamine Reagent)

Nitrite Test Reagents

Solution A: Dissolve 8 g sulfanilic acid in 1000 ml 5N acetic acid (1 part glacial acetic acid to 2.5 parts water).
Solution B: Dissolve 5 g dimethyl-alpha-naphthylamine in 1000 ml 5N acetic acid. Do not mix solutions.
Caution: Although at this time it is not known for sure, there is a possibility that dimethyl-alpha-naphthylamine in solution B may be carcinogenic. For reasons of safety, avoid all contact with tissues.

Oxidase Test Reagent

Mix 1.0 g of dimethyl-ρ-phenylenediamine hydrochloride in 100 ml of distilled water.

Preferably, the reagent should be made up fresh, daily. It should not be stored longer than one week in the refrigerator. Tetramethyl-ρ-phenylenediamine dihydrochloride (1%) is even more sensitive, but is considerably more expensive and more difficult to obtain.

Phenolized Saline

Dissolve 8.5 g sodium chloride and 5.0 g phenol in 1 liter distilled water.

Physiological Saline

Dissolve 8.5 g sodium chloride in 1 liter distilled water.

Potassium Permanganate

(for fluorochrome staining)

KMnO$_4$ . . . . 2.5 g
Distilled water . . . . 500.0 ml

Safranin (for Gram staining)

Safranin O (2.5% sol'n in 95%
    ethyl alcohol) . . . . 10.0 ml
Distilled water . . . . 100.0 ml

Trommsdorf's Reagent (nitrite test)

Add slowly, with constant stirring, 100 ml of a 20% aqueous zinc chloride solution to a mixture of 4.0 g of starch in water. Continue heating until the starch is dissolved as much as possible, and the solution is nearly clear. Dilute with water and add 2 g of potassium iodide. Dilute to 1 liter, filter, and store in amber bottle.

Vaspar

Melt together 1 pound of Vaseline and 1 pound of paraffin. Store in small bottles for student use.

Voges-Proskauer Test Reagent

(see Barritt's Reagent)

Appendix C—Media

Conventional Media   The following media are used in the experiments of this manual. All of these media are available in dehydrated form from either Difco Laboratories, Detroit, Michigan; or Baltimore Biological Laboratory (BBL), a division of Becton, Dickinson & Co., Cockeysville, Maryland. Compositions, methods of preparation, and usage will be found in their manuals, which are supplied upon request at no cost. The source of each medium is designated as (B) for BBL and (D) for Difco.

Bile esculin (D)
Brewer's anaerobic agar (D)
Desoxycholate citrate agar (B,D)
Desoxycholate lactose agar (B,D)
DNase test agar (B,D)
Endo agar (B,D)
Eugon agar (B,D)
Fluid thioglycollate medium (B,D)
Heart infusion agar (D)
Hektoen enteric agar (B,D)
Kligler iron agar (B,D)
Lead acetate agar (D)
Levine EMB agar (B,D)
Lipase reagent (D)
Lowenstein-Jensen medium (B,D)
MacConkey agar (B,D)
Mannitol salt agar (B,D)
MR-VP medium (D)
m-Staphylococcus broth (D)
Mueller-Hinton medium (B,D)
Nitrate broth (D)
Nutrient agar (B,D)
Nutrient broth (B,D)

Nutrient gelatin (B,D)
Phenol red sucrose broth (B,D)
Phenylalanine agar (D)
Phenylethyl alcohol medium (B)
Russell double sugar agar (B,D)
Sabouraud's glucose (dextrose)
   agar (D)
Semisolid medium (B)
Simmons citrate agar (B,D)
Snyder test agar (D)
Sodium hippurate (D)
Spirit blue agar (D)
SS agar (B,D)
Staphylococcus medium 110 (D)
Starch agar (D)
Trypticase soy agar (B)
Trypticase soy broth (B)
Tryptone glucose extract agar (B,D)
Urea (urease test) broth (B,D)
Veal infusion agar (B,D)
Xylose lysine desoxycholate agar
   (B,D)

Special Media   The following media are not included in the manuals that are supplied by Difco and BBL; therefore, methods of preparation are presented here.

Bile Esculin Slants (Ex. 53)

Heart infusion agar . . . . 40.0 g
Esculin. . . . . 1.0 g
Ferric chloride. . . . . 0.5 g
Distilled water . . . . 1000.0 ml

Dispense into sterile 15 × 125 mm screw-capped tubes, sterilize in autoclave at 121°C for 15 minutes, and slant during cooling.

Blood Agar

Trypticase soy agar powder. . . . . .40 g
Distilled water .1000 ml
   Final pH of 7.3
Defibrinated sheep or rabbit blood . . . . . .50 ml

Liquefy and sterilize 1000 ml of trypticase soy agar in a large Erlenmeyer flask. While the TSA is being sterilized, warm up 50 ml of defibrinated blood to

50°C. After cooling the TSA to 50°C, aseptically transfer the blood to the flask and mix by gently rotating the flask (cold blood may cause lumpiness).

Pour 10–12 ml of the mixture into sterile petri plates. If bubbles form on the surface of the medium, flame the surface gently with a Bunsen burner before the medium solidifies. It is best to have an assistant to lift off the petri plate lids while pouring the medium into the plates. A full flask of blood agar is somewhat cumbersome to handle with one hand.

Emmons' Culture Medium for Fungi

C. W. Emmons developed the following recipe as an improvement over Sabouraud's glucose agar for the cultivation of fungi. Its principal advantage is that a neutral pH does not inhibit certain molds that have difficulty growing on Sabouraud's agar (pH 5.6). Instead of relying on a low pH to inhibit bacteria, it contains chloramphenicol, which does not adversely affect the fungi.

Glucose . . . . 20 g
Neopeptone . . . . 10 g
Agar . . . . 20 g
Chloramphenicol . . . . 40 mg
Distilled water . . . . 1000 ml

After the glucose, peptone, and agar are dissolved, heat to boiling, add the chloramphenicol, which has been suspended in 10 ml of 95% alcohol, and remove quickly from the heat. Autoclave for only 10 minutes.

Glucose Peptone Acid Agar

Glucose . . . . 10 g
Peptone . . . . 5 g
Monopotassium phosphate . . . . 1 g
Magnesium sulfate (MgSO$_4$ · 7H$_2$O) . . . . . 0.5 g
Agar . . . . 15 g
Water . . . . 1000 ml

While still liquid after sterilization, add sufficient sulfuric acid to bring the pH down to 4.0.

Glycerol Yeast Extract Agar

Glycerol . . . . 5 ml
Yeast extract . . . . 2 g
Dipotassium phosphate . . . . 1 g
Agar . . . . 15 g
Water . . . . 1000 ml

LB Broth (Ex. 51)

Tryptone . . . . 10 g
Yeast extract . . . . 5 g

NaCl . . . . 5 g
Distilled water . . . . 1000 ml

Add 1 M NaOH to adjust the pH to 7. Autoclave at standard conditions.

m Endo MF Broth (Ex. 45)

This medium is extremely hygroscopic in the dehydrated form and oxidizes quickly to cause deterioration of the medium after the bottle has been opened. Once a bottle has been opened it should be dated and discarded after one year. If the medium becomes hardened within that time it should be discarded. Storage of the bottle inside a larger bottle that contains silica gel will extend shelf life.

Failure of Exercise 45 can often be attributed to faulty preparation of the medium. It is best to make up the medium the day it is to be used. It should not be stored over 96 hours prior to use. The Millipore Corporation recommends the following method for preparing this medium. (These steps are not exactly as stated in the Millipore Application Manual AM302.)

1. Into a 250 ml screw-cap Erlenmeyer flask place the following:

   Distilled water . . . . 50 ml
   95% ethyl alcohol . . . . 2 ml
   Dehydrated medium (*m* Endo MF broth) . . 4.8 g

   Shake the above mixture by swirling the flask until the medium is dissolved and then add another 50 ml of distilled water.
2. Cap the flask loosely and immerse it into a pan of boiling water. As soon as the medium begins to simmer, remove the flask from the water bath. Do not boil the medium any further.
3. Cool the medium to 45°C, and adjust the pH to between 7.1 and 7.3.
4. If the medium must be stored for a few days, place it in the refrigerator at 2–10°C, with screw-cap tightened securely.

Milk Salt Agar (15% NaCl)

Prepare three separate beakers of the following ingredients:

1. Beaker containing 200 g of sodium chloride.
2. Large beaker (2000 ml size) containing 50 g of skim milk powder in 500 ml of distilled water.
3. Glycerol-peptone agar medium:

   MgSO$_4$ · 7H$_2$O . . . . 5.0 g
   MgNO$_3$ · 6H$_2$O . . . . 1.0 g
   FeCl$_3$ · 7H$_2$O . . . . 0.025 g
   Difco proteose-peptone #3 . . . . 5.0 g
   Glycerol . . . . 10.0 g

Agar. . . . . 30.0 g
Distilled water . . . . 500.0 ml

Sterilize the above three beakers separately. The milk solution should be sterilized at 113–115°C (8 lb pressure) in autoclave for 20 minutes. The salt and glycerol-peptone agar can be sterilized at conventional pressure and temperature. After the milk solution has cooled to 55°C, add the sterile salt, which should also be cooled down to a moderate temperature. If the salt is too hot, coagulation may occur. Combine the milk salt and glycerol-peptone agar solutions by gently swirling with a glass rod. Dispense aseptically into petri plates.

Nitrate Agar

Beef extract . . . . .3 g
Peptone . . . . .5 g
Potassium nitrate .1 g
Agar. . . . . .12 g
Distilled water . . . . .1000 ml
Final pH 6.8 at 25°C

Nitrate Broth

Beef extract . . . . .3 g
Peptone . . . . .5 g
Potassium nitrate .1 g
Distilled water . . . . .1000 ml
Final pH 7.0 at 25°C

Phage Growth Medium (Ex. 22)

KH_2PO_4 . . . . 1.5 g
Na_2HPO_4 . . . . 3.0 g
NH_4Cl . . . . 1.0 g
$MgSO_4 \cdot 7H_2O$. . . . . 0.2 g
Glycerol. . . . . 10.0 g
Acid-hydrolyzed casein. . . . . 5.0 g
dl-Tryptophan . . . . 0.01 g
Gelatin. . . . . 0.02 g
Tween-80. . . . . 0.2 g
Distilled water . . . . 1000.0 ml
Sterilize in autoclave at 121°C for 20 minutes.

Phage Lysing Medium (Ex. 22)

Add sufficient sodium cyanide (NaCN) to the above growth medium to bring the concentration up to 0.02 M. For 1 liter of lysing medium this will amount to about 1 g (actually 0.98 g) of NaCN. When an equal amount of this lysing medium is added to the growth medium during the last 6 hours of incubation, the concentration of NaCN in the combined medium is 0.01 M.

Kligler's Iron Agar (Ex. 38 and 54)

Beef extract . . . . .3 g
Proteose Peptone No. 3 (Difco). . . . . .5 g
Lactose. . . . . .10 g
Dextrose. . . . . .1 g
Sodium chloride . . . . .5 g
Agar. . . . . .12 g
Phenol red (Difco). . . . . 0.024 g
Peptone . . . . 15.0 g
Yeast extract . . . . .3 g
Sodium thiosulfate. . . . . 0.3 g
Ferrous sulfate. . . . . 0.2 g
Deionized water. . . . . .1000 ml
Final pH 7.4 +/− 0.2 at 25°C

Dissolve ingredients in water, and bring to boiling. Cool to 50–60°C, and dispense about 8 ml per tube (16 mm dia tubes). Slant tubes to cool. Butt depth should be about $\frac{1}{2}$ ".

Skim Milk Agar

Skim milk powder. . . . . .100 g
Agar. . . . . .15 g
Distilled water . . . . .1000 ml

Dissolve the 15 g of agar into 700 ml of distilled water by boiling. Pour into a large flask and sterilize at 121°C, 15 lb pressure.

In a separate container, dissolve the 100 g of skim milk powder into 300 ml of water heated to 50°C. Sterilize this milk solution at 113–115°C (8 lb pressure) for 20 minutes.

After the two solutions have been sterilized, cool to 55°C and combine in one flask, swirling gently to avoid bubbles. Dispense into sterile petri plates.

Sodium Chloride (6.5%) Tolerance Broth (Ex. 53)

Heart infusion broth . . . . .25 g
NaCl. . . . . .60 g
Indicator (1.6 g bromcresol purple in 100 ml
    95% ethanol) .1 ml
Dextrose. . . . . .1 g
Distilled water .1000 ml

Add all reagents together up to 1000 ml (final volume). Dispense in 15 × 125 mm screw-capped tubes and sterilize in an autoclave 15 minutes at 121°C.

A positive reaction is recorded when the indicator changes from purple to yellow or when growth is obvious even though the indicator does not change.

Sodium Hippurate Broth

Heart infusion broth . . . . 25 g
Sodium hippurate . . . . 10 g
Distilled water . . . . 1000 ml
Sterilize in autoclave at 121°C for 15 minutes after dispensing in 15 × 125 mm screw-capped tubes. Tighten caps to prevent evaporation.

Soft Nutrient Agar (for bacteriophage)

Dehydrated nutrient broth . . . . 8 g
Agar. . . . . 7 g
Distilled water . . . . 1000 ml
Sterilize in autoclave at 121°C for 20 minutes.

Spirit Blue Agar (Ex. 37)

This medium is used to detect lipase production by bacteria. Lipolytic bacteria cause the medium to change from pale lavender to deep blue.
Spirit blue agar (Difco) . . . . 35 g
Lipase reagent (Difco). . . . . 35 ml
Distilled water . . . . 1000 ml

Dissolve the spirit blue agar in 1000 ml of water by boiling. Sterilize in autoclave for 15 minutes at 15 psi (121°C). Cool to 55°C and slowly add the 35 ml of lipase reagent, agitating to obtain even distribution. Dispense into sterile petri plates.

Tryptone Agar

Tryptone . . . . 10 g
Agar. . . . . 15 g
Distilled water . . . . 1000 ml

Tryptone Broth

Tryptone . . . . 10 g
Distilled water . . . . 1000 ml

Tryptone Yeast Extract Agar

Tryptone . . . . 10 g
Yeast extract . . . . 5 g
Dipotassium phosphate . . . . 3 g
Sucrose . . . . 50 g
Agar. . . . . 15 g
Water . . . . 1000 ml
   pH 7.4

Appendix D—Identification Charts

Chart I **Symbol Interpretation of API® 20E System**

READING TABLE

TESTS	ACTIVE INGREDIENTS	QTY (mg/cup.)	REACTIONS/ENZYMES	RESULTS	
				NEGATIVE	POSITIVE
ONPG	2-nitrophenyl-ßD-galactopyranoside	0.223	ß-galactosidase (Ortho NitroPhenyl-ßD-Galactopyranosidase)	colorless	yellow (1)
ADH	L-arginine	1.9	Arginine DiHydrolase	yellow	red / orange (2)
LDC	L-lysine	1.9	Lysine DeCarboxylase	yellow	red / orange (2)
ODC	L-ornithine	1.9	Ornithine DeCarboxylase	yellow	red / orange (2)
CIT	trisodium citrate	0.756	CITrate utilization	pale green / yellow	blue-green / blue (3)
H2S	sodium thiosulfate	0.075	H2S production	colorless / greyish	black deposit / thin line
URE	urea	0.76	UREase	yellow	red / orange (2)
TDA	L-tryptophane	0.38	Tryptophane DeAminase	*TDA / immediate* yellow	reddish brown
IND	L-tryptophane	0.19	INDole production	*JAMES / immediate* colorless pale green / yellow	pink
VP	sodium pyruvate	1.9	acetoin production (Voges-Proskauer)	*VP 1 + VP 2 / 10 min* colorless	pink / red (5)
GEL	Gelatin (bovine origin)	0.6	GELatinase	no diffusion	diffusion of black pigment
GLU	D-glucose	1.9	fermentation / oxidation (GLUcose) (4)	blue / blue-green	yellow / greyish yellow
MAN	D-mannitol	1.9	fermentation / oxidation (MANnitol) (4)	blue / blue-green	yellow
INO	inositol	1.9	fermentation / oxidation (INOsitol) (4)	blue / blue-green	yellow
SOR	D-sorbitol	1.9	fermentation / oxidation (SORbitol) (4)	blue / blue-green	yellow
RHA	L-rhamnose	1.9	fermentation / oxidation (RHAmnose) (4)	blue / blue-green	yellow
SAC	D-sucrose	1.9	fermentation / oxidation (SACcharose) (4)	blue / blue-green	yellow
MEL	D-melibiose	1.9	fermentation / oxidation (MELibiose) (4)	blue / blue-green	yellow
AMY	amygdalin	0.57	fermentation / oxidation (AMYgdalin) (4)	blue / blue-green	yellow
ARA	L-arabinose	1.9	fermentation / oxidation (ARAbinose) (4)	blue / blue-green	yellow
OX	(see oxidase test package insert)		cytochrome-OXidase	(see oxidase test package insert)	

SUPPLEMENTARY TESTS

TESTS	ACTIVE INGREDIENTS	QTY (mg/cup.)	REACTIONS/ENZYMES	RESULTS	
				NEGATIVE	POSITIVE
Nitrate reduction GLU tube	potassium nitrate	0.076	NO2 production	*NIT 1 + NIT 2 / 2–5 min* yellow	red
			reduction to N2 gas	*Zn / 5 min* orange-red	yellow
MOB	API M Medium or microscope		motility	nonmotile	motile
McC	MacConkey medium		growth	absence	presence
OF-F	glucose (API OF Medium)		fermentation : under mineral oil	green	yellow
OF-O			oxidation : exposed to the air	green	yellow

(1) A very pale yellow should also be considered positive.
(2) An orange color after 36-48 hours incubation must be considered negative.
(3) Reading made in the cupule (aerobic).
(4) Fermentation begins in the lower portion of the tubes, oxidation begins in the cupule.
(5) A slightly pink color after 10 minutes should be considered negative.
• The quantities indicated may be adjusted depending on the titer of the raw materials used.
• Certain cupules contain products of animal origin, notably peptones.

CHART II Characterization of Gram-Negative Rods—The API® 20E System

IDENTIFICATION TABLE

% of positive reactions after 18–24/48 hrs. at 36°C ± 2°C

API 20 E V4.1	ONPG	ADH	LDC	ODC	CIT	H2S	URE	TDA	IND	VP	GEL	GLU	MAN	INO	SOR	RHA	SAC	MEL	AMY	ARA	OX	NO2	N2	MOB	McC	OF/O	OFF
Buttiauxella agrestis	100	0	0	85	25	0	0	0	0	0	0	100	100	0	1	99	0	92	99	100	0	100	0	100	100	100	100
Cedecea davisae	99	89	0	99	75	0	0	0	0	89	0	100	100	10	0	0	100	1	100	1	0	99	0	87	100	100	100
Cedecea lapagei	99	99	0	0	75	0	0	0	0	90	0	100	100	0	0	0	0	0	100	1	0	99	0	87	100	100	100
Citrobacter braakii	50	45	0	75	75	81	1	0	4	0	0	100	100	1	100	99	1	91	99	99	0	100	0	95	100	100	100
Citrobacter freundii	90	24	0	75	75	75	1	0	1	0	0	100	99	1	99	99	99	82	40	99	0	98	0	95	100	100	100
Citrobacter koseri/amalonaticus	99	75	0	99	75	0	1	0	99	0	0	100	100	25	99	99	1	99	98	99	0	100	0	95	100	100	100
Citrobacter koseri/farmeri	99	2	0	100	25	0	1	0	99	0	0	100	100	25	1	99	99	80	99	99	0	100	0	95	100	100	100
Citrobacter youngae	100	50	0	1	80	80	1	0	1	0	0	100	100	0	95	100	1	80	25	100	0	85	0	100	100	100	100
Edwardsiella hoshinae	0	0	100	99	50	94	0	0	99	0	0	100	100	0	0	0	100	0	0	1	0	100	0	98	100	100	100
Edwardsiella tarda	0	0	100	99	1	75	0	0	99	0	0	100	100	0	0	0	0	0	0	1	0	100	0	98	100	100	100
Enterobacter aerogenes	99	25	99	98	82	0	1	0	0	85	0	99	99	99	99	99	99	99	99	99	0	100	0	97	100	100	100
Enterobacter amnigenus 1	99	25	0	40	40	0	1	0	0	75	0	100	100	0	1	100	99	99	99	99	0	100	0	92	100	100	100
Enterobacter amnigenus 2	99	80	0	99	80	0	0	0	0	75	0	100	100	0	1	100	99	99	99	99	0	100	0	92	100	100	100
Enterobacter asburiae	100	25	0	80	80	0	0	0	0	10	0	100	99	25	100	99	99	1	99	99	0	100	0	0	100	100	100
Enterobacter cancerogenus	100	75	0	99	99	0	0	0	0	89	0	100	100	0	0	100	0	1	100	99	0	100	0	99	100	100	100
Enterobacter cloacae	98	82	1	92	90	0	1	0	0	85	0	99	99	12	90	85	96	90	99	99	0	100	0	95	100	100	100
Enterobacter gergoviae	99	0	32	90	75	0	99	0	0	90	0	100	99	23	1	100	100	100	100	99	0	100	0	90	100	100	100
Enterobacter intermedius	99	0	0	99	1	0	0	0	0	2	0	100	97	0	88	99	40	99	99	99	0	100	0	92	100	100	100
Enterobacter sakazakii	100	96	0	94	94	0	1	0	0	91	10	100	100	75	1	99	99	99	99	99	0	100	0	96	100	100	100
Escherichia coli 1	90	1	74	70	0	1	3	0	25	0	0	99	98	1	8	82	36	90	3	99	0	100	0	95	100	100	100
Escherichia coli 2	26	45	45	20	0	1	1	0	89	0	0	99	98	1	91	30	3	75	3	70	0	98	0	5	100	100	100
Escherichia fergusonii	96	18	99	100	18	0	1	0	99	0	0	99	99	1	0	87	3	3	1	99	0	100	0	93	100	100	100
Escherichia hermannii	100	0	1	100	86	0	1	0	99	0	0	100	100	0	0	75	25	1	99	99	0	100	0	99	100	100	100
Escherichia vulneris	100	30	50	0	1	0	0	0	0	0	0	99	100	0	1	95	7	95	95	99	0	95	0	60	100	100	100
Ewingella americana	98	0	0	0	75	0	0	0	0	95	1	99	99	0	0	1	0	1	50	1	0	100	0	60	100	100	100
Hafnia alvei 1	75	0	99	98	50	0	0	0	0	85	0	99	99	0	0	1	0	90	1	99	0	100	0	85	100	100	100
Hafnia alvei 2	50	1	99	99	1	0	1	0	0	10	0	99	99	1	1	99	1	99	25	99	0	100	0	85	100	100	100
Klebsiella oxytoca	99	0	99	0	89	0	78	0	99	80	0	100	100	99	100	99	99	100	100	100	0	100	0	0	100	100	100
Klebsiella pneumoniae ssp ozaenae	94	25	25	1	18	0	1	0	0	0	0	99	96	57	66	58	20	80	97	85	0	92	0	0	100	100	100
Klebsiella pneumoniae ssp pneumoniae	99	0	99	0	86	0	75	0	0	90	0	99	99	99	99	99	98	99	99	99	0	100	0	0	100	100	100
Klebsiella pneumoniae ssp rhinoscleromatis	1	0	0	0	0	0	0	0	0	0	0	99	100	90	90	75	72	1	99	10	0	100	0	0	100	100	100
Kluyvera spp	95	0	25	99	60	0	0	0	80	0	0	99	99	0	25	93	75	99	99	99	0	95	0	94	100	100	100
Leclercia adecarboxylata	99	0	0	0	12	0	0	0	99	0	0	100	99	0	2	99	66	99	99	100	0	100	0	100	100	100	100
Moellerella wisconsensis	97	0	0	0	40	0	0	0	15	0	0	100	1	0	0	0	100	0	100	100	0	90	0	0	100	100	100
Morganella morganii	1	0	10	98	1	1	99	93	99	0	0	99	1	0	0	0	1	0	1	1	0	88	0	95	100	100	100
Pantoea spp 1	85	1	0	0	13	0	1	0	0	9	1	100	99	1	26	1	98	26	59	61	0	85	0	85	100	100	100
Pantoea spp 2	99	1	0	0	99	0	1	0	53	62	4	100	99	36	82	90	98	81	99	99	0	85	0	85	100	100	100
Pantoea spp 3	99	1	0	0	21	0	1	0	0	86	15	100	99	34	1	97	93	23	65	97	0	85	0	85	100	100	100
Pantoea spp 4	86	1	0	0	29	0	1	0	59	1	0	100	100	10	1	75	72	89	99	99	0	85	0	85	100	100	100
Proteus mirabilis	1	0	0	99	50	75	99	98	1	1	50	98	0	0	0	0	72	0	0	0	0	93	0	85	100	100	100
Proteus penneri	1	0	0	0	0	20	100	99	0	0	50	99	0	0	0	0	100	0	0	0	0	99	0	85	100	100	100
Proteus vulgaris group	1	0	0	0	12	83	99	99	92	0	74	99	1	0	0	1	89	1	66	1	0	99	0	85	100	100	100
Providencia alcalifaciens/rustigianii	0	0	0	0	80	0	0	100	99	0	0	99	0	1	0	0	0	0	0	0	0	100	0	94	100	100	100
Providencia rettgeri	1	0	0	0	74	0	99	98	90	0	0	100	82	78	0	97	93	0	65	0	0	98	0	96	100	100	100
Providencia stuartii	1	0	0	0	85	0	30	98	95	0	0	100	3	80	0	0	15	0	40	0	0	98	0	85	100	100	100
Rahnella aquatilis	100	0	0	50	50	0	0	0	0	99	0	100	100	0	98	99	100	97	100	98	0	100	0	6	100	100	100
Raoultella ornithinolytica	100	0	99	99	99	0	85	0	100	65	0	99	100	99	100	100	89	100	100	100	0	100	0	0	100	100	100
Raoultella terrigena	100	0	99	6	52	0	0	0	1	75	0	99	99	99	99	99	100	99	100	100	0	100	0	0	100	100	100
Salmonella choleraesuis ssp arizonae	98	75	97	99	75	99	0	0	1	0	0	100	100	0	98	99	1	78	100	99	0	99	0	99	100	100	100
Salmonella choleraesuis ssp choleraesuis	0	15	99	99	64	64	0	0	0	0	0	100	100	0	98	99	0	0	0	100	0	100	0	95	100	100	100
Salmonella ser Gallinarum	0	1	100	1	0	25	0	0	0	0	0	100	100	0	98	1	0	100	100	100	0	100	0	0	100	100	100

CHART II Characterization of Gram-Negative Rods—The API® 20E System (continued)

API® 20E V4.1	ONPG	ADH	LDC	ODC	CIT	H2S	URE	TDA	IND	VP	GEL	GLU	MAN	INO	SOR	RHA	SAC	MEL	AMY	ARA	OX	NO2	N2	MOB	McC	OF/O	OF/F
Salmonella ser Paratyphi A	0	5	0	99	0	1	0	0	0	0	0	100	99	0	99	98	0	96	0	99	0	100	0	95	100	100	100
Salmonella ser Pullorum	0	1	75	100	0	85	0	0	0	0	0	100	100	0	100	100	0	0	0	75	0	100	0	0	100	100	100
Salmonella typhi	0	1	99	0	0	8	0	0	0	0	0	100	99	0	99	0	1	90	0	0	0	100	0	97	100	100	100
Salmonella spp	1	56	82	93	65	83	0	0	1	0	1	99	100	40	99	86	1	90	1	99	1	92	0	95	100	100	100
Serratia ficaria	99	0	0	0	100	0	0	0	0	40	90	100	100	50	100	74	99	99	100	99	0	99	0	100	100	100	100
Serratia fonticola	99	0	73	99	75	0	0	0	0	0	65	100	100	97	100	99	30	72	97	99	0	99	0	91	100	100	100
Serratia liquefaciens	95	1	78	98	80	0	2	0	0	59	87	100	99	80	98	2	99	68	97	97	0	100	0	95	100	100	100
Serratia marcescens	94	0	95	95	96	0	0	0	0	70	99	100	99	85	99	1	99	1	99	25	0	95	0	97	100	100	100
Serratia odorifera 1	95	0	95	99	95	0	0	0	99	50	99	100	99	99	99	99	1	99	99	99	0	99	0	100	100	100	100
Serratia odorifera 2	95	0	96	0	95	0	0	0	99	50	50	100	99	70	99	1	99	85	98	95	0	99	0	50	100	100	100
Serratia plymuthica	99	0	0	0	65	0	1	0	0	65	50	100	90	70	70	3	99	95	99	98	0	100	0	85	100	100	100
Serratia rubidaea	99	0	30	0	92	0	0	0	0	71	82	99	99	75	1	1	99	95	99	99	0	100	0	0	100	100	100
Shigella spp	1	0	0	1	0	0	0	0	29	0	0	99	63	0	7	7	0	20	0	50	0	100	0	0	100	100	100
Shigella sonnei	96	0	0	93	0	0	0	0	0	0	0	99	99	0	1	75	0	1	0	99	0	98	0	2	100	100	99
Yersinia enterocolitica	80	0	0	90	0	0	98	0	50	5	0	99	99	25	98	1	99	4	75	75	0	98	0	5	100	99	99
Yersinia frederiksenii/intermedia	99	0	0	75	1	0	99	0	99	0	0	100	99	25	99	1	99	99	99	99	0	98	0	5	100	99	99
Yersinia kristensenii	80	0	0	80	0	0	99	0	97	0	0	100	99	10	99	0	0	0	99	99	0	98	0	5	99	99	99
Yersinia pestis	68	0	0	0	0	0	0	0	0	0	0	99	99	0	70	0	0	30	30	50	0	47	0	0	99	99	94
Yersinia pseudotuberculosis	98	0	0	0	1	0	99	0	0	1	0	99	99	0	1	75	0	50	75	75	0	95	0	95	99	99	99
Aeromonas hydrophila gr. 1	98	90	25	1	25	0	0	0	85	25	90	99	99	1	3	5	97	1	75	75	100	97	0	95	99	99	99
Aeromonas hydrophila gr. 2	99	97	80	1	80	0	0	0	85	80	97	97	99	9	9	1	80	1	75	5	100	98	1	95	99	99	99
Aeromonas salmonicida ssp salmonicida	1	60	1	0	0	0	0	0	0	0	75	50	54	0	0	0	0	1	0	0	100	100	0	1	95	99	99
Grimontia hollisae	1	0	75	0	0	0	0	0	94	0	0	99	0	0	0	0	0	0	0	99	100	100	0	25	2	23	23
Photobacterium damselae	1	99	75	0	1	0	98	0	0	10	1	50	0	0	0	0	1	0	0	0	99	99	0	95	2	23	23
Plesiomonas shigelloides	95	99	100	100	0	0	0	0	100	0	0	99	0	99	0	0	0	75	10	0	89	99	0	95	9	33	33
Vibrio alginolyticus	0	1	94	97	75	0	1	0	100	58	92	98	98	0	0	0	94	0	5	0	100	47	0	100	90	98	0
Vibrio cholerae	98	1	97	99	75	0	0	0	99	80	75	75	80	1	0	0	75	75	36	0	100	96	0	100	96	97	99
Vibrio fluvialis	95	99	0	0	50	0	0	0	0	1	75	99	99	0	0	0	1	1	75	75	100	100	0	100	95	99	6
Vibrio mimicus	99	0	99	99	50	0	1	0	80	1	75	99	99	0	0	0	1	1	12	0	99	95	0	100	98	99	49
Vibrio parahaemolyticus	0	0	100	99	25	0	1	0	100	1	75	100	75	0	0	1	1	1	90	75	100	63	0	100	100	100	2
Vibrio vulnificus	99	0	91	90	25	0	0	0	99	0	99	99	99	0	1	1	99	1	90	0	99	54	0	100	100	100	49
Pasteurella aerogenes	4	0	0	80	0	0	99	0	99	0	0	99	1	0	0	1	99	1	0	0	99	90	0	0	2	23	23
Pasteurella multocida 1	7	0	25	45	0	0	25	0	15	0	75	29	1	1	99	0	75	10	1	10	99	90	1	0	2	23	23
Pasteurella multocida 2	7	0	1	10	0	0	25	0	99	7	3	44	12	0	99	0	99	25	2	0	89	90	0	0	9	33	33
Pasteurella pneumotropica/ Mannheimia haemolytica	60	0	0	0	0	0	14	0	15	5	5	35	12	0	12	1	35	1	1	1	80	99	1	0	90	98	33
Acinetobacter baumannii/calcoaceticus	0	0	0	0	51	0	1	0	0	0	1	0	0	0	0	0	1	99	0	99	0	3	0	0	90	98	0
Bordetella/Alcaligenes/Moraxella spp *	0	0	0	0	52	0	14	1	0	5	43	60	1	0	0	0	13	0	7	20	95	62	1	75	75	97	0
Burkholderia cepacia	50	0	25	0	78	0	0	0	0	25	1	99	99	0	0	0	10	0	0	0	90	40	0	99	88	99	99
Chromobacterium violaceum	5	99	0	0	12	0	0	0	14	99	99	99	0	0	0	0	10	0	0	0	99	75	0	99	57	90	10
Chryseobacterium indologenes	77	0	0	0	20	0	1	0	75	0	80	0	0	0	0	0	0	0	0	0	99	20	0	1	48	93	6
Chryseobacterium meningosepticum	77	0	0	0	20	0	1	0	85	0	90	1	0	0	0	0	1	0	0	0	100	6	0	1	48	93	6
Eikenella corrodens	0	0	75	99	0	0	0	0	0	1	75	0	0	0	0	0	0	0	0	0	99	95	0	1	84	49	49
Myroides /Chryseobacterium indologenes	15	0	0	0	50	0	75	0	0	15	75	0	0	0	0	0	0	0	0	10	99	42	60	0	99	47	2
Ochrobactrum anthropi	15	89	0	0	30	0	25	0	0	0	75	50	0	0	1	0	0	10	1	25	97	12	56	97	100	98	0
Pseudomonas aeruginosa	0	75	0	0	92	0	25	0	0	10	75	84	0	0	0	0	0	25	1	20	99	26	0	100	96	93	0
Pseudomonas fluorescens/putida	86	75	0	0	94	0	25	0	0	25	94	84	0	0	0	0	0	15	0	85	0	30	0	100	91	94	0
Pseudomonas luteola	0	1	0	0	89	0	1	0	0	25	13	10	0	0	0	1	1	10	0	45	0	7	0	99	99	99	0
Pseudomonas oryzihabitans	1	1	0	0	37	0	1	0	0	15	9	10	0	0	0	1	1	0	0	1	93	48	35	99	85	49	0
Non-fermenter spp	1	0	0	0	75	75	1	0	0	0	75	1	1	0	0	1	1	0	0	2	99	96	0	100	96	9	0
Shewanella putrefaciens group	0	0	80	1	75	75	1	0	0	0	75	1	0	0	0	1	1	0	0	0	99	96	0	100	85	9	0
Stenotrophomonas maltophilia	70	0	75	1	75	1	0	0	0	0	90	1	0	0	0	1	1	0	0	0	1	26	1	100	91	49	0

* *Brucella* spp possible

Courtesy of bioMérieux, Inc.

Chart III Characterization of *Enterobacteriaceae*—The Entero*Pluri* System

Group	Genus / species	GLUCOSE	GAS PRODUCTION	LYSINE	ORNITHINE	H₂S	INDOLE	ADONITOL	LACTOSE	ARABINOSE	SORBITOL	VOGES-PROSKAUER	DULCITOL	PHENYLALANINE DEAMINASE	UREA	CITRATE
ESCHERICHIEAE	Escherichia	+ / 100.0	+J / 92.0	d / 80.6	d / 57.8	−K / 4.0	+ / 96.3	− / 5.2	+J / 91.6	+ / 91.3	± / 80.3	0.0	d / 49.3	0.0	0.1	0.2
ESCHERICHIEAE	Shigella	+ / 100.0	−A / 2.1	− / 0.0	∓B / 20.0	− / 0.0	∓ / 37.8	− / 0.0	−B / 0.3	± / 67.8	∓ / 29.1	0.0	d / 5.4	0.0	0.0	0.0
EDWARDSIELLEAE	Edwardsiella	+ / 100.0	+ / 99.4	+ / 100.0	+ / 99.0	+ / 99.6	+ / 99.0	− / 0.0	− / 0.0	∓ / 10.7	0.2	0.0	0.0	0.0	0.0	0.0
SALMONELLEAE	Salmonella	+ / 100.0	+C / 91.9	+H / 94.6	+I / 92.7	+E / 91.6	1.1	0.0	0.8	± / 89.2	94.1	0.0	dD / 86.5	0.0	0.0	dF / 80.1
SALMONELLEAE	Arizona	+ / 100.0	+ / 99.7	+ / 99.4	+ / 100.0	+ / 98.7	− / 2.0	− / 0.0	d / 69.8	+ / 99.1	+ / 97.1	0.0	0.0	0.0	0.0	+ / 96.8
SALMONELLEAE / CITROBACTER	freundii	+ / 100.0	+ / 91.4	− / 0.0	d / 17.2	± / 81.6	− / 6.7	− / 0.0	d / 39.3	+ / 100.0	+ / 98.2	0.0	d / 59.8	0.0	dw / 89.4	+ / 90.4
SALMONELLEAE / CITROBACTER	amalonaticus	+ / 100.0	+ / 97.0	− / 0.0	+ / 97.0	− / 0.0	+ / 99.0	− / 0.0	± / 70.0	+ / 99.0	+ / 97.0	0.0	∓ / 11.0	0.0	± / 81.0	+ / 94.0
SALMONELLEAE / CITROBACTER	diversus	+ / 100.0	+ / 97.3	− / 0.0	+ / 99.8	− / 0.0	+ / 100.0	+ / 100.0	d / 40.3	+ / 98.0	+ / 98.2	0.0	± / 52.2	0.0	dw / 85.8	+ / 99.7
PROTEEAE / PROTEUS	vulgaris	+ / 100.0	±G / 86.0	− / 0.0	− / 0.0	+ / 95.0	+ / 91.4	− / 0.0	− / 0.0	− / 0.0	− / 0.0	0.0	− / 0.0	+ / 100.0	+ / 95.0	d / 10.5
PROTEEAE / PROTEUS	mirabilis	+ / 100.0	+G / 96.0	− / 0.0	+ / 99.0	+ / 94.5	− / 3.2	− / 0.0	− / 2.0	− / 0.0	− / 0.0	0.0	∓ / 16.0	+ / 99.6	± / 89.3	± / 58.7
PROTEEAE / MORGANELLA	morganii	+ / 100.0	±G / 86.0	− / 0.0	+ / 97.0	− / 0.0	+ / 99.5	− / 0.0	− / 0.0	− / 0.0	− / 0.0	0.0	− / 0.0	+ / 95.0	+ / 97.1	−L / 0.0
PROTEEAE / PROVIDENCIA	alcalifaciens	+ / 100.0	dG / 85.2	− / 0.0	− / 1.2	− / 0.0	+ / 99.4	+ / 94.3	− / 0.3	− / 0.7	− / 0.6	0.0	− / 0.0	+ / 97.4	− / 0.0	+ / 97.9
PROTEEAE / PROVIDENCIA	stuartii	+ / 100.0	− / 0.0	− / 0.0	− / 0.0	− / 0.0	+ / 98.6	∓ / 12.4	− / 3.6	− / 4.0	− / 3.4	0.0	− / 0.0	+ / 94.5	∓ / 20.0	+ / 93.7
PROTEEAE / PROVIDENCIA	rettgeri	+ / 100.0	∓G / 12.2	− / 0.0	− / 0.0	− / 0.0	+ / 95.9	+ / 99.0	d / 10.0	− / 0.0	− / 1.0	0.0	− / 0.0	+ / 98.0	+ / 100.0	+ / 96.0
KLEBSIELLEAE / ENTEROBACTER	cloacae	+ / 100.0	+ / 99.3	− / 0.0	+ / 93.7	− / 0.0	∓ / 0.0	∓ / 28.0	± / 94.0	+ / 99.4	+ / 100.0	+ / 100.0	d / 15.2	− / 0.0	± / 74.6	+ / 98.9
KLEBSIELLEAE / ENTEROBACTER	sakazakii	+ / 100.0	+ / 97.0	− / 0.0	+ / 97.0	− / 0.0	∓ / 16.0	− / 0.0	+ / 100.0	+ / 100.0	− / 0.0	+ / 97.0	6.0	− / 0.0	− / 0.0	+ / 94.0
KLEBSIELLEAE / ENTEROBACTER	gergoviae	+ / 100.0	+ / 93.0	± / 64.0	+ / 100.0	− / 0.0	− / 0.0	− / 0.0	∓ / 42.0	+ / 100.0	− / 0.0	+ / 100.0	0.0	− / 0.0	+ / 100.0	+ / 96.0
KLEBSIELLEAE / ENTEROBACTER	aerogenes	+ / 100.0	+ / 95.9	+ / 97.5	+ / 95.9	− / 0.0	− / 0.8	+ / 97.5	+ / 92.5	+ / 100.0	+ / 98.3	+ / 100.0	4.1	− / 0.0	− / 0.0	+ / 92.6
KLEBSIELLEAE / ENTEROBACTER	agglomerans	+ / 100.0	∓ / 24.1	− / 0.0	− / 0.0	− / 0.0	∓ / 19.7	− / 7.5	d / 52.9	+ / 97.5	d / 26.3	± / 64.8	d / 12.9	∓ / 27.6	d / 34.1	d / 84.2
KLEBSIELLEAE / HAFNIA	alvei	+ / 100.0	+ / 98.9	+ / 99.6	+ / 98.6	− / 0.0	− / 0.0	− / 0.0	d / 2.8	+ / 99.3	− / 0.0	± / 65.0	− / 2.4	− / 0.0	− / 3.0	d / 5.6
KLEBSIELLEAE / SERRATIA	marcescens	+ / 100.0	±G / 52.6	+ / 99.6	+ / 99.6	− / 0.0	−w / 0.1	∓ / 56.0	− / 1.3	− / 0.0	+ / 99.1	+ / 98.7	− / 0.0	− / 0.0	dw / 39.7	+ / 97.6
KLEBSIELLEAE / SERRATIA	liquefaciens	+ / 100.0	d / 72.5	± / 64.2	+ / 100.0	− / 0.0	−w / 1.8	− / 8.3	d / 15.6	+ / 97.3	+ / 97.3	∓ / 49.5	− / 0.0	− / 0.9	dw / 3.7	+ / 93.6
KLEBSIELLEAE / SERRATIA	rubidaea	+ / 100.0	dG / 35.0	± / 61.0	− / 0.0	− / 0.0	−w / 2.0	± / 88.0	+ / 100.0	+ / 100.0	− / 8.0	+ / 92.0	− / 0.0	− / 0.0	dw / 4.0	± / 88.0
KLEBSIELLEAE / KLEBSIELLA	pneumoniae	+ / 100.0	+ / 96.0	+ / 97.2	− / 0.0	− / 0.0	− / 0.0	± / 89.0	+ / 98.7	+ / 99.9	+ / 99.4	+ / 93.7	∓ / 33.0	− / 0.0	+ / 95.4	+ / 96.8
KLEBSIELLEAE / KLEBSIELLA	oxytoca	+ / 100.0	+ / 96.0	+ / 97.2	− / 0.0	− / 0.0	+ / 100.0	± / 89.0	∓ / 98.7	+ / 100.0	+ / 98.0	+ / 93.7	∓ / 33.0	− / 0.0	∓ / 95.4	∓ / 96.8
KLEBSIELLEAE / KLEBSIELLA	ozaenae	+ / 100.0	d / 55.0	∓ / 35.8	1.0	− / 0.0	− / 0.0	+ / 91.8	d / 26.2	+ / 100.0	± / 78.0	− / 0.0	− / 0.0	− / 0.0	d / 14.8	d / 28.1
KLEBSIELLEAE / KLEBSIELLA	rhinoscleromatis	+ / 100.0	− / 0.0	− / 0.0	− / 0.0	− / 0.0	− / 0.0	+ / 98.0	d / 6.0	+ / 100.0	+ / 98.0	− / 0.0	− / 0.0	− / 0.0	− / 0.0	− / 0.0
YERSINIAE / YERSINIA	enterocolitica	+ / 100.0	− / 0.0	− / 0.0	+ / 90.7	− / 0.0	∓ / 26.7	− / 0.0	− / 0.0	+ / 98.7	+ / 98.7	− / 0.1	− / 0.0	− / 0.0	+ / 90.7	− / 0.0
YERSINIAE / YERSINIA	pseudotuberculosis	+ / 100.0	− / 0.0	− / 0.0	− / 0.0	− / 0.0	− / 0.0	− / 0.0	− / 0.0	± / 55.0	− / 0.0	− / 0.0	− / 0.0	− / 0.0	+ / 100.0	− / 0.0

E. *S. enteritidis* bioserotype Paratyphi A and some rare biotypes may be H₂S negative.

F. *S. typhi*, *S. enteritidis* bioserotype Paratyphi A and some rare biotypes are citrate-negative and *S. cholerae-suis* is usually delayed positive.

G. The amount of gas produced by *Serratia*, *Proteus* and *Providencia alcalifaciens* is slight; therefore, gas production may not be evident in the ENTEROTUBE II.

H. *S. enteritidis* bioserotype Paratyphi A is negative for lysine decarboxylase.

I. *S. typhi* and *S. gallinarum* are ornithine decarboxylase-negative.

J. The Alkalescens-Dispar (A-D) group is included as a biotype of *E. coli*. Members of the A-D group are generally anaerogenic, nonmotile and do not ferment lactose.

K. An occasional strain may produce hydrogen sulfide.

L. An occasional strain may appear to utilize citrate.

Courtesy and © Becton, Dickinson and Company

Chart IV Biochemistry of API® Staph Tests

READING TABLE

TESTS	ACTIVE INGREDIENTS	QTY (mg/cup.)	REACTIONS / ENZYMES	RESULT	
				NEGATIVE	POSITIVE
0	No substrate		Negative control	red	—
GLU	D-glucose	1.56	(Positive control) (D-GLUcose)		
FRU	D-fructose	1.4	acidification (D-FRUctose)		
MNE	D-mannose	1.4	acidification (D-ManNosE)		
MAL	D-maltose	1.4	acidification (MALtose)		
LAC	D-lactose (bovine origin)	1.4	acidification (LACtose)	red *	yellow
TRE	D-trehalose	1.32	acidification (D-TREhalose)		
MAN	D-mannitol	1.36	acidification (D-MANnitol)		
XLT	xylitol	1.4	acidification (XyLiTol)		
MEL	D-melibiose	1.32	acidification (D-MELibiose)		
NIT	potassium nitrate	0.08	Reduction of NITrates to nitrites	NIT 1 + NIT 2 / 10 min	
				colorless-light pink	red
PAL	ß-naphthyl phosphate	0.0244	ALkaline Phosphatase	ZYM A + ZYM B / 10 min	
				yellow	violet
VP	sodium pyruvate	1.904	Acetyl-methyl-carbinol production (Voges-Proskauer)	VP 1 + VP 2 / 10 min	
				colorless-light pink	violet-pink
RAF	D-raffinose	1.56	acidification (RAFfinose)		
XYL	D-xylose	1.4	acidification (XYLose)		
SAC	D-saccharose (sucrose)	1.32	acidification (SACcharose)	red	yellow
MDG	methyl-αD-glucopyranoside	1.28	acidification (Methyl-αD-Glucopyranoside)		
NAG	N-acetyl-glucosamine	1.28	acidification (N-Acetyl-Glucosamine)		
ADH	L-arginine	1.904	Arginine DiHydrolase	yellow	orange-red
URE	urea	0.76	UREase	yellow	red-violet

The acidification tests should be compared to the negative (0) and positive (GLU) controls.

* When MNE and XLT are preceded or followed by positive tests, then an orange test should be considered negative.

- The quantities indicated may be adjusted depending on the titer of the raw materials used.
- Certain cupules contain products of animal origin, notably peptones.

Lysostaphin resistance test

Determine resistance to lysostaphin on P agar, as indicated in the following procedure or according to the manufacturer's recommendations.

To perform the test, inoculate the surface of a P agar plate by flooding it with a bacterial suspension (approximately 10^7 organisms/ml).

Leave to dry for 10−20 minutes at $36°C \pm 2°C$.

Place a drop of lysostaphin solution (200 µg/ml) on the surface of the agar.

Incubate for 18−24 hrs. at 35−37°C.

Total or partial lysis of the bacterial culture indicates susceptibility to the enzyme.

This test constitutes the 21st test of the strip. It is considered positive if resistance to lysostaphin is determined.

Courtesy of bioMérieux, Inc.

Chart V   API® Staph Profile Register

IDENTIFICATION TABLE

% of reactions positive after 18–24 hrs. at 36°C ± 2°C

API STAPH V4.0	0	GLU	FRU	MNE	MAL	LAC	TRE	MAN	XLT	MEL	NIT	PAL	VP	RAF	XYL	SAC	MDG	NAG	ADH	URE	LSTR
Staphylococcus aureus	0	100	100	95	96	88	91	80	0	0	83	97	78	1	0	97	2	90	80	80	0
Staphylococcus auricularis	0	100	99	36	72	10	90	9	0	0	81	0	1	0	0	40	0	15	90	1	0
Staphylococcus capitis	0	100	99	80	43	22	2	36	0	0	86	23	90	0	0	50	0	1	85	35	0
Staphylococcus caprae	0	100	99	70	10	75	74	10	0	0	99	95	99	0	0	0	0	1	99	60	0
Staphylococcus carnosus	0	100	100	99	0	99	99	99	0	0	99	83	83	0	0	0	0	100	100	0	0
Staphylococcus chromogenes	0	100	100	99	79	100	100	13	0	0	96	96	1	0	1	100	0	31	89	95	0
Staphylococcus cohnii ssp cohnii	0	100	99	66	99	2	97	88	33	0	21	66	94	0	0	2	0	9	2	1	0
Staph.cohnii ssp urealyticum	0	100	100	99	98	98	100	94	64	0	1	94	87	0	0	0	0	98	0	99	0
Staphylococcus epidermidis	0	100	99	70	99	81	2	0	0	1	80	84	68	1	0	97	4	18	73	88	0
Staphylococcus haemolyticus	0	99	75	5	99	80	91	60	0	1	78	3	57	0	0	98	13	83	85	1	0
Staphylococcus hominis	0	98	94	41	97	50	86	28	0	1	82	27	70	1	0	97	4	50	43	84	0
Staphylococcus hyicus	0	100	99	99	0	87	99	0	0	0	90	90	15	0	0	99	2	93	100	68	0
Staphylococcus lentus	0	100	100	100	100	100	100	100	7	99	92	21	57	100	100	100	28	100	0	1	0
Staphylococcus lugdunensis	0	100	89	88	99	66	99	0	0	0	99	16	99	0	0	100	0	90	1	50	0
Staphylococcus saprophyticus	0	100	99	2	97	90	99	88	22	0	35	14	79	1	0	96	1	70	30	65	0
Staphylococcus schleiferi	0	100	80	100	0	1	71	0	0	0	99	97	99	0	0	0	0	94	99	0	0
Staphylococcus sciuri	0	99	99	99	99	70	93	98	0	0	83	67	30	0	16	95	7	68	0	0	0
Staphylococcus simulans	0	100	100	57	11	95	92	73	4	0	83	27	38	0	4	97	2	90	97	84	0
Staphylococcus warneri	0	99	99	50	98	19	96	70	0	0	23	16	90	0	0	99	0	6	77	97	0
Staphylococcus xylosus	0	100	100	92	81	85	95	90	30	9	82	75	67	11	82	87	10	80	5	90	0
Kocuria kristinae	0	99	96	99	90	9	84	3	0	0	6	3	93	0	0	90	12	0	0	0	97
Kocuria varians/rosea	0	91	92	8	1	1	8	1	0	0	75	4	8	4	8	4	0	1	1	29	95
Micrococcus spp	0	2	4	0	1	0	1	0	0	0	8	15	1	0	0	1	0	1	11	11	91

Courtesy of bioMérieux, Inc.

Appendix E—Metric and English Measurement Equivalents

THE METRIC SYSTEM

The metric system comprises three basic units of measurement: distance measured in meters, volume measured in liters, and mass measured in grams. In order to designate larger and smaller measures, a system of prefixes based on multiples of ten is used in conjunction with the basic unit of measurement. The most common prefixes are

$$\text{kilo} = 10^3 = 1{,}000$$
$$\text{centi} = 10^{-2} = 0.01 = 1/100$$
$$\text{milli} = 10^{-3} = 0.001 = 1/1{,}000$$
$$\text{micro} = 10^{-6} = 0.000001 = 1/1{,}000{,}000$$
$$\text{nano} = 10^{-9} = 0.000000001 = 1/1{,}000{,}000{,}000$$

THE ENGLISH SYSTEM

The measurements of the English system used in the United States unfortunately are not systematically related. For example, there are 12 inches in a foot and 3 feet in a yard. Quick conversion tables for the metric and English systems are listed below.

UNITS OF LENGTH

Metric to English

1 centimeter (cm) or 10 mm = 0.394 in or 0.0328 ft
1 meter (m) = 100 cm or 1,000 mm = 39.4 in or 3.28 ft or 1.09 yd
1 kilometer (km) = 1,000 m = 3,281 ft or 0.621 mile (mi)

THE NUMBER OF:	MULTIPLIED BY:	EQUALS:
millimeters	0.04	inches
centimeters	0.4	inches
meters	3.3	feet
meters	1.1	yards
kilometers	0.6	miles

English to Metric

1 in = 2.54 cm
1 ft or 12 in = 30.48 cm
1 yd or 3 ft or 36 in = 91.44 cm or 0.9144 m
1 mi or 5,280 ft or 1,760 yd = 1,609 m or 1.609 km

THE NUMBER OF:	MULTIPLIED BY:	EQUALS:
inches	2.5	centimeters
feet	30.5	centimeters
yards	0.9	meters
miles	1.6	kilometers

UNITS OF AREA

Metric to English

1 square centimeter (cm^2) or 100 mm^2 = 0.155 in^2
1 square meter (m^2) = 1,550 in^2 or 1.196 yd^2
1 hectare (ha) or 10,000 m^2 = 107,600 ft^2 or 2.471 acres (A)
1 square kilometer (km^2) or 1,000,000 m^2 or 100 ha = 247 A or 0.3861 mi^2

THE NUMBER OF:	MULTIPLIED BY:	EQUALS:
square centimeters	0.16	square inches
square meters	1.2	square yards
square kilometers	0.4	square miles

English to Metric

1 square foot (ft^2) or 144 in^2 = 929 cm^2
1 square yard (yd^2) or 9 ft^2 = 8,361 cm^2 or 0.836 m^2
1 acre or 43,560 ft^2 or 4,840 yd^2 = 4,047 m^2 = 0.405 ha
1 square mile (mi^2) or 27,878 ft^2 or 640 A = 259 ha or 2.59 km^2

THE NUMBER OF:	MULTIPLIED BY:	EQUALS:
square inches	6.5	square centimeters
square feet	0.09	square meters
square yards	0.8	square meters
square miles	2.6	square kilometers
acres	0.4	hectares

UNITS OF VOLUME

Metric to English

1 cubic centimeter (cm^3 or cc) or 1,000 mm^3 = 0.061 in^3

1 cubic meter (m^3) or 1,000,000 cm^3 = 61,024 in^3 or 35.31 ft^3 or 1.308 yd^3

THE NUMBER OF:	MULTIPLIED BY:	EQUALS:
cubic meters	35	cubic feet
cubic meters	1.3	cubic yards

English to Metric

1 cubic ft (ft^3) or 1,728 in^3 = 28,317 cm^3 or 0.02832 m^3

1 cubic yard (yd^3) or 27 ft^3 = 0.7646 m^3

THE NUMBER OF:	MULTIPLIED BY:	EQUALS:
cubic feet	0.03	cubic meters
cubic yards	0.76	cubic meters

UNITS OF LIQUID CAPACITY

Metric to English

1 milliliter (ml) or 1 cm^3 = 0.06125 in^3 or 0.03 fl oz

1 liter or 1,000 ml = 2.113 pt or 1.06 qt or 0.264 U.S. gal

THE NUMBER OF:	MULTIPLIED BY:	EQUALS:
milliliters	0.03	fluid ounces
liters	2.10	pints
liters	1.06	quarts

English to Metric

1 fluid ounce or 1.81 in^3 = 29.57 ml

1 pint or 16 fl oz = 473.2 ml

1 qt or 32 fl oz or 2 pt = 946.4 ml

1 gal or 128 fl oz or 4 qt or 8 pt = 3,785 ml or 3.785 liters

THE NUMBER OF:	MULTIPLIED BY:	EQUALS:
teaspoons	5	milliliters
tablespoons	15	milliliters
fluid ounces[a]	30	milliliters
cups	0.24	liters
pints	0.47	liters
quarts	0.95	liters
gallons	3.8	liters

UNITS OF MASS

Metric to English

1 gram (g) or 1,000 mg = 0.0353 oz

1 kilogram (kg) or 1,000 g = 35.2802 oz or 2.205 lb

1 metric or long ton or 1,000 kg = 2,205 lb or 1.102 short tons

THE NUMBER OF:	MULTIPLIED BY:	EQUALS:
grams	0.035	ounces
kilograms	2.2	pounds

English to Metric

1 ounce (oz) or 437.5 grains (gr) = 28.35 g

1 pound (1b) or 16 ounces = 453.6 g or 0.454 kg

1 ton (short ton) or 2,000 lb = 907.2 kg or 0.9072 metric ton

THE NUMBER OF:	MULTIPLIED BY:	EQUALS:
ounces	28	grams
pounds	0.45	kilograms
tons	0.9	metric ton

[a]1 British fluid ounce = 0.961 U.S. fluid ounces or, conversely, 1 U.S. fluid ounce = 1.041 British fluid ounces. The British pint, quart, and gallon = 1.2 U.S. pints, quarts, and gallons, respectively. To convert these U.S. fluid measures, multiply by 0.8327.

Index

INDEX

Wine production, 347–48
Work area disinfection, 63
Working stock culture, 243–44

Xanthophyll, 36
Xeromyces, 197
Xylene, for cleaning microscopes, 6

Yeasts, 52–55
    molds *vs.*, 51, 53
    in normal flora of skin, 234
    pathogenic, 52
    reproduction of, 52
    Saccharomyces cerevisiae study, 55

Yersinia, 395–96
Yersinia pestis, 396, 404

Z

Zeiss microscopes, 20
Zernike, Frits, 17–18
Zernike microscope, 17–19
Ziehl-Neelsen staining method, 117
Zinc test, 263
Zone of inhibition, 216, 236–41
Zoospores, 54
Zoothamnium, 35, 36
Zygnema, 39
Zygomycetes, 54
Zygospores, 54